© Bob Willoughby

About the Author

William R. Polk taught Middle Eastern history and politics and Arabic at Harvard until 1961, when he became a member of the Policy Planning Council of the Department of State, responsible for the Middle East and North Africa. In 1965, he resigned to become Professor of History and Founding Director of the Center for Middle Eastern Studies at the University of Chicago. He is the author of *Understanding Iraq, The United States and the Arab World,* and *The Elusive Peace*, among many other books.

Also by William R. Polk

Understanding Iraq

The United States and the Arab World

The Elusive Peace: The Middle East in the Twentieth Century

Neighbors and Strangers: The Fundamentals of Foreign Affairs

Polk's Folly: An American Family History

THE BIRTH OF AMERICA

From Before Columbus to the Revolution

WILLIAM R. POLK

NEW YORK • LONDON • TORONTO • SYDNEY

For William Polk Carey—financier, statesman, humanitarian, friend

HARPER PERENNIAL

A hardcover edition of this book was published in 2006 by HarperCollins Publishers.

HarperCollins books may be purchased for educational, business, or sales promotional use. For information please write: Special Markets Department, HarperCollins Publishers, 10 East 53rd Street, New York, NY 10022.

FIRST HARPER PERENNIAL EDITION PUBLISHED 2007.

Designed by C. Linda Dingler

Maps by Paul J. Pugliese

Library of Congress Cataloging-in-Publication Data is available upon request.

ISBN: 978-0-06-075090-9
ISBN-10: 0-06-075090-1

ISBN: 978-0-06-075093-0 (pbk.)
ISBN-10: 0-06-075093-6 (pbk.)

07 08 09 10 11 ❖/RRD 10 9 8 7 6 5 4 3 2 1

Acknowledgments

Researching and writing *The Birth of America* has stretched over several years, interrupted by writing another book on the Middle East, *Understanding Iraq,* and writing a long introduction to the diary of President James K. Polk. I could not have undertaken it without the support of the W. P. Carey Foundation and William Polk Carey. I am grateful to the foundation and have manifested my affection and thanks to Bill Carey by dedicating this volume to him.

As I make clear in my introduction, I have approached this period of American history from the outside. My own academic work has been primarily concerned with Asia and Africa. But over the years in which I have "read myself into" the topic of this book, I have so immersed myself in the works of specialists in the many fields that I feel that they have become not only colleagues but friends. As the reader will see from the notes, I have relied heavily on them, even though, since I long ago left my professorship at the University of Chicago and moved to the south of France, I have had opportunities to meet only a few of them.

I am grateful to my literary agent, Sterling Lord, for support and advice, and my editor at HarperCollins, Cass Canfield Jr., for his good editorial eye. And, finally, I particularly want to thank my wife, Elisabeth, for her forbearance and my daughter, Milbry, for many useful suggestions and encouragement.

William R. Polk
Vence, France

Contents

PART III: BREAKDOWN OF THE IMPERIAL SYSTEM

CONCLUSION: STUMBLING TOWARD WAR

Maps and Illustrations

INTRODUCTION

Revising the American Past

There was a time when people thought "history" was "the past," uniform, unchanging, finished; but as with styles in clothing, personal appearance, and deportment, we now know our view of the past to be affected by shifting tastes and values. What one age thinks insignificant is thrust to the center by another, and what another age regarded as crucial may later seem trivial. New information is discovered, influences are traced to different causes, and new light is cast on the actors. History is caught in a cycle of rebirth, never completely to come to a final rest, always open to revision.

The first American historians began, understandably, to write about what was at hand. Each one set out "the beginnings" by drawing on materials in the language he knew: Englishmen, on what was available in English; Spaniards, in Spanish; Frenchmen in French; Dutchmen, in Dutch.

Sadly, no literate person knew enough of any Native American language to record the histories of the many societies of early America north of Mexico. To judge by what we know of similar societies elsewhere, myths, legends, and deeds were recited or sung. Like Homer in his account of the Trojan War, orators often "froze" their accounts by putting them into poetry or rhymed prose; probably American Indians did too. Others, the first American historians, used mnemonic devices, often patterned wampum belts or notched sticks, to help them remember. Some of the belts survive, but we cannot capture what they helped Indian historians recall. The best we can do is to "reconstruct" what might have been from the information we get from contemporary visitors, accounts from later times, comparative studies from other tribal peoples, and on occasion archaeology.

For the blacks who began to arrive in the New World in the sixteenth century, information is even more tenuous or distorted: this is not only because few Europeans were interested in the lives of people they treated as domesticated animals but also because by the time blacks settled in America, their diverse African cultural heritages were eroded or overlaid and the web of their traditional social relations had been shattered.

So what we now have, like all history, is imperfect, incomplete, subject to revision. We keep struggling to get what to us seems a better vision. This effort involves not only digging up more information and revising what we have but also seeking different angles of vision. It is in the angle of vision that I offer a modest contribution because, having spent most of my scholarly career studying Asia, Africa, and Europe, I approach American history from an "external" perspective. That is, of course, how most of those who came to the New World approached America, either driven by events in the Old World or conditioned by experiences there. I believe that this approach enriches the insights to be gained from a study of what happened in the New World. To illustrate this point, consider how colonists thought of the native peoples they encountered.

The initial attitudes of southern Europeans toward native peoples were shaped by their first ventures in colonialism, particularly in the Canary Islands. There, the Spaniards encountered a people known as the Guanche. We would describe them as having a more or less Neolithic culture, but so primitive did they appear that the Spaniards regarded them as mere animals. When the Guanche tried to prevent the Spaniards from seizing their lands, the Spaniards enslaved or exterminated them. Then, on reaching the Caribbean island of Hispaniola, where a score of primitive societies lived, the Spaniards treated these natives as they had the Guanche. And, as they spread out into Mexico and what they called La Florida, it was the Canaries-Hispaniola colonial model they followed. Ironically, had Columbus actually reached Cathay (China) or Chipango (Japan) as he had hoped, the model the Spaniards would probably have followed was not what they learned dealing with the Guanche but what they had been taught in their centuries-long complex relationship with the highly cultured Muslim and Jewish peoples of Muslim Spain, al-Andalus—that is, diplomacy interspersed with warfare but not genocide. Of course, they did not reach Cathay or Chipango, so it was the Guanche model they applied to Hispaniola, Mexico, La Florida, and, eventually, the North American West.

Meanwhile, the northern Europeans, particularly the English and those people who were to play such a crucial role in their colonies, the so-called Scotch-Irish, were learning the craft of colonialism in fighting the Irish. Since shortly after the Norman invasion of England in the eleventh century, the English had been trying to subdue, segregate, or exterminate the Irish. At the beginning of the seventeenth century, James I encouraged a large-scale migration of Scots to assist them. The Scots took to that task with ruthless vigor. Then, toward the end of the century, when many of them were ejected or frightened into moving across the Atlantic, the Scotch-Irish found the Native Americans much like the Irish in dress, housing, and lifestyle and began to treat them as they, and the English, had treated the Irish.

In these and many other ways, the habits of the Old World were formative in the New World. It follows that understanding them is crucial to an appreciation of early American history; so I begin my account of the "birth" of America at conception rather than delivery. What I particularly want to emphasize is that this book attempts to appreciate what came "before" in Europe and Africa; what role the Spaniards, French, and Dutch played before the "British" became the dominant white Americans; and, insofar as it is now possible to understand, how all these groups impacted upon and interacted with the Native Americans.

What of the Native Americans? Were they really comparable to the Guanche or the Irish? The answer is both obscure and complex: complex because the Native Americans were divided into hundreds of societies at various levels of cultural florescence, and obscure because none of those living in the areas that ultimately became the United States left written records. Consequently, we must derive what we can know of them from the observations of Europeans who were usually ignorant of their languages and hostile to their beliefs. Two exceptions are the Virginia planter Robert Beverley in the late seventeenth century and the surveyor-general of North Carolina, John Lawson, who visited Indians, saw how they lived, inquired about their beliefs, and then recorded them. Other than they, the most observant of the European reporters visited the native societies after their ways of life had been affected by difficult-to-estimate degrees of contact.

Thus even the best accounts are seriously flawed, and flawed in ways we sometimes cannot judge.

From their earliest accounts it is clear that the British colonists, unlike the Spaniards, found the Indians worthy of respect. They observed Indians living in well-organized societies, comfortably adapted to their environment, healthy, and physically impressive. It was only as the immigrants learned how to feed themselves, grew in number, and overwhelmed their immediate neighbors that they began to hate and despise the Indians. Within a few generations, as they spread inland and increasingly took over Indian lands, they came to share the belief that "the only good Indian is a dead Indian." So most of their observations can be used by historians only with extreme caution.

Indians were not the only neighbors of the British colonists. They would have found our neglect of their French and Spanish contemporaries curious indeed. Yet it was not until the 1920s that Spain came to be treated as an integral part of the story of "America." It was not just that the Spaniards had led the European advance into the New World but also that much of what they did set patterns that were, unconsciously and with significant variations, followed by others. Their accounts naturally focused on the areas where they were most active, Nueva España (mainly modern Mexico), which was much more important to Spain than La Florida (the North American southeast); and Nueva Andalucia (the North American west). Although what happened in these areas had important consequences for what ultimately became the United States, few American historians were interested.

We can now see, however, that Spanish explorers not only "opened" the interior of our continent but changed it in significant ways. Perhaps the most powerful change was brought about by their unwitting importation of European diseases to peoples without immunities to them. Thus they set in motion a demographic revolution in which many Native American societies would be wiped out. Indian population uniformly shows catastrophic decline in size, often as much as 90 percent. At the time of "contact," the Native American population east of the Mississippi had reached somewhere around 2 million; and by roughly 1750 it had dropped to approximately 250,000.

With effects similar to disease, Spaniards imposed upon the New World the thriving new business of sugar plantations. They did so partly because the contemporary African states prevented the Portuguese and them from creating sugar plantations, then a major source of wealth, in Africa. Some of those states would not even allow Europeans to establish trading stations in their territories. Thwarted there, the Spaniards (and eventually the French and English) created plantations in the Caribbean where they formed economies, legal systems, and a political order profoundly different from what had existed before. It was this new system, originally worked by Indian slaves, which they carried onto the mainland.

Bartolomé de las Casas, who began his career at age eighteen in 1502 as a *conquistador* and became the first man to be ordained a priest in the New World, recounted the dreadful fate of the the Indians of the Caribbean and Nueva España, which presaged the tragedy that awaited all Indians. *The Devastation of the Indies* shocked many of his contemporaries, but it did not halt a flow of events that, time after time, in various guises and degrees, was to be played out in the conflicts that make up so large a part of North American history.

While Spain was a vigorous imperial power, it was less interested than either France or England in trade and made only limited attempts to colonize North America. Land, which exercised such a profound attraction on British colonists, was far less important to the Spaniards (or the French) than "reducing" the Native Americans and incorporating them into the Catholic fold. The ways in which they attempted to do this illuminate the psychological dimension of colonization.

The ventures of sixteenth-century Spain were tightly organized and centrally controlled. Consequently, we find Spanish state or church documents more prolific and important than personal records. But earlier generations of English-speaking historians used these "quarries" very little, so the integration of Spanish experiences into American history is relatively recent. It effectively began with the work of Herbert Eugene Bolton whose 1921 book, *The Spanish Borderlands*, inspired a new school of American historians.

The French role in American history came earlier to the attention of American historians primarily because the French were active closer to the main population centers of the British colonists and because their European rivalry spilled over into the New World where it also drew in Indian allies and ultimately exploded in the French and Indian War. In addition to these stirring events, French officials and explorers also left several works of vivid description that were easier to access than dry official reports. Notable among these was the account of Sammuel (as he spelled his name) de Champlain, who tells us how French imperial strategy emerged and sets out the impact of firearms on Indian societies. While little remains of the engravings of Jacque Le Moyne de Morgues, they offer a contemporary view of Indian villages. Canada naturally figured prominently because that was where most of the French efforts focused, but the French role on the southern Atlantic coast has now begun to attract the attention of scholars.

Most of the raw materials and finished products (the primary and secondary sources) we have, however, are by English-speaking colonists. The first materials that came to hand were accounts of the voyages of discovery. Historians treat these tales of derring-do, many of which were collected by Richard Hakluyt, as raw materials; but Hakluyt, as he freely acknowledged, thought of them as propaganda for English imperialism. The adventurers who wrote them were reporting to their sponsors, glorifying their achievements, justifying their failures, or encouraging further investment in their projects. Some of their reports, then considered confidential government documents, were suppressed, expurgated, lost, or destroyed. Others were not read until long after they were written, but in the aggregate they were the "mine" that earlier American historians worked.

English ventures into the northeastern coastal areas, Virginia and New England, were initially "privatized" and more freewheeling than those of the Spanish, so the early reports were more personal and less official. Captain John Smith was as accomplished and flamboyant an autobiographer as he was a soldier; his account of the Jamestown settlement was published in London in 1624 to great acclaim. Among modern historians, David Beers Quinn was particularly diligent and able in his efforts to compile records of the early English, Spanish, and French explorers such as *The Roanoke Voyages, 1584–1590* and *England and the Discovery of America, 1481–1620.*

Just as the early colonists were sensitive to the activities of Spaniards to the south of them, they were acutely aware of—indeed, often respectful of—

their Native American neighbors, on whom they depended for food and instruction. Because they often did not understand the nature of Indian society or government, they explained it in the only terms they knew, European titles and offices. They found "kings" everywhere they went. They were also struck by how much the Indian villages reminded them of English villages. Later historians, whose contemporary Native Americans were the semi-nomads of the Great Plains, forgot these early accounts. It was "politically correct" to portray all Indians as nomads to whom land had little value; this explained and partly justified colonial land-grabbing. It is only recently that we have revised our vision of the northeastern Indians to see them, as indeed they were, as settled villagers and farmers to whom land was supremely important. In some ways, our most valuable legacy of the very early times is the collection of what survived of the drawings made in 1585 by John White. They enable us to come as close as we ever will to seeing how the native population actually looked and in what they lived.

For the growing Atlantic coastal colonies of Englishmen, Scots, Irish, Germans, and others who were absorbed into what modern writers call British America, relatively abundant records exist. From the very early days, William Bradford is one of the most important chroniclers. His and other early accounts were concerned primarily with their tiny societies: how communities ruled themselves; drew up laws; developed legislatures, courts, and administrations; and, above all, worked their lands. While they focused on local events and papers, they had also to deal with their British patrons and crown authorities. It was not until nearly the end of the nineteenth century that the first historian tried to make use of the British records. Alexander Brown's *The First Republic in America* and *The Genesis of the United States* challenged scholars to go back to original documents, because, he argued, the British had suppressed some of the important information in the published materials. Partly stimulated by him, a virtual industry began to assemble and make available all the papers that survived.

Until fairly recently, these and other colonial records focused attention on the creation of institutions and the promulgation of laws and regulations by the settlers and away from relations with the native peoples. "Serious" historians paid little attention to the way people lived in the new environment.

The first ventures into social history—what people ate, how they were housed, the clothing they wore, how they kept warm, how they moved about—these and other things were stimulated by the writings of the colonists about themselves. One of the fundamental problems they faced was their identity: at first all of them would have said that they were Englishmen; but as years passed, we begin to hear three words that suggest a blurring of this identity. Increasingly, England was referred to as the "mother"; the new lands, originally thought of as "plantations," came to be identified as discrete "colonies" that were the "daughters" of England. In them we can discern a common historical experience of growing up and growing apart. That is everywhere the nature of colonialism.

Unveiling the minds of the men and women who lived not only long ago but in circumstances very different from ours is one of the most difficult challenges of history. When people write for others to read, they often magnify or diminish or even invent or suppress, so what they really thought or did is often elusive. Private diaries are therefore especially valuable. One of the more bizarre pieces of scholarship was the deciphering of the code used by William Byrd to record his frank and sometimes salacious account of his life. Landon Carter's diary, which is more accessible, has more recently become available and is the basis of an excellent account of eighteenth-century Virginia. From them we get rare and intimate views of the thoughts of prominent members of the Virginia aristocracy in the early eighteenth century. Letters and papers become increasingly rich and abundant later in the eighteenth century. Benjamin Franklin provided us with a gold mine; George Washington's diary is now available on a compact disc; most of the figures who played key roles in the Revolution—except for Samuel Adams, who destroyed his—have left us their voluminous correspondence. These sources can be mined both for accounts of the major events and also for intimate details of the writers' lives.

Such materials virtually forced an enlargement of the scope of American history. It was no longer enough to know that institutions existed; they had to be seen in action, and the purposes of the men who manipulated them had to be explored. Then overarching themes could be announced to attempt to bring order into the myriad details.

Frederick Jackson Turner in 1893 provided one of the most stimulating of these overarching themes in his interpretation of the role of the frontier in American history. He turned our attention away from the towns along the Atlantic coast toward the interior in the seventeenth century. For Turner, the actors who counted were the white settlers and those who urged them on or tried to control them; but inevitably those with whom they interacted, the Native Americans, had to be understood. That understanding was long in coming.

Until the twentieth century, historians were content to portray what Richard White called *The White Man's Indian.* "Colonial and early American historians have made Indians marginal to the periods they describe," wrote White. "They have treated them as curiosities in a world that Indians also helped create." Having rediscovered them, some historians echoed Bartolomé de las Casas in treating them as tragic victims, as merely the objects of white people's action. This was perhaps inevitable, since most of what we have to draw upon was recorded by whites who not only did not fully understand Indian culture but regarded Indians only as competitors or even terrorists. Progress toward regarding Indians as people with discernible interests, policies, and both successes and failures came only recently. Carl Bridenbaugh rediscovered perhaps the first of the anticolonial Indian statesmen in late seventeenth-century Spanish America and speculated on his possible role in Jamestown in the early eighteenth century, thus tying together three early American themes: the Spanish, English, and Indian. Many scholars have now joined a wave of revisionism regarding the Indians. Stimulated by a new generation of historians, we are beginning to see them vigorously interacting with one another and with white societies; recognizing them as actors rather than as objects; and seeking to comprehend their often complex attempts to protect their way of life through resistance, accommodation, diplomacy, and unification. By the middle of the eighteenth century, Indian "seers" were trying to formulate religiopolitical doctrines to stave off the religiopolitical attack of the Europeans. We know something about a few of them but virtually nothing about what were probably many more. The most imaginative and interesting speculation on what an Indian history might have been is Daniel K. Richter's *Facing East from Indian Country: A Native History of Early America.*

Inspired by their propaganda, infuriated by European disdain, and terrified by the destruction of their societies, Indians in Spanish-controlled

Pueblo country and in the Ohio Valley rose against and almost destroyed the Europeans. Theirs were the great rebellions against colonialism of the seventeenth and eighteenth centuries. They failed. What is interesting, and only now beginning to be appreciated, is how sophisticated and subtle were the diplomatic, military, and religious forces at work and how astute were some of the Indian leaders. For the first time since the very early colonial period we come up against true Indian statesmen like Shingas, Pontiac, and Joseph Brant who tried to formulate a way to reconcile the Indian desire for autonomy with European ambitions. Despite recent progress, we still look through this glass darkly.

Until recently, black people were seen both in the singular, one category rather than peoples of diverse languages, religions, societies, and even races, and only when they came "onstage" in the New World. Attempts are being made to understand better the often violent but frequently also subtle ways in which blacks were integrated into American history. What now emerges is that we have to deal with the African component of our history in three interlocking but different categories: who the Africans were, how the blacks got to the New World, and what happened to them when they arrived.

Little by little, they are now being traced back to their complex roots. African societies were as diverse as Indian societies, with some large and centralized empires commanding relatively huge armies while hundreds of village-based states had evolved not only legal systems but impressive public institutions; agriculture and long-distance trade were even more highly developed than those practiced in North America; some societies had developed impressive metallurgy and textile industries; and literacy, particularly in the Muslim areas, was widespread. Much of the African part of this story is still obscure.

Relatively speaking great progress has been made on understanding the Middle Passage, the slave routes to the New World. A number of accounts by white slave traders describe in horrifying detail what the voyages were like. Randy Sparks followed the path of two African slave traders who were themselves enslaved to put together a fascinating account of their experience in *The Two Princes of Calabar.* Robert Harms took the log of one vessel, the *Diligent,* to give a detailed but relatively benign picture of the passage.

Another account by a former slave, Olaudah Equiano (Gustavus Vassa), is unique.

We have almost nothing by blacks offering real insights into slave life in America during the seventeenth and eighteenth centuries and probably never will. We are forced to rely on accounts by contemporary whites and some later accounts by blacks. We are now able, however, to move out onto the fringes, to see how runaway communities (of which some were mixtures of blacks, Indians, and whites, known as maroons or *cimarones*) established themselves; how transracial ties were created; how religious thought permeated the hundreds of small and often isolated communities; and how visions of freedom survived to be expressed when opportunity afforded at the outset of the Revolution. A major start has been made, but this remains still almost uncharted territory in American history.

Analogous to but less tragic than the story of the blacks who came unwillingly to America was the migration of the many thousands of frightened, condemned, indentured, or exiled Scots, Englishmen, Irish, French, and German people. Abbot Emerson Smith led the way in the 1930s, and now scores of studies take up each ethnic and social group.

Once the immigrants from Europe and Africa arrived, it seemed possible for American historians to stop at the water's edge. What happened *here* was American history; what happened *there* was European, English, French, or Spanish. Contemporaries did not share this view. They were well aware of what was going on in and beyond Europe; already in the seventeenth century they were creating markets, exporting and importing. Later, in the eighteenth century, popular movements in England, Scotland, Ireland, France, Corsica, and even distant Poland would attract their attention. Only recently have we rediscovered the "outreach" of our history. Franco Venturi, Pauline Maier, and Bernard Bailyn have each contributed to broadening this view.

Economic affairs were already being discussed in the seventeenth and eighteenth centuries. Adam Smith wrote perceptively about colonialism in *The Wealth of Nations,* and he was by no means the first or only one to do so. Mercantilism, against which he inveighed, was in part an attempt to figure out the proper relationship of the "mother" country to the "daughter" colonies.

Religion as expressed in cultural orientation was at the forefront of Spanish, French, English, and—as we now know—Native American thought. It was what brought many of the Europeans to the New World and was one means through which Indians in the eighteenth and nineteenth centuries sought salvation. Each major group sought to protect itself from outside intrusion, and particularly from attempts by the Spanish, French, and English to destroy Indian beliefs. Catholic missions or *reducciones* in La Florida, sometime around the middle of the seventeenth century, contained perhaps as many as 30,000 Indians; New England Protestant "praying towns" were smaller partly because the people who might have lived in them had been killed or chased away, but at their height, they may have reached over 2,000. The French efforts fell somewhere in between and were scattered more diffusely across a far larger expanse around the Great Lakes region.

The most interesting, longest-lasting, and largest-scale field of this cross-cultural influence is the experience of the American black community. Black people's modification and adoption of Christianity is particularly fascinating because, on the surface, it appears so illogical that blacks would adopt what they and the Indians regarded as "white men's religion." How this happened, who promoted it, why they did, and what the blacks found in Christianity are among the most subtle problems of American history. They have echoes today in the attempts by large numbers of blacks to find a different religious and cultural orientation in their interpretation of Islam. But we are learning that these ventures also have African roots: large numbers of Africans had become Muslim and even larger numbers were certainly influenced by Islamic ideas. Moreover, at least some African religions were monotheistic and lent themselves to being influenced by either Islam or Christianity.

In the period that led up to the Revolution, books, articles, and collections of papers are uncountable. Several, however, merit special attention. Pauline Maier's, noted above, is one. Another of Bernard Bailyn's studies, in addition to his *Ideological Origins,* is his focus on the conflict between "radicals" and "conservatives" in Boston in *The Ordeal of Thomas Hutchinson.* Arthur M. Schlesinger studies the "newspaper war" in *Prelude to Independence;* and Hiller B. Zobel gives an almost blow-by-blow account of the last days before the Revolution.

Among the newer tools available to historians is archaeology. Archaeology not only enables us to understand what sort of buildings early peoples built and what sort of tools and weapons they used, but also what they ate and how hard they worked.

Climatic change has also shed new light on events: in the "little ice age," the migration of cod caused the first European fishermen to come to America. On the North American landmass, it caused a decline in agricultural productivity; consequently, Indian societies scattered from previous urban concentrations like the surprisingly large town of Cahokia, where perhaps as many as 20,000 people lived. From this new knowledge, we have had to revise the earlier and more comfortable notion that, since the Indians were nomads, taking Indian land did not much matter.

Certainly no society has ever expended more time, talent, and treasure on learning about itself than the American. The task is far from complete. And now we are finding that in almost every way, we have had to revise, enrich, extend, and internalize our view of the American past. My purpose in writing this account is to put together an overall view of the process during which the America we have so fortunately inherited was born.

PART I

Europe and Africa Come to America

CHAPTER 1

The Native Americans

Who were the Native Americans? The Spanish, French, and English explorers were perplexed by that question. Their first assumption that the natives were Chinese was soon abandoned; the natives obviously were not European and did not seem to be African either. The explorers could not think of any other possibilities. William Strachey spoke for them in 1612 in his *Historie of Travell into Virginia Britania:* "It were not perhappes too curyous a thing to demaund, how these people might come first, and from whome, and whence, having no entercourse with *Africa, Asia* nor *Europe,* and considering the whole world, so many years, by all knowledg receaved, was supposed to be only conteyned and circumscrybed in the discovered and travelled Bowndes of those three."

The Indian societies he saw, Strachey would have been astonished to learn, were formed by thousands of years of migration, splitting apart, rejoining, exchanging mates, settling, and adapting—essentially the same process that shaped European lives and culture. Just as Europeans were products of the migrations of western Asians, so the Native Americans were descendants of migrants from eastern Asia. And just as the Europeans' languages give a view of their history, so American Indians' languages illustrate their background.

The first Indians the Spaniards encountered in what they named La Florida spoke dialects of a language known as Muskhogean. It was one of 583 languages that have so far been identified as spoken by natives in North and South America. Linguists trace it back to a tongue they call Amerind. Linguistic evidence points toward northeastern Asia as their "origin." What the spread of language indicates has now been confirmed by genetic studies. Together they suggest that ancestors of the American Indians probably began crossing to North America roughly 30,000 years ago. Climatologists now believe that from about 60,000 years ago, Asia and North America were joined at what is now the Bering Strait and archaeologists have found evidence of human settlements in northeastern Siberia from about 40,000 years ago. So it was possible for humans and animals to walk across a land bridge, which geologists call Beringia. They began to do so because, although much of North America was covered by huge glaciers and sheets of ice, parts of Alaska enjoyed a relatively mild climate. Even in the coldest times, there was a corridor of relatively open countryside that channeled movement of animals and men to the south. Then, about 10,000 years ago, with the coming of what geologists term the Holocene, a warmer epoch, so much ice melted that the sea rose as much as 120 meters and submerged the land bridge. Those people who had already made the passage from Asia profited from the melting of the vast sheets of ice to move inland and further south. By about 14,000 years ago, some had reached Patagonia and others had spread over both continents.

After their arrival in the New World, the speakers of Amerind spread out across almost the whole of North and South America. Pockets of other languages remained in the American Southwest and the Canadian Northwest. These were derivatives of an Old World language now called Na-Dene and were spoken in the far north of the continent where what is known as Eskimo-Aleut was the common tongue. Then, for thousands of years as families and small clans moved apart from one another, they acquired different habits, adapted to different environments, and made changes in the way they spoke. We can see how this process works by delving back into the past of our own language. Shakespeare's English is intelligible to us although it contains expressions we no longer understand. Middle English, spoken a few centuries earlier, is arcane. Farther back and farther away, English's close cousins—Spanish, Italian, French, and Portuguese—although sharing some vocabulary and much syntax and grammar, were already largely foreign. If we move yet farther afield to languages in our same Indo-European family, Russian, Persian, Greek, Armenian, and Sanskrit appear almost totally alien. So it was with the

Indian languages. Over thousands of years and a large stretch of geography, each society elaborated from the common ancestor its own way of thought and speech.

When the French explorer Sammuel Champlain landed on the Saint Lawrence in 1608, he encountered a people speaking Algonquian, a language related to the language spoken far to the south in Virginia. What that seemingly unlikely fact tells us is that the two groups must have originally been one people; as one or both migrated, they first became neighbors and finally strangers, just the way our European ancestors did.

The largest, most sophisticated, and most warlike of the northeastern Indian societies were speakers of Iroquoian. When the explorers first encountered them, they had divided into five nations but were still linked by a confederation which they called Haudenosaunee (Iroquoian for "the Long House"). The 6,000 or so members of the confederation were tribes known as the Kaniengebaga (or, as their enemies called them, the Mohawks), the Oneidas, Onondagas, Cayugas, and Senecas. Related to them and speaking dialects of Iroquois were the Cherokee and Tuscarora, who had earlier migrated southward.

Another family of languages was Siouan, spoken by peoples who dominated the Piedmont southward from Maryland. They included the Catawba, Saponi, Tutelo, Occaneechee, and Cheraw of the Carolinas and the Creek of what became Georgia, as well as scores of smaller, now mainly forgotten groups.

A fourth collection of societies, the North American native people the Spaniards first knew, spoke varieties of Muskhogean and inhabited the Gulf coastal region from Georgia to the Mississippi.

Speakers of these four groups of languages, the Native Americans living east of the Mississippi, probably numbered about 2 million on the eve of first contact with Europeans.

As they spread out over North America, the Indians developed such distinctive characteristics as to seem alien to one another, just as the European and African nations did. Frequently clashing with one another as they sought to defend or enlarge territories, each emphasized its uniqueness. Many of the Indian names for themselves meant "*the* people" or "the *real* people"; that is, each group asserted that it was fundamentally unlike its neighbors, who were not "real" people. As the Spanish, French, and English invaders quickly realized, these differences had given rise to bitter and long-standing hostilities that made it virtually impossible for the Indians to combine against Europeans. With their eyes firmly fixed on their neighbors, time after time, group after group, Indians would welcome foreign invaders and attempt to use them in local struggles. This was also the experience during the European conquest of Africa and Asia: European military force often provided only "stiffening" to the mainly indigenous armies that established European rule. Moreover, individual tribal societies could never match the manpower that national consolidation in Europe gave Britain, France, and Spain. Fragmented as the vast areas of Africa, Asia, and the Americas in fact were, each was outnumbered as well as outgunned by the invaders.

While diverse and often mutually hostile, the societies also retained many common features. Since no Native American society north of Mexico produced or bequeathed to us its own record, we glimpse them only in the blurred picture presented by European visitors. None of the first Europeans, of course, spoke native languages; when a few did learn some dialect, they rarely thought it worthwhile to try to capture the thoughts, fears, or hopes of any Native Americans.

The most sophisticated of the observers, the Jesuit priests in the areas controlled by Spain, learned more but also were more hostile to Indian culture. A striking example is given in the account of the Jesuit missionary Juan Nentuig:

> The ceremonies of their heathenish weddings are not fit to be described in detail. I shall only mention the more decent. They gather together, old and young, and the young men and marriageable women are placed in two files. At a given signal the latter begin to run, and at another signal the former to follow them. When the young men overtake the young women each one must take his mate by the left nipple and the marriage is made and confirmed. After this preliminary ceremony they devote themselves to dancing. . . . Then all at once they take mats of palm tree leaves, which are prepared beforehand, and without further ceremony each couple is placed on a mat, and the rest of the people go on rejoicing.

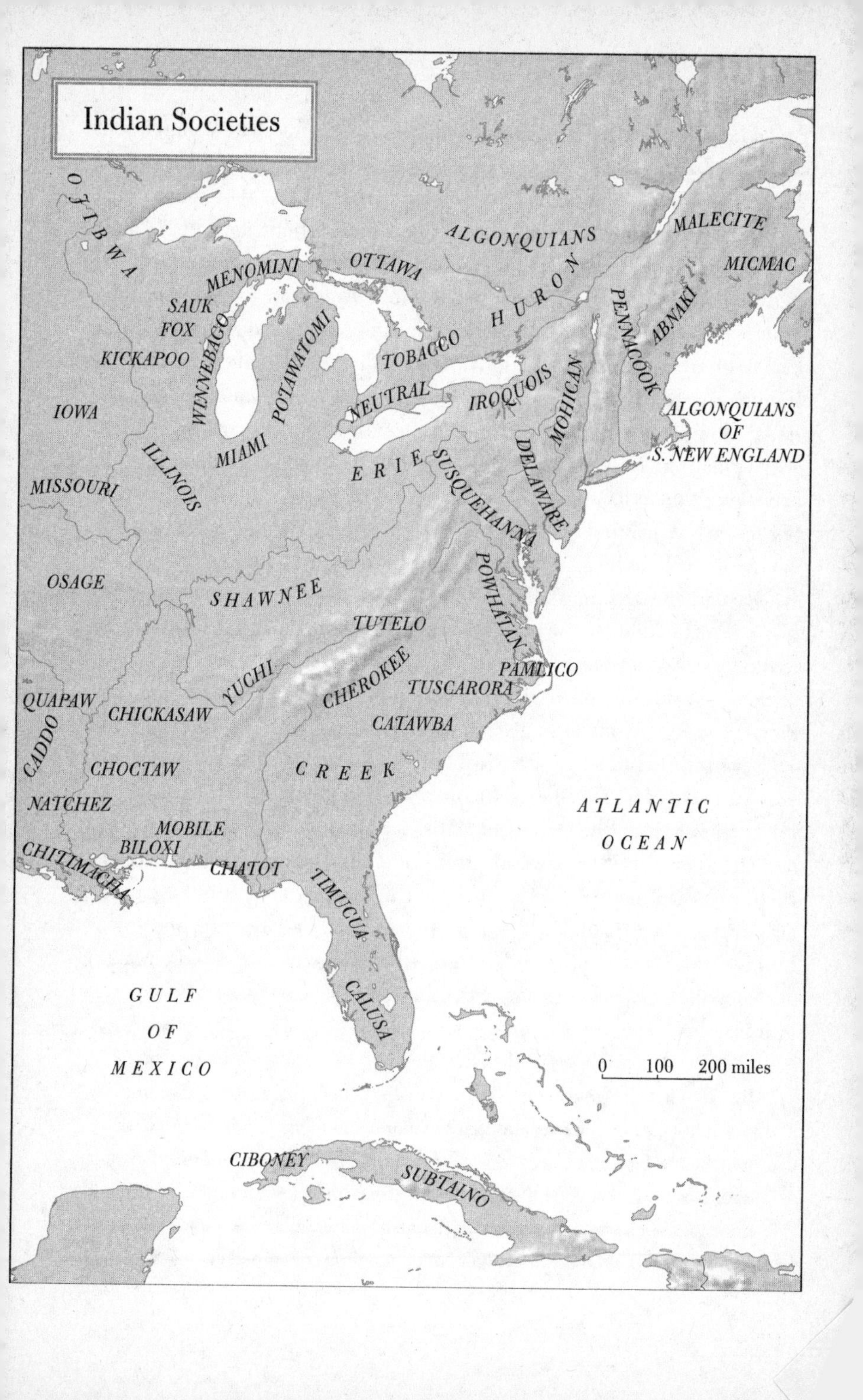
Indian Societies
OJIBWA
MENOMINI
OTTAWA
ALGONQUIANS
MALECITE
MICMAC
SAUK
FOX
WINNEBAGO
POTAWATOMI
HURON
PENNACOOK
ABNAKI
KICKAPOO
TOBACCO
MOHICAN
IOWA
NEUTRAL
IROQUOIS
ALGONQUIANS
OF
S. NEW ENGLAND
ILLINOIS
MIAMI
ERIE
SUSQUEHANNA
DELAWARE
MISSOURI
OSAGE
SHAWNEE
POWHATAN
TUTELO
PAMLICO
YUCHI
CHEROKEE
TUSCARORA
QUAPAW
CHICKASAW
CATAWBA
CADDO
CHOCTAW
CREEK
NATCHEZ
ATLANTIC
OCEAN
MOBILE
BILOXI
CHITIMACHA
CHATOT
TIMUCUA
CALUSA
GULF
OF
MEXICO
0 100 200 miles
CIBONEY
SUBTAINO

The priests considered dances, feasts, and even athletic competitions almost as bad as sex: to them all ceremonies except their own were satanic. Even an outburst of joy when rain fell on the parched desert seemed sinful.

Most of those who could have told us much about the Indians were simply not interested. Perhaps the longest and most intimate contact between a group of Europeans and a Native American was at Plymouth. There, in 1620, William Bradford met a man he called Squanto. This man had visited England and had, apparently, a considerable command of English. As Bradford describes it, Squanto had a warm and friendly relationship with the Pilgrims. By teaching them how to plant corn and how to survive in cold New England, he quite literally saved their lives; and he stayed with them for over thirty years—until about 1653. But Bradford, in his otherwise instructive writings, gives no hint of what Squanto might have related about his people. Obviously, Bradford did not care.

Remarkably different—indeed, for the late seventeenth century, probably uniquely different—was the Virginia colonist Robert Beverley. He wrote, "I have been at several of the Indian towns and conversed with some of the most sensible of them in that country, but I could learn little from them, it being reckoned sacrilege to divulge the principles of their religion." So Beverley did not sit passively by. He took an opportunity to break into a *quioccasan* (shrine) to see what it contained. He found it almost as bare as a Puritan church, but he examined an ossuary and an idol (variously known as Okee, Quioccos, or Kiwasa). From other Indians he had learned that each town had its own shrine. Unsatisfied, Beverley sought out a particularly intelligent Indian, and "seating him close by a large fire, and giving him plenty of strong cider which I hoped would make him good company and openhearted," plied him with questions. The response he got was remarkable:

> [his people] believed God was universally beneficent, that his dwelling was in the heavens above, and that the influences of his goodness reached to the earth beneath. . . . God is the giver of all good things, but they flow naturally and promiscuously from him; that they are showered down upon all men indifferently without distinction; that God does not trouble himself with the impertinent affairs of men, nor is concerned at what they

> do, but leaves them to make the most of their free will and to secure as many as they can of the good things that flow from him. That therefore it was to no purpose either to fear or worship him, but on the contrary if they did not pacify the evil spirit and make him propitious, he would take away or spoil all those good things that God had given and ruin their health, their peace, and their plenty by sending war, plague, and famine among them. For, said he, this evil spirit is always busying himself with our affairs and frequently visiting us, being present in the air, in the thunder, and in the storms.

I know of no other early account in which a white colonist tried so carefully to understand Indian belief or, indeed, other aspects of Indian life. Much of what we know of Indian life comes from later centuries. But by then, Indian societies and culture had been violently transformed. The older people, who were the carriers and disseminators of tradition, died before they could perpetuate their lore; societies imploded and mingled with survivors of alien groups so that the sense of being "a people" withered. And the whites, having filled the lands with their own kind, moved ever westward. It is what white people later saw or heard about in the West that, largely unconsciously, we see as "Indian" in our mind's eye. What our more recent ancestors learned about the nineteenth-century nomads of the Great Plains has been further programmed in our minds by the cinema. Sitting Bull, Geronimo, and Crazy Horse are the quintessential Indian warriors, flashing across the screen on their Spanish ponies, hot after buffalo or white settlers. Even after stripping away the stereotype, we can see that the actual way of life of the hunters of the Great Plains was very different from that of the farmers met by the colonists on the Atlantic coast in the sixteenth, seventeenth, and eighteenth centuries. So I begin with a basic feature of their lives, agriculture.

Native Americans began the transition from hunting and gathering to agriculture at roughly the same time as Old World farmers but faced a more difficult task. Some of the plant species with which they experimented, notably the potato (*Solanum tuberosum*), of the nightshade family, contained poisons that had to be extracted before the plants could be eaten. It must have taken generations of experimentation and selective cultivation to turn the potato into a vegetable that was safe to eat. All the Europeans who met the Indians were impressed by their knowledge of edible and medicinal plants. In plants they were relatively rich, but in animals they were poor. They had no animals comparable to the goats, sheep, horses, donkeys, wild asses, camels, or elephants of the Old World. Their only domesticated animal was the dog.

The most distinctive Indian food crop, corn (in Algonquian, *maize*), was first domesticated in Mexico about 3000 B.C.E. and slowly made its way all over both North and South America, growing in an astonishing variety of climates and altitudes. The English marveled at its fecundity: it produced a far higher yield than European food grains. In 1587, in "A Briefe and True Report of the New Found Land of Viriginia," Thomas Hariot exclaimed, "It is a graine of marvellous great increase: of a thousand, fifteene hundred, and some two thousand folde." For the East Coast Indians, corn was life: they pounded it to make bread, boiled it to make gruel, and treated it with wood ash to make hominy.

They also ate a variety of tubers, gourds, pumpkins, and squashes (in the Narragansett dialect of Algonquian, *askútasquash*); they demonstrated their feel for agronomy by planting the lima bean and the kidney bean in the same fields with squash and corn to enrich the soil. And just as they combined these plants in the fields, they also did in the cooking pot, making mixtures of vegetables like succotash (Narragansett, *msiquatash*). Farther north, Indians harvested the misnamed "wild rice" (*Zizania aquatica*) from marshes and ponds. Wild rice had the great advantages of being highly nutritious, growing in very cold areas, and being sown simply by scattering some of each crop back into the water.

Having no draft animals, Indians did not use the plow. Neither did they have iron with which to make alternative implements. Their common tool was a pointed stick with which they punched in the earth holes into which seeds were dropped; they weeded furrows with a sort of hoe and dug with a tool comparable to a spade. With these wooden tools, it was easy to plant around trees or stumps, so although they cleared considerable stretches of agricultural land, much of their produce came from wooded terrain. Rather than trying to "fight" the trees as the white colonists would do, the Indians made a virtue out of forested land: they had learned that some shade made planting easier because the soil remained moist and soft. As they taught the Pilgrims, they fertilized land (and got rid of weeds) by burning brush and by burying fish heads in the hillocks in which they placed seed. Like contemporary European farmers, they let exhausted fields lie fallow temporarily.

Among the East Coast Indians, as among many tribal peoples, agriculture was primarily woman's work. While the women planted, weeded, harvested, and processed vegetables, the men hunted and fished. Wild turkey, deer, bear, and a variety of smaller animals they found in the forests; and beavers, otters, and fish from the rivers provided a high-protein, nutritious, and abundant diet.

It followed that Indians were generally healthy. As Helen C. Rountree and Thomas E. Davidson comment, "Indian people of both sexes were healthier than early seventeenth century English people . . . [who] were almost continually ailing: their diet was poor in fruits and vegetables for most of the year." Indian men, at about 5 feet 7½ inches, were taller than Englishmen by an average of 1 inch; Indian women on average were over 2 inches taller than Englishwomen. The early English visitors and colonists marveled at the size and strength of the Indians. For example, William Strachey described a meeting in what became Maryland where "inhabite a people called the *Sasquesahonougs*. . . . Such great and well proportioned men are seldome seene, for they seemed like Gyantes to the English [with voices] . . . sounding from them (as yt were) a great voyce in a vault or caue as an Eccoe."

Along the Saint Lawrence, Champlain similarly found all the people

> well-formed and proportioned in body, some of the men being very strong and robust. And there are also women and girls who are very beautiful and attractive in figure, coloring (although it is olive) and in features, all in proportion; and their breasts hang down hardly at all, unless they are old. Some of them are very powerful and of extraordinary height. The women are equally well-formed, plump, and of a tawny complexion.

Particularly striking to the early observers was the strength of the Indians and their skill with their primitive weapons. George Percy described a test of the Indians' strength and skill with a bow and arrow in these words:

> One of our gentlemen having a target [a shield] which he trusted in, thinking it would bear out a slight shot, he set it up against a tree, willing one of the savages to shoot, who took from his back an arrow of an ell [45 inches] long, drew it strongly in his bow, shoots the target a foot through or better; which was strange, being that a pistol could not pierce it.

Even toward the end of the seventeenth century, when Indians had suffered much from the effects of imported disease and alcohol, Robert Beverley waxed lyrical when he described Indians who, he said, have the

> cleanest and most exact Limbs in the World: They are so perfect in their outward frame, that I never heard of one single *Indian,* that was either dwarfish, crooked, bandy-legg'd, or otherwise misshapen. . . . Their women are generally beautiful, possessing an uncommon delicacy of shape and features and wanting no charm but that of a fair complexion. . . . They are remarkable for having small round Breasts, and so firm, that they are hardly ever observ'd to hang down, even in old women.

After years of travel among Indians in the Carolinas around 1700, John Lawson found that women's "Breaths are as sweet as the Air they breathe in, and the Woman seems to be of that tender Composition, as if they were design'd rather for the Bed then Bondage."

About men, the opinions were less lyrical. When Thomas Jefferson circulated the manuscript of his *Notes on the State of Virginia,* he received a comment by a French naturalist, the comte de Buffon, that although the Indians are generally larger and taller than Englishmen, "their organs of generation are smaller and weaker than those of Europeans." (The opposite belief about black males would later enter the mythology of race believed by southern whites.) Machismo aside, John Lawson in 1701 and James Adair in 1765 agreed that "they seldom saw a crippled Indian, never a blind one."

It was not only their diet that gave the Indians their health; it was also the simplicity and economy of their mode of living. We can single out certain of their advantages. First was cleanliness. Indians, at least those close to the myriad waterways of the Virginia coast and the Chesapeake, are recorded as bathing every morning, whereas sixteenth- and seventeenth-century Englishmen, who considered bathing unhealthy, generally bathed only about once a year. In drafty English houses, they had a point; but filth provoked a variety of skin diseases. The Virginia Indians, the Delaware, the Creek, and other societies enjoyed steam baths in buildings rather like Finnish saunas where groups of neighbors gathered to gossip and refresh themselves. Robert Beverley wrote in 1705 that "in every town they have a sweating house, and a doctor is paid by the public to attend it."

Since the English went weeks or months without taking off their clothes, even to sleep, they had to resign themselves to hordes of bodily pests. In contrast, the Indians provided a poor habitat for vermin, since they had little body hair and wore few clothes. Champlain saw 500 or 600 Indians on the New England coast "who were all naked, except for their private parts, which they cover with a little piece of doe-skin or seal-skin. The women also cover theirs with skins, or white leaves, and all have the hair well combed, and braided in various ways.

The common dress of a man was a fur mantle that the English called a "matchcoat" (from the Odjibwa dialect of Algonquian, *matchigode*). This cloak made the Indians appear to the English like the despised Irish in their *falaing*. In everyday working situations Indian men normally wore just a loincloth, and children wore nothing. Women were usually bare-breasted but normally wore a sort of loincloth, which the English called a flap. However, while picking fruit, working in a cornfield, or bathing, they dispensed with all clothing, as the traveler William Bartram rather closely observed of a group of young Cherokee women, "disclosing their beauties to the fluttering breeze, and bathing their limbs in the cool, flitting streams." Another somewhat less poetic but even more observant Englishman, William Fyffe, remarked in 1761 that both sexes shave pubic hair "off their privities." Travelers were fascinated by what they saw. John Lawson remarked: "As for their Privities, since they wore Tail-Clouts, to cover their Nakedness, several of the Men have a deal of Hair thereon. It is to be observ'd that the Head of the *Penis* is cover'd (throughout all the Nations of the *Indians* I ever saw)."

What was true of clothing was also true of housing: simplicity was the rule. In a pamphlet commissioned by Lord Baltimore for the use of colonists going to Maryland in 1635, the author wrote that Indian "houses are made like our Arboures, covered with matts, others with barke of trees, which defend them from the injury of the weather." So adept at defending against "injury of the weather" were they that Captain John Smith reported that the colonists often preferred these houses (Algonquian, *wigwangs*) to English-style houses.

Even in the relatively warm Southeast, as an early Spanish visitor wrote, when "They shut the very small door at night and build a fire inside the house . . . it gets as hot as an oven, and stays so all night long so that there is no need of clothing." In contrast to the houses in which the colonists froze, the surveyor and presumed expert on Indians, John Lawson, found them "as hot as Stoves." Chickasaw round houses reminded James Adair of Dutch stoves. They were, he wrote, woven around upright posts and daubed "all over about six or seven inches thick with tough clay, well mixt with withered grass."

Houses were also intended, like the garrison houses of the New England frontier, for defense. Winter was generally a time of truce, but summer was a time of raids, so, as James Adair commented, summer houses also served as "a savage philosopher's castle, the side and gables of which are bullet proof."

Indians' materials and designs, like their clothing, were similar to those of the Irish. To build their houses, Indians first set stout poles at the corners. They then cut tall bamboo-like poles 15 or 20 feet long and set them along the line where they wanted a wall. Then they wove into that line lighter saplings or reeds, lashing each piece into position with leather thongs or strips of wood. In effect, they wove the walls and roof like a basket. Walls could be built to almost any length, since each section was self-supporting. When the wall reached the desired height, two or more builders would climb up on each facing wall until it bent inward under their weight. Where the poles met, they were tied together and the roof was interlaced like the sides. This formed a building somewhat similar to the frame of a Mongol or Turkman yurt. And, like the yurt, the house was then sealed against rain and cold by a covering of skins, bark, or—as in the Irish post-and-wattle house—with thatch. Robert Beverley observed that the

"Chimney, as among the true Born *Irish,* is a little hole in the top of the House" placed directly above the fire pit where the cooking was done. Houses like this were built up and down the East Coast. As with Irish houses, the English visitors remarked, building was relatively simple and quick, taking just a few days of labor and lasting for about ten years. After that time, the family probably needed to find a new neighborhood, as it would have used up all the convenient firewood and might also have outgrown the house or separated from the relatives with whom the house was shared. Also, despite the smoke, the furs or thatch would have been invaded by rodents and other pests.

Generally, Indian houses were not scattered across the landscape as they were among the colonists; the East Coast Indians from the Gulf of Mexico to Newfoundland were townspeople. Although later British colonists insisted that Indians were nomads (and so not really attached to the land the colonists wanted to appropriate), their eyes told them otherwise. As the pamphlet commissioned by Lord Baltimore for the use of colonists in 1635 admitted, "They live for the most parts in Townes, like Countrey Villages in England." Earlier, in 1539, the country of the Florida Indians was described as "greatly inhabited with many great towns and many sown fields which reached from one to the other." On the site of modern East St. Louis before the arrival of the Spaniards, the city we call Cahokia had a population of "upwards of 20,000." Unlike the Great Plains nomads whom the white men would meet centuries later, East Coast Indians were settled farmers and villagers.

Villages were usually surrounded by a pale or stockade of upright logs, as we can see in a drawing made in 1585 by John White. Indian villages struck some observers as miniature counterparts of the city-states of medieval Italy and were as frequently drawn to war by their governments. The author of *A Relation of Maryland* described the government of the Indian societies known to the English as "Monarchiacall, he that governs in chiefe, is called the Werowance, and is assisted by some that consult with him of the common affaires, who are called Wisoes: They have no Lawes, but the Law of Nature and discretion."

Discretion exercised a powerful restraint on Indian statecraft. The power of the werowance (or, as most whites called him, from the

Narragansett, the sachem; in the Delaware dialect, *sakiam;* in Micmac, *sakumow;* and in Pnobscot, *sagamo*) was limited in four ways. First, "the Werowance himselfe plants Corne, makes his owne Bow and Arrowes, his Canoo, his Mantle, Shooes, and what ever else belongs unto him, as any other common Indian." There was little distinction of wealth; if a "chief" had more, he was also expected to give more away. As one modern historian has said, "Chiefs were men with large responsibilities and few resources."

Second, the leader made no decisions without the approval of his counselors. As William Penn wrote, "nothing of Moment is undertaken, be it War, Peace, Selling of Land or Traffick, without advising with them." Major decisions required the unanimous approval of the whole community, after hours or even days of discussion and speechmaking that seem like the accounts of Greek assemblies in Thucydides. From this we get the concept and probably the word "caucus." The Algonquian *cau'-cau-as'u* became under Captain John Smith's pen *caw-cawassough* with roughly the meaning we give it. As William Penn poetically wrote of the Shawnee chiefs, "they move by the breath of their people." In 1755 Colonel James Smith was captured by the Indians and had a chance to observe their organization under rather intense circumstances. He reported that the Indian chief was:

> neither a supreme ruler, monarch or potentate—he can neither make war or peace, leagues or treaties—He cannot impress soldiers, or dispose of magazines.—He cannot adjourn, prorogue or dissolve a general assembly, nor can he refuse his assent to their conclusions, or in any manner controul them. . . . The chief of a nation has to hunt for his living, as any other citizen.

Perhaps the best observation on the power (or lack of it) of Indian chiefs, was made among the Delawares in the eighteenth century by Reverend David Zeisberger. A chief, he wrote, "may not presume to rule over the people, as in that case he would immediately be forsaken by the whole tribe, and his counsellors would refuse to assist him. He must ingratiate himself with the people and stand by his counsellors. Hence, it is that chiefs are generally friendly, gracious, hospitable, communicative, affable and their house is open to every Indian." If a chief violated that custom, he paid heavily, perhaps even with his life. When Pontiac, the great hero of the revolt of 1763 by the Indian peoples of inner America against the British, made the mistake of acting like an autocratic ruler, his followers killed him. As Richard White has written, his assassination was "a monument to the limits of chieftainship."

Third, unlike European monarchs, werowances were not war leaders. That role was vested in a man known on the Chesapeake as a *cawcawaassough* or, as the English called him, a cockoroose. Among the Seneca he was the "great war soldier" (Iroquoian, *hos-gä-ä-geh'-da-go-wä*). The attributes of the two positions of leadership shaped their holders: the war leaders were expected to be violent, impulsive, brave, and occasionally cruel, whereas the political leader was to be moderate and wise. Promotion from war leader to political leader was expected to effect this transition. How such men were chosen can be reasonably guessed. A successful hunter was always sought out by the aged, the sick, and the indigent. Moreover, the less capable tended to gravitate toward men with reputations as good shots and skilled trackers. As Harold Driver commented, "This was the lowest level of leadership and political organization."

From time to time among the Indians, as among Europeans, a leader arose who transcended these restraints. Such a man, if we can believe William Strachey, was the first ruler encountered by the English, Powhatan. "Powhatan" was probably not a given name but a title—the English often referred to him as "the Powhatan," as they were accustomed to refer to Irish clan leaders. Both Strachey and Captain John Smith thought of him as an emperor (Algonquian, *mamantowick*). Powhatan did, indeed, rule over twenty-eight villages containing perhaps 13,000 people in an area about a quarter the size of what became the state of Virginia. He was obviously keen to impress the English, and he certainly did. They portrayed him as the contemporary English imagined the Ottoman sultans, Persian shahs, or Mughal emperors, inflicting "Oriental" tortures and having "as many women as he will, and hath (as is supposed) many more than one hundred. All which he doth not keepe, yet as the Turke in one Saraglia or howse." If Strachey's portrait was accurate, Powhatan's despotic government was certainly an exception among the East Coast Indians.

The fourth difference between European and Indian leaders was that, unlike the European monarchs with whose powers the settlers were familiar, the werowances did not "own" even a part of their kingdoms. No matter how powerful or popular they were, they had no right to dispose of land, which was regarded as the inalienable possession of the whole people. But it was convenient, indeed crucial, for the English settlers to believe that a "chief" had this right, since they, and the Americans after the Revolution, bribed or forced chiefs to sell them the communal lands.

Rights over land among most Indian groups divided into two separate but compatible categories: usually by decision of a clan or village, an individual acquired, often for his lifetime, *use* of a given piece of land. This right was more or less what in English law is known as usufruct rather than ownership. As Lewis H. Morgan wrote, "No person in Indian life could obtain the absolute title to land, since it was vested by custom in the tribe as one body, and they had no conception of what is implied by a legal title in severalty with power to sell and convey the fee." For Indians, "ownership" of a given plot amounted to a sort of trusteeship.

Even the Dutch and Swedish colonists who recognized Indian titles did not really comprehend the Indian concept; nor did William Penn. Penn, who has rightly been considered the most humane of the English proprietors, did not grasp, and certainly did not want to grasp, the Indian idea of collective ownership. Other English and colonial authorities completely disregarded it and ultimately forced the Indians to accept, even if they did not understand, something like the English legal definition—so that between 1630 and 1767, the one society of the Delawares turned over nearly 800 deeds of lands to incoming whites. With the Indians, at least at first, allowing whites to use their land did not entail alienating it but more or less corresponded to hospitality.

How did the Indians' concept of ownership fit into their concept of their role on earth? This was not a question most whites asked, but at least some of the missionaries did. The Indians, wrote the Moravian missionary John Heckewelder, who lived among them for fifteen years from 1771 to 1786, believed that:

> the Great Spirit . . . made the earth and all that it contains for the common good of mankind; when he stocked the country that he gave them

> with plenty of game, it was not for the benefit of a few, but of all. Everything was given in common to the sons of men. Whatever liveth on the land, whatsoever groweth out of the earth, and all that is in the rivers and waters flowing through the same, was given jointly to all, and every one is entitled to his share. From this principle hospitality flows as from its source. With them it is not a virtue, but a strict duty.

It was a dangerous duty.

Had the Arawak Indians not salvaged the cargo of Columbus's *Santa María* and taken in the nearly drowned crew, Spain's venture in the New World might have turned out differently or at least been delayed. Elsewhere, Indians' hospitality to Europeans would be repeated time after time, and often with results painfully destructive to the Indians. "Upon the advent of the European race among them," Lewis Morgan wrote, "it was also extended to them." That certainly was the experience of the early colonists. Captains Philip Amadas and Arthur Barlowe lovingly detail the generosity of the first Indians they encountered in 1584. A generation later the settlers at Jamestown would have starved without the Indians' generosity. No lover of the Indians, Captain John Smith remarks that in 1607, when the settlers' fortunes were at a low ebb, "it pleased God (in our extremity) to move the Indians to bring us Corne, ere it was halfe ripe, to refresh us, when we rather expected that they would destroy us." On his tour of the Virginia hinterland, when he and his little group were virtually defenseless, Smith remarks on "the people in all places kindely intreating us, daunsing and feasting us with strawberries, Mulberies, Bread, Fish, and other of their Countrie provisions."

Hospitality was universal and enduring. Even Hernando de Soto experienced it during his rampaging tour through La Florida in 1539; there the Indians turned inhabitants out to give their houses to the visitors. A century later, in Maryland, Lord Baltimore's incoming colonists benefited from a similar gesture by the Yoacomaco Indians, who turned over to them what became St. Mary's. Accounts everywhere echo comparable acts of generosity right down to the end of the pre-Revolutionary era. During his captivity by the Indians in the 1750s, James Smith was taught that "when strangers come to our camp, we ought always to give them the best that we have." Not only the English but also some of the French Jesuits saw

> in these savages the fine roots of human nature, which are entirely corrupted in civilized nations. . . . Living in common, without disputes, content with little, guiltless of avarice . . . it is impossible to find people more patient, more hospitable, more affable, more liberal, more moderate in their language. In fine, all our fathers and the French who have lived with the savages consider that life flows on more gently among them than with us.

And far to the west, in 1766, when Jonathan Carver visited the Dakota tribes of the Mississippi, he observed, "No people are more hospitable, kind, and free than the Indians." About the same time, James Adair remarked of the southern tribes, the Cherokees, Choctaws, Chickasaws, and confederated Creek tribes, "They are so hospitable, kind-hearted, and free, that they would share with those of their own tribe the last part of their own provisions, even to a single ear of corn. . . . To be narrow-hearted, especially to those in want, to any of their own family, is accounted a great crime, and to reflect scandal on the rest of the tribe."

With such widespread sharing of attitudes and customs, there must have been much movement among Indian societies. Indians, as Paul A. W. Wallace has written, "laced our hills and valleys with a complex system of paths which not only drew local communities together but spread out into a vast continental network." On the waterways, Indians used dugout canoes to move from one settlement to another. Their lack of wheeled vehicles, animal-borne transport, or watercraft larger than canoes caused Indian societies to be economically localized, but despite this limitation they certainly had extensive contacts.

One of the things eastern Indians shared among themselves, in a way similar to West African societies, was money in the form of seashells. Having access to the "mines," the beds along the coast, gave some Indians the kind of advantage that access to gold or silver gave the Spaniards. And just as Europeans divided their money into higher-value gold and lesser-value silver and copper, so the Indians divided theirs: *wompompeag* was worth three times as much as the more common *roanoke.* So widespread was the use that colonists employed it in their trade with Native Americans, as European traders did with African states. *Wampumpeag,* or *peag* as it was sometimes known, was also legal tender among whites in New England for small sums during the seventeenth century. The Dutch in New Amsterdam quoted prices in silver, beaver skins, and "wampum." But wampum seems to have had, at least initially, more symbolic or prestige value than what we consider money; most trade was by barter.

Despite what they shared and how they moved, rarely could Indians manage to create supra-village organizations that might have been capable of withstanding the invasions by whites. Some steps in the direction of confederation were being taken—Lewis Morgan counts three among the Ottawa, six among the Creek, and the "Seven Council Fires" of the Dakota, in addition to the best-known of all, the Iroquois Five Nations (later called the Six Nations when they were joined by the southern Tuscarora). Each of these, however, was within a local language group; none managed to transcend the language barrier until nearly the end of the eighteenth century, when the Indians were driven by desperation and no longer had coherent, single-language communities.

Meanwhile, Indian societies were frequently at war with one another. Before they acquired firearms, they fought wars that were largely ceremonial but with hard and painful elements. The first such element was that booty consisted in part of slaves. Slavery was as common among Indian societies in the New World as among Africans and Europeans. Practices were equally cruel. As among some African groups, slaves who attempted to run away were disabled by having their Achilles tendons severed or their feet mutilated, as John Lawson observed among Indians in the Carolinas. The difference between Indians enslaving Indians and whites enslaving Indians was the scale. The accounts we have of slavery before the coming of the whites are not explicit, but they suggest that the numbers involved were small; under white supremacy, they became vast.

Scalping and torture of captured prisoners were both widely practiced by American Indians before Columbus arrived in the New World. Some Indians, notably the Mohawk, Algonquians, and Hurons—like the Caribs whom the Spaniards met in the Caribbean—also practiced cannibalism. These, to us ghastly, practices had separate and distinct meaning for the tribes that engaged in them. In part, both scalping and cannibalism were believed to transfer the strength and "persona" of the fallen warrior to the victor and to allow the victor to demonstrate his prowess.

More complex were what came to be called "mourning wars." Many Indian societies, particularly the Iroquois group, believed that a fallen warrior must be both mourned and replaced. By all reports, the Indians mourned the fallen more deeply, indeed more bitterly, than perhaps any other society. Friends and relatives made every attempt to lessen the pain with condolence ceremonies, feasts, and the giving of presents. If these did not cover the loss, the society raided another Indian or white group for captives. Some captives were adopted but others were cruelly tortured to assuage the pain of the bereaved. Captives were first made to run a gauntlet that tested their morale and physical well-being. "Men—but usually not women or young children—received heavy blows designed to inflict pain without serious injury. Then they were stripped and led to a raised platform . . . where old women led the community in further abuse, tearing out fingernails and poking sensitive body parts with sticks and firebrands." Then the prisoners were fed and allowed to rest while the grieving families among whom they were apportioned decided whether to adopt them or have them killed. Those who were to be killed were decorated, politely treated until the appointed time, and then tortured and (among the Mohawks and some other peoples) eaten.

Many prisoners did not suffer this fate. White captives, whose reports are the only eyewitness accounts we have, frequently mention the kindness with which they were treated, as one member of my family did. Her account is borne out by those of other whites. Captives who were destined to be adopted were quickly received into their new families, bathed, given new clothes and started on a road of acculturation. As Philip Mazzei wrote in 1788, "A father who has lost his son adopts a young prisoner in his place. An orphan takes a father or mother; a widow a husband; one man takes a sister and another a brother." As difficult as it was for whites in the nineteenth century to believe, life among the Indians had great attractions for many whites. Benjamin Franklin, who was as ready as anyone to exploit them, remarked that

> when white persons of either sex have been taken prisoners young by the Indians and lived a while among them, tho' ransomed by the Friends, and treated with all imaginable tenderness to prevail with them to stay among

> the English, yet in a Short time they become disgusted with our manner of life, and the care and pains that are necessary to support it, and take the first good Opportunity of escaping again into the Woods, from whence there is no reclaiming them.

Many would be bound and dragged back, weeping, to white society.

In the days when warfare was largely ceremonial and casualties were light, few people were involved; but later, as warfare became endemic and casualties resulting from firearms and disease mounted alarmingly, some Indian societies faced the collapse of their world and engaged in virtually perpetual hunts for replacements. How these events appeared to the Indians, we can only imagine.

CHAPTER 2

The Fearsome Atlantic

Looking west, the people of the gentle Mediterranean saw the Atlantic as the great unknown, obscured by storm clouds, without welcoming shores, and given to violence. Their ships were too light to withstand its waves; their navigation instruments were baffled by its immensity; their experience stopped short of its rocks and reefs. So fearful were they of losing touch with land that they coined for us the concept of being "disoriented"—to have lost one's east, the European shore, in the watery wastes of the western sea.

Columbus inherited a long tradition of the search for the far reaches of this apparently limitless sea. We do not know about most of his predecessors, because many were drowned, marooned, murdered, or eaten by sharks or cannibals; but at least as early as the time of the great Greek traveler and gossip Herodotus, we know that sailors were already venturing out into the ocean and had visited at least the Azores and Canaries. In the fifth century B.C.E., a Phoenician navigator wrote a sailing guide for Africa, later translated into Greek, that must have been based on experience. Two centuries before Columbus, a group of Genoese set sail, bound for India. They disappeared into the sunset, never to return.

Also shrouded in mystery are the Portuguese expeditions in the generation before Columbus sailed. Fearing rivals, the Portuguese treated their records as state secrets; but, in their time, the secrets "leaked." Hearing of the successes of the Portuguese with sugar, slaves, and gold, the Spanish wanted to break into their monopoly. Partly because of this, Portugal and Castile fought a bitter four-year war which ended just thirteen years before Columbus sailed. The key provision of the peace treaty, judged by its importance for the discovery of America, was that Spain acquired the Canary Islands. Also important for Columbus's voyage, the war hastened the growth of shipbuilding technology: so much of the fighting was at sea, both states had to strengthen the hulls of their Mediterranean ships to cope with the Atlantic. Since the galleys used in the Mediterranean were powered partly by oarsmen, they were lightly constructed; and because they often had to tack across or against the wind, their sails were triangular, or lateen. To venture out into the Atlantic, where food and potable water had to be rationed, oarsmen were an expensive luxury. But if oars were given up, a different sail was needed. While well adapted to the Mediterranean, the triangular sail proved to be inefficient where a ship might sail for days or even weeks on a single reach. The square sail was used on the newly designed, longer, heavier carrack, which was possibly inspired by Turkish or North African ships. The Portuguese led the way with this ship, which they called a *não,* half a century before Columbus sailed. Recognizing the efficiency of its sails, he stopped in the Canaries on his way across the Atlantic and rerigged the little *Niña,*

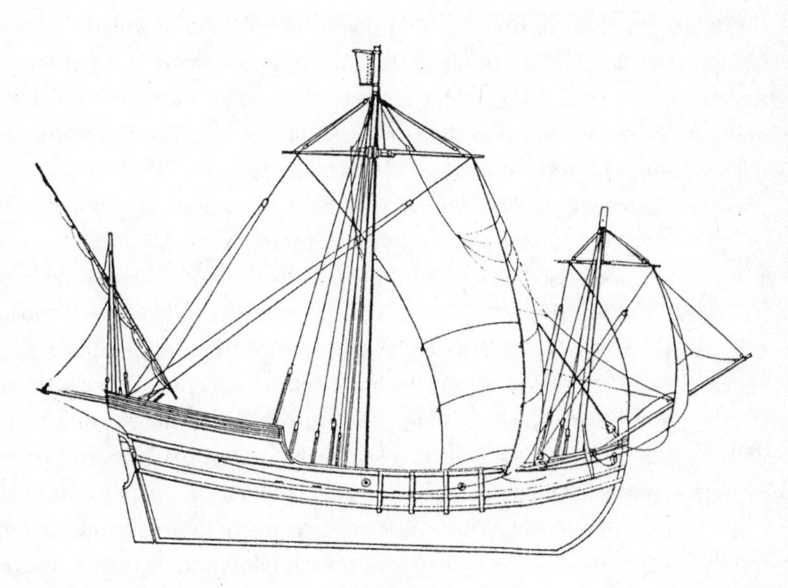

The Niña

transforming it from a "Mediterranean" *caravela latina,* with *lateen* sails, to an "Atlantic" *caravela redonda,* with square sails. The war had given Atlantic explorers their vehicle.

Technical innovation, so often a by-product of war, was important, but what really shaped the Spanish and Portuguese empires for centuries was wind. In the age of sailing ships, the wind pattern of the Atlantic governed where and how fast ships could sail. Columbus was lucky. Had he been sailing on behalf of Portugal, to whose king he first applied for patronage, he would have had to sail from the Azores, into a headwind. From the Azores, the Portuguese had to go south; but from the Canaries, the Spaniards could go west. Columbus—who had lived in the Azores, where he was married to the daughter of a sugar planter, and had sailed for the Portuguese—must have known this. He almost certainly would have seen flotsam, coming from somewhere to the west, washed up on the shores of the Azores, and as an experienced sailor, he could not have missed sensing the prevailing winds.

Geographical knowledge and information on winds and currents had slowly accumulated in the two centuries before Columbus sailed. For the Mediterranean, the Black Sea, and the Atlantic coast, pilots' instructions—*routiers* or *portolani*—began to be assembled around the thirteenth century into a single guide known as the *Compasso da Navigare.* The *Compasso* is the ancestor of such modern guides as the British Admiralty *Sailing Directions.* Some of the *portolani* were put into visual form as charts drawn on ox hides by Italian craftsmen. But none existed for the Atlantic.

The Atlantic posed challenges quite different from the Mediterranean. In the vast and relatively empty ocean, sailors could not check their position by observing islands and coasts. Once out of sight of land, as noted above, they "lost their east"; they became "disoriented." Sailing south, as the Portuguese were doing, posed a different risk, losing the North Star by dipping below the equator. So in Portuguese to lose one's bearings was expressed not as losing the east but as being "unnorthed," *desnorteado.*

Being disoriented or *desnorteado* was a concern of ship captains, but armchair navigators were governed less by what pilots told them than by cartographers' speculations and the inherited wisdom of Greek and Arab scientists. For them navigation was not a practice but an art. And the first task was to get some notion of the size of the Earth. Since Earth was known to be a round ball, the key measurement was the size of a degree of longitude, that is, 1/360 of its circumference. The first (and as it turned out the best) estimate was given in the third century B.C.E. by the Greek geographer and mathematician Eratosthenes. He arrived at his estimate by measuring the angle of shadows cast at noon both in Alexandria, where he was director of the great library, and also about 800 kilometers (500 miles) to the south in what is today called Aswan. With simple geometry, he estimated a degree of longitude at 59½ miles. Much of the scientific speculation of the Greeks was lost as a result of the collapse of the western part of the classical world, but their tradition was carried on by Muslim scholars. So it was that a medieval Muslim geographer, al-Faraghani, recalculated Eratosthenes' experiment and arrived at 56¾ miles for a degree of longitude.

Given the tools that Eratosthenes and al-Faraghani had to work with, these measurements were remarkably precise, but when passed down to later readers, they were distorted by one crucial factor: the length of a mile. That measure varied from place to place. When Columbus read of al-Faraghani's measurement, he transposed al-Faraghani's mile into the shorter Italian mile and came up with a degree nearly 12 miles shorter than al-Faraghani had meant. For Columbus, therefore, the world was thus about 25 percent smaller than the real world. He was in good company. In the influential work *Imago Mundi,* Cardinal Pierre d'Ailly argued (without objective reasons) that the "Ocean Sea" was not immense; and in 1474 the Florentine cosmographer Paolo Toscanelli (again with no objective reason) put a number on it: he pontificated that only about 5,000 miles separated Europe and Cathay (China), instead of 15,000 miles.

Reaching Cathay was what Columbus desperately wanted to do. He was inspired by Marco Polo's account, as we can see from the fact that his own annotated copy of *The Travels* (*Il Milione*) has survived. Marco Polo's Cathay was the magnet that drew Columbus's mind to the Atlantic. So, with the blessing and finances of Queen Isabella of Castile, he set off in three little carracks from Palos, on the southern Atlantic coast of Spain, on August 3, 1492. A week later they were in the Canaries. On September 6, with sails flapping in the calm air, his ships set off again. Then, picking up the "westerlies" offshore, they sailed steadily to the west for a month of nearly perfect days until, on October 12, they caught sight of an island in the Bahamas.

However much he excited the navigators who followed him, Columbus inadvertently terribly misled them. His own trip from the Canaries to the Caribbean was virtually a pleasure cruise. With a following wind, at the right season of the year, and not reaching the brutal latitudes along the North American coast, he sailed steadily and serenely across a calm sea. What most of those who followed him experienced can best be described as horror. Even three centuries later, one of them told of the experience in these words:

> . . . terrible misery, stench, fumes, horror, vomiting, many kinds of seasickness, fever dysentery, headache, heat constipation, boils, scurvy, cancer, mouthrot, and the like, all of which comes from old and sharply salted food and meat, also from very bad and foul water, so that many die miserably.

That horror was what Spanish, French, and English voyagers in the sixteenth and seventeenth centuries had to expect. As a contemporary poem describes them, the fragile, miniature immigrant ships were "freighted with fools." To embark upon one, a person had to be desperate, a condemned felon, a captive, mad—or perhaps just ignorant. Written accounts of the voyages probably reached few, although as would-be travelers gathered in seaside inns, they overheard sailors' accounts. These tales must have been so lurid and arresting that they would have been told and retold often and widely. But the sailors whose stories could be heard were the survivors. Many others were not there to tell their tales.

Many ships simply disappeared. The death rate was appalling. When the great Spanish treasure fleet, the *flota,* set out from Havana in 1591, it was a convoy of seventy-seven ships. When the ships reached the latitude of Cape Hatteras, as survivors later reported, they were beset by a violent wind and high seas. The largest of their ships, with 500 men aboard, was lost; and a few days later, hit by another storm, "five or sixe other of the biggest shippes cast away with all their men, together with their Vice-Admirall." The survivors kept sailing north to the Chesapeake, where they turned due east to sail across the Atlantic to the westernmost islands of the Azores. There they were hit by another storm. In that year only 25 ships of a total of 123 survived

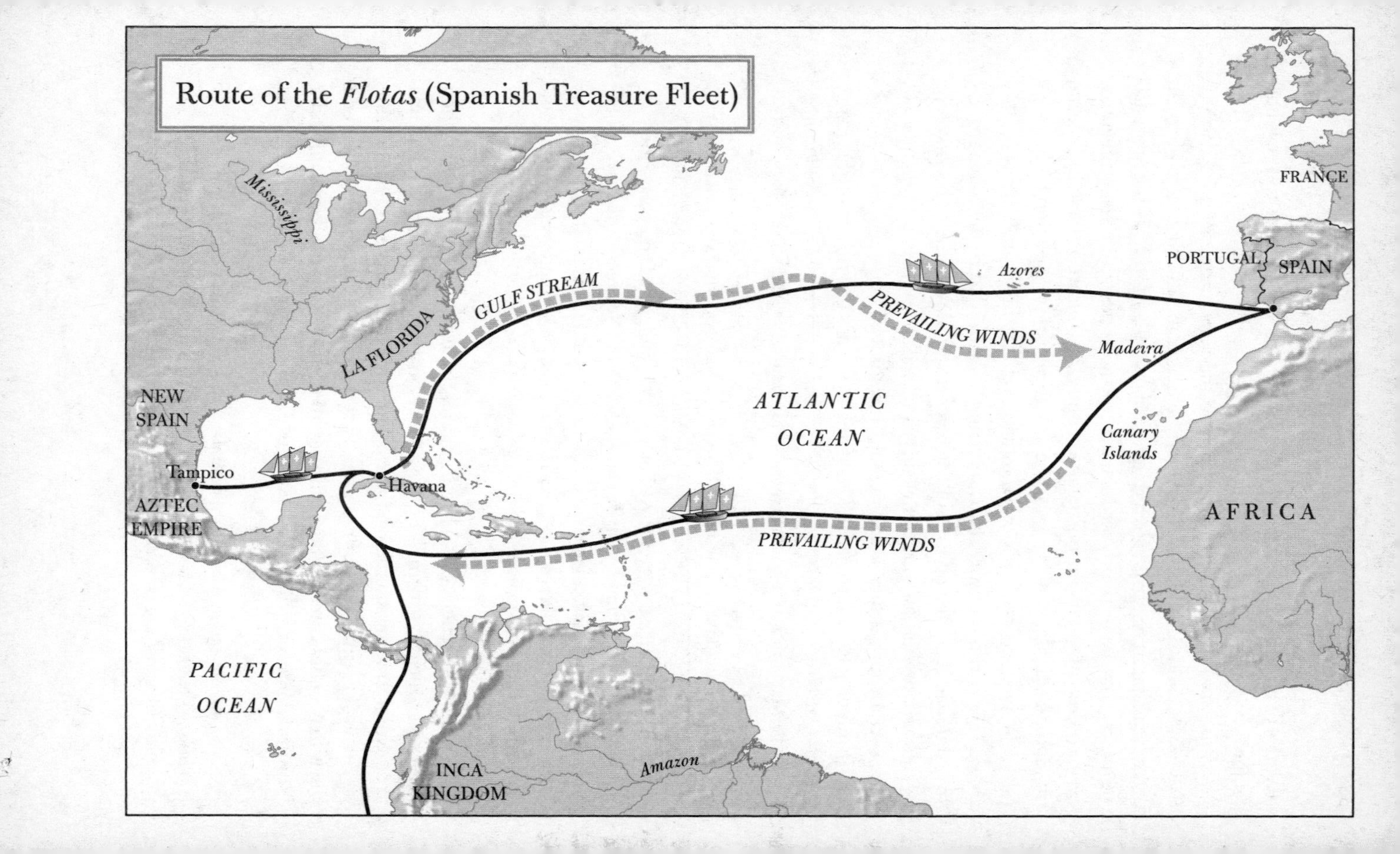
Route of the *Flotas* (Spanish Treasure Fleet)
Mississippi
FRANCE
PORTUGAL
SPAIN
Azores
GULF STREAM
PREVAILING WINDS
Madeira
LA FLORIDA
NEW SPAIN
ATLANTIC OCEAN
Canary Islands
Tampico
Havana
AZTEC EMPIRE
AFRICA
PREVAILING WINDS
PACIFIC OCEAN
INCA KINGDOM
Amazon

the trip. The zephyr that had carried Columbus had turned to howling tempest.

Since no one could predict how long the passage across the Atlantic would take, and since the owners of private ships and the chandlers of government ships were obviously trying to make as much profit as they could, vessels were usually poorly provisioned in the sixteenth and seventeenth centuries, and well into the nineteenth century. Biscuits were the staple; meat was a luxury. Soaking in brine was the only means of preservation for beef or pork, and passengers often found it rotten. Even if food began by being palatable or even edible, it was almost certain to be invaded by seawater while stored in barrels in the hold. And wherever it was, it was sure to be gnawed by rats and infested by worms. Biscuits soon became moldy and soggy, or as one seventeenth-century sufferer put it, "most beastly rotten."

Prudent passengers packed their own rations. Cognizant of the high rate of death from bad food or a lack of food, Lord Baltimore advised immigrants to Maryland to take their own and provided an ideal inventory:

> Fine Wheate-flower, close and well packed, to make puddings, etc. Clarret-wine burnt. Canary Sacke. Conserves, Marmalades, Suckets, and Spices. Sallet Oyle. Prunes to stew. Live Poultry. Rice, Butter, Holland-cheese, or old Cheshire, gammons of Bacon, Porke, dried Neates-tongues, Beefe packed up in Vinegar, some Weather-sheepe, meats baked in earthen potts, Leggs of Mutton minced, and stewed, and close packed up in tried Sewet, or Butter, in earthen pots: Juyce of Limons, etc.

Ideal but not practical. Few passengers were rich enough to buy even simple food sufficient for the long voyage; most were completely dependent upon the shipowners and crew for the little they got. When food ran out, as was likely if a ship was becalmed, passengers sometimes ate one another or, when they could catch any, the rats that infested all wooden vessels. On the *Virginia Merchant*, where the passengers were starving,

> The infinite number of rats that all the voyage had been our plague, we now were glad to make our prey to feed on, and as they were insnared and taken, a well grown rat was sold for sixteen shillings as a market rate.

> Nay, before the voyage did end, a woman great with child offered twenty shillings for a rat, which the proprietor refusing, the woman died.

One voyage in 1731 took twenty-four weeks, and of the 150 passengers on this ship more than 100 starved to death. As ghastly as the trip was for adults, it was worse for children. Few lived to tell the tale.

Apart from being poorly provisioned by owners and chandlers, ships had little space for food and drink. Few seventeenth-century ships displaced as much as 150 tons. Columbus's flagship, the *Santa María,* is thought to have displaced about 100 tons and to have been about 75 feet long. The *Niña* and the *Pinta* were about half that size. Many of the remarkable voyages of exploration by the Portuguese, Spanish, and Genoese in the fifteenth and sixteenth centuries were made in vessels as small as modern sailboats—30 tons and 25 or 30 feet in length. In his remarkable 1524 voyage on behalf of King François I of France, Giovanni Verrazzano reconnoitered almost the whole North American coast in a ship that displaced only about 70 tons and was perhaps 65 or 70 feet long. The *Mathew* of the Genoese John Cabot (Cabotto, Italian for "the coaster") was little more than half that size.

Big ships were rare. In 1582, just before the Spanish Armada sailed against England, the English owned only twenty ships of more than 200 tons. Two out of three of the rest were less than 80 tons. Among the ships of its time, the 90-foot-long *Mayflower* was a relatively ample 180 tons; and the *Ark,* which took the first settlers to Maryland, was a nearly gigantic 350 tons. The *Dove,* which accompanied the *Ark,* was only about 50 tons. The most common ship of the seventeenth century was the Dutch flat-bottomed *fluyt* or fly boat, which usually displaced only about 50 tons. The *Tygre,* on which people and supplies were sent to Virginia in 1621, was a pinnace or brigantine rated at only 45 tons. Even smaller was *Sparrow,* 30 tons, which went to Plymouth in 1624. We do not know of many of the seventeenth-century ships, but eighteenth-century ships were also as small as modern yachts: the *William and Mary,* the *William and Elizabeth,* and the *Beginning* each displaced only about 30 tons and were only 30 to 45 feet long.

Packed with people, their belongings, their equipment, and their animals, such diminutive vessels had scant room for anything else. Even when, as on the *Mayflower,* the passengers rather than avaricious owners bought the provisions and so made sure the food was reasonably palatable, the little ships could carry only a small amount of food and water.

Water for the trip had to be procured from rivers which flowed down to the ports of departure and to which the inhabitants of all the riverside towns had contributed their garbage and excrement. Put into barrels, river water was sure to become stagnant and covered with slime. During his crossing in 1612, George Percy found the water "so stencheous thatt onely washeinge my hands there wth I cold nott endure the sentt thereof." Even ashore prudent people in the seventeenth century hardly touched water except to bathe, which they did rarely. Their favorite drink was beer.

Beer was safer, but even in stout kegs it could not be completely protected aboard a ship; worse, only a limited amount could be afforded or carried, so the supply would be exhausted on very long trips. The crew and passengers of an average-size ship on a three-month trip were expected to drink 3,500 gallons of beer. Much as they hated to drink water, the ship also carried about 1,500 gallons of it. Ships could hardly carry more. In fact, the Pilgrims landed at Plymouth Rock because they had run out of beer.

Running out of beer or water was lethal but could not necessarily be avoided, since the length of a trip was decided by weather. Just getting started was completely unpredictable. A ship might spend weeks swaying at anchor while waiting for favorable winds. Once under way, it faced months at sea—how many, no one could foresee. The *Dove* took only two months plus layovers at Barbados and Saint Christopher to reach Maryland; but in 1670 another ship, fitted out in part by the young philosopher John Locke, took seven months.

Three or four or even seven months—that is, if the voyagers were lucky. The last part of the trip, where the warm waters of the Gulf Stream collided with colder waters off Cape Hatteras, was known as a ship killer. Storms there were often so violent as to dismast ships, tear their rudders from the moorings, rip the sails apart, smash lifeboats and cabins, and cave in decks and hulls. William Strachey described "a most terrible and vehement storme, which was the taile of the West Indian Huracano . . . so violent that men could scarce stand upon the Deckes, neither could any man heare another speake." It was that Atlantic which inspired Shakespeare's play *The Tempest*.

A fluyt *(flyboat), the most common ship of the seventeenth century*

That also was the Atlantic which the Spanish *flota* endured on the return leg of its treasure-gathering trip to the Caribbean. But the Spanish soldiers and colonists going out to the New World had an easier passage than northern Europeans. Like Columbus, they sailed downwind first to the Canaries. There, already a third of the way across the Atlantic, they rested a few days and took on wine, fresh water, and *gofia*—a bread, made of barley and goat milk, which the Bristol merchant Nicolas Thorne described in 1526 as "exceedingly holesome." Then, because they stopped in the Caribbean or on the Florida coast, they missed the terror of Cape Hatteras.

Because Europeans attempting to reach the northeastern shore of America could not avoid the terror of Cape Hatteras, the southern route fell out of favor in the eighteenth century. Thereafter, their ships tended to follow what we know as the great circle course, northwest from the British Isles or northern France nearly to Iceland, then southwest past Nova Scotia and Newfoundland and south along the New England coast. That route more or less followed the fishermen who in their little two-masted doggers or dog boats had been chasing the codfish for centuries. The northern route was shorter, optimistically put at thirty to forty days east to west and about twenty-four days west to east; so it was cheaper. It also had the advantage of shielding the passengers from tropical diseases, but of course it was even more susceptible to violent storms than most of the southern route.

As sailors' tales portrayed them, storms could last for days or even weeks during which time a ship would be nearly totally out of control and would be blown far, and unpredictably, off course. Then, in the ominous phrase of the sailors, the ship might be "cast away." Sailors' tales are filled with accounts of what "cast away" really meant: broken to pieces and sunk without a trace.

Even ships built for the Atlantic were clumsy and often unseaworthy. The *fluyt,* the mainstay of Dutch commerce, had been designed for the relatively protected North Sea and shallow coasts of Holland; for economy, it was very lightly built. Its timbers were ill matched to the Atlantic swells and storms. But it became the most common ship of the seventeenth century because about 1,000 *fluyts* were seized by England in its wars with Holland and sold cheaply to English merchants. Opposite the little *fluyt* were the relative huge Spanish galleons. Far heavier and more sturdy they were, but their towering hulls and high rigging made an even better target for wind and wave.

Light or heavy, humble or majestic, *fluyt* or galleon, ships sailed nearly blind. "Pilots" or "ruttiers," the traditional logbooks in which for centuries Mediterranean navigators had recorded their accumulated experience, were only beginning to focus on the Atlantic. No maps yet showed coasts with sufficient accuracy, much less the positions of reefs, rocks, or sandbars. Even if they had had sailing pilots and reasonably precise maps, navigators could not have fixed their positions accurately with the crude instruments, hardly changed from the Middle Ages, upon which they had to rely. To calculate latitude they used the astrolabe, cross staff, and quadrant; but having no reliable clock, they had no means to do more than roughly estimate longitude. The best they could do was to work out dead reckonings with a compass, a sand glass, and a log line. Or, like the Vikings and early Portuguese and Spanish sailors, they could watch birds and pick up flotsam. Finding an exact position was not so critical in the open ocean, but once ships got close to land, particularly during a fog or storm, they were in mortal danger. Running aground on a reef, striking a rock, or slamming into a coast caused many a ship to be "cast away." That is how Columbus lost the *Santa María* and was forced to establish the first Spanish colony in the New World.

Even if the passengers, mercifully, knew little of these dangers, hunger, thirst and sickness were inescapable realities. For much of a voyage, passengers were jammed into the area known as 'tween-decks, that is, the space between the hold and the deck. Below them, they could hear bilgewater sloshing back and forth and timbers groaning as the ship rocked or plunged; above, where the ship's gear and boats were stored, the rigging was fixed, and the crew worked, the wind and sea spray drowned out all other sound. 'Tween-decks could be ventilated only in fair weather; in heavy seas the hatch was battened down to prevent seawater from flooding in as each wave hit. On many of the boats in the seventeenth century, particularly the common *fluyt,* 'tween-decks was lower than the height of a man, sometimes only 4 feet high, so passengers spent days or even weeks hunched over, squatting, or lying down—and always hanging on for dear life as the ship reeled and tossed.

As they got their sea legs, the passengers would begin to try to eat. During storms there was no way to do more than gnaw on biscuits, because the only kitchen was what was called a caboose. On the larger boats, the caboose was a small hut on the upper deck where a brick fireplace was set up. On the smaller and more common boats, the caboose was simply a barrel, sawed in half and filled with sand on which a fire could more or less safely be made under a pot to boil mush or stew or to roast salted meat.

On even the larger ships, there were virtually no means available for sanitation. Fresh water was far too precious to be used for bathing or washing clothes. Seawater left clothes full of salt, which irritated the skin and caused boils and lesions. In the frequent periods of fog and rain, clothes could not be dried and so mildewed; and bedding, such as it was, became fusty. In this luxuriant environment, lice and other vermin proliferated and had to be picked off one's body by hand. As a French traveler remembered, "Each time we left the 'tween-decks we found ourselves covered with vermin. I found them even in my shoes." Combating vermin on themselves and on one another—what in monkeys is called grooming—was the passengers' main occupation throughout the voyage. And, bad as the inability to ever get clean must have been, defecating was not only messy but dangerous. Toilets were either buckets, which must often have overturned in rough seas, or openings in the railings where, as waves slammed into the ship's hull, the brave or desperate precariously perched, fearing to be washed overboard and getting frequent baths of seawater while they were relieving themselves.

Even on relatively calm days, everything loose—boxes, bedding, utensils, and people—skidded unceasingly, like a pendulum or a metronome, back and forth, port to starboard, starboard to port, across the deck, slamming into other bales and boxes, upending people, overturning buckets of slops. There was no escape, since every inch of 'tween-decks was packed with people, goods, and animals. The stench of excrement and vomit, the bellows of the terrified animals, the moaning of the sick, the screams of the children, the creaking of the ship's timbers, and the crash of the sea can scarcely be imagined. This bedlam was enveloped in an almost pitch-black gloom. No candles or lanterns could be allowed between decks because a fire at sea was too frightening to contemplate. Passengers and crew alike thought of 'tween-decks as a dungeon and spoke of the inhabitants getting what some of them, probably from their English penal days, called jail fever.

Jail fever was probably typhus. Other diseases were just as lethal. Scurvy was the most frightening. In sixteenth- and seventeenth-century Europe, no one understood its cause or prevention. Any traveler who had read the already popular accounts of Richard Hakluyt, the intellectual father of colonization in British America, would know its terrible symptoms—"swolne euery ioint withall." With scurvy, people were lamed, their teeth fell out, and bleeding sores erupted. It was not until 1753, when passengers were advised to provide themselves with prunes and "Juyce of Limons," that Europeans caught up with the Arabs and the Indians, who had long understood scurvy. Smallpox also was dreaded as a killer both ashore and at sea, but at sea there could be no quarantine—which was the only way then known to contain it. Of the 100 passengers on William Penn's *Welcome* in 1682, thirty died of smallpox.

In addition to storms, starvation, and sickness, passengers and crew had to reckon with the Atlantic as a sort of "no man's sea," contested among the British, French, Dutch, and Spanish navies in times of hot and cold war and at all times among privateers and pirates. Not only the richly laden Spanish treasure ships that made up the *flota* but even humble barks were targets. The 45-ton *Tygre* was captured for its sails, anchors, hourglass, compass, and provisions on its way to Virginia in 1621.

Within thirty years after Columbus had arrived, French privateers were savaging Spanish fleets—and not only the fleets. As these privateers grew bolder, both they and the English began to sack cities. The most spectacular raid was Jacques de Sores's 1555 sack of the great Spanish city of Havana. Even relatively peaceful trade sometimes faded into piracy when better terms might be extorted by threat. The Spanish word *rescates* summed up the transaction: it meant both barter and ransom. Even in times when England and Spain or France and Spain were at peace, their agents, both privateers and pirates, engaged in dirty tricks and casual mayhem. The hand that held out goods for trade easily strayed to the hilt of the sword.

Many hands held those swords. No reliable estimates exist for the numbers of pirates in the seventeenth century, but during the first half of the eighteenth century about 5,000 were active in the Atlantic. Some of these outlaws were corsairs sailing out of North African ports. They are usually called "Turks," but their crews and even their commanders were often European outcasts, then called *renegadoes.* Nearly half were English or colonial American. The pirate world was polyglot and truly international: its frontiers were not fixed geographically, ethnically, socially, or legally. Even law-abiding masters of merchant ships frequently augmented their incomes by a little piracy on the side. As David Beers Quinn remarked, "Coasters were continually being robbed when they were not taking their turn to rob. . . . Every ship's weapons [were] being turned against all others—Scots, French, English, Hansards, Flemings, each attacking and being attacked indiscriminately." The large number of pirates can be explained, in part, because as navies contracted after wars ended, discharged veterans often drifted into piracy. So numerous and brazen were they that in the last years of the seventeenth century and the early years of the eighteenth, they set up "republics" on islands scattered around the world from the Caribbean to Madagascar.

The Spaniards and the British, even when they fought one another, were engaged in a bloody and unending war against pirates. When pirates were caught by the Spaniards, as an English pirate recounted, they were enslaved or, as Protestants, turned over to the Inquisition. At the hands of the Inquisitors, they "were all rackt" and given a last breakfast of wine and a slice of bread fried in honey; "every man alone in his yellow coat, and a rope about his neck, and a great greene Waxe candle in his hand unlighted" was marched to a scaffold, some to be burned as "English heretickes, Lutherans" while others were whipped and sent to the galleys as slaves.

At the same time, the newly organized British navy deployed about 13,000 men mainly to catch pirates. When they succeeded, they either killed the pirates immediately or brought them back to London where wholesale executions became a popular spectacle. About 600 Anglo-American pirates were hanged between 1716 and 1726. These hangings entertained Londoners but did not stop piracy.

Most pirates got away. When the navy appeared in force, the pirates scattered to prearranged hideouts where shallow waters or reefs prevented the larger warships from following. Some even attacked naval warships, as Captain Kidd did at the mouth of the Chesapeake in 1699. No matter how savage the punishments, piracy remained a terrifying danger throughout the sixteenth, seventeenth, and eighteenth centuries—and not just for ordinary passengers; even an English royal governor of the Carolinas was kidnapped.

The danger of piracy and the commonplace reality of hunger, thirst, and storm were bad enough for normal passengers, but far worse for convicts or prisoners of war. Often chained, thousands of prisoners of war were packed off as virtual slaves to Virginia and the West Indies after Oliver Cromwell crushed the Irish revolt of 1651. The treatment of prisoners of war after the battle of Sedgemoor, where James II defeated the Monmouth rebels in 1685, became the paradigm for suppression of the Scottish and Irish revolts in the eighteenth century. When James II set up what came to be known as the Bloody Assizes to stamp out opposition to his reign, his courtiers urged him not to kill the prisoners but to sell them as slaves. He did, and nearly 1,000 were parceled out to be jammed into available boats and set off across the Atlantic to be sold into servile exile.

Not surprisingly, exiled white prisoners suffered casualties comparable to those that took place on the ships transporting black slaves. Even among free passengers, one death among each ten passengers was not considered unusual. The rates of survival among the Spanish and French were apparently far higher, but of the "first wave" of English reaching Virginia around 1609, less than half survived.

Powerful forces must have driven or lured Europeans out into the stormy Atlantic to suffer great danger and constant misery. We must now ask what those forces were.

CHAPTER 3

Sugar, Slaves, and Souls

Throughout the Middle Ages, western Europeans had bought sugar from the lands of the eastern Mediterranean. Added to purchases of other luxuries from Byzantium and spices from the Indian Ocean islands, sugar put a heavy burden on their economies. Europeans had little to offer in exchange for it except gold, and by the end of the eleventh century they had used up nearly all the gold they had. So, beginning early in the thirteenth century, Europeans set about trying to find a cheaper way to satisfy their sweet tooth. Every step they could bring production closer to the markets would cut costs. So, first on the island of Cyprus, then on Crete, and next on Sicily, they invested in projects to grow sugarcane and refine sugar. By 1404, they had reached Portugal. Sugar was still an expensive luxury beyond the reach of any but the rich. Even Queen Isabella of Spain treated it as a special Christmas gift for the royal children. If sugar was to become a popular condiment, cheaper means of production had to be found.

Meanwhile, the Europeans were also trying to find new supplies of gold. Africa was thought to be the most promising source, but no one in Europe knew exactly where in Africa gold came from. Europeans knew only that gold arrived by camel caravan at Moroccan ports, where it was exchanged for cloth, timber, and glass beads, and that it arrived in sufficient quantities to enable the Florentines, in 1254, to produce the first gold coin minted in centuries. Rumors of mines or gold-bearing rivers far in the interior of the vast Sahara spurred Italian explorers to try to find them. Many died or were murdered on the way, but at least one Florentine merchant had reached Timbuktu by 1470. The quest for fabled cities would ever lure

Europeans onward, but they realized that the desert route would never become economic or even safe because the Berber middlemen jealously guarded it.

The Portuguese decided that the sea was the better way. It was partly in quest of gold that Prince Henry of Portugal, known as Henry the Navigator, encouraged expeditions along the African coast. Finding gold was a dream; but more practically, Henry, who had invested in the first Portuguese sugar refinery in 1452, looked for places where he could grow sugar cheaply. Overcoming the conflicting winds and currents on the uncharted African coast, his captains plunged south. Strong African societies prevented them from establishing colonies on the mainland, but they discovered islands that were well adapted to sugarcane. After Henry died in 1460, the advance continued under a private charter, and in 1471 the Portuguese set up the first significant trading station on what came to be known as the Gold Coast. From São Jorge da Mina (later known as Elmina), they were soon importing much of Africa's gold.

Growing and refining sugarcane was hard work and of a kind not congenial to the merchants of Genoa, the warriors of Spain, or the sailors of Portugal. Thus, to the earlier search for gold and sugar a third quest was added: for slaves to hoe and cut the cane, carry it to mills, and grind and boil it; and for slaves themselves, to become a commodity to be sold to others. Within a few years, the Portuguese were shipping about 2,000 black slaves a year to work their new plantations in the Azores, Madeira, and the Cape Verde islands. Half a century later, Lisbon itself had a population of 10,000 black slaves.

In his explorations, Henry had made use of the Genoese, who were among the most aggressive of the Europeans. The Genoese had to be aggressive, as they had little to keep them at home. As the French explorer Sammuel de Champlain later memorably noted, Genoa was "built in a region surrounded by mountains, very wild, and so sterile that the inhabitants were obliged to have soil brought from outside to cultivate their garden plots, and their sea is without fish." So the Genoese became Europe's "sailors of fortune." By the fifteenth century, they had arrived in Portugal, Spain, France, and England, where their experience with the sea fitted them to be sailors, navigators, and mapmakers. Columbus is the example we know best, but he was just one of many who freighted sugar from the Atlantic islands to western Europe.

Of the Atlantic islands, the most important were the Canaries, the Azores, Madeira, and, farther down the African coast, São Tomé. By 1480, Madeira was loading about sixty or seventy ships a year with sugar for Spain and Portugal. No longer a rare luxury item, sugar had become an important commodity. Producing it had become the major commercial enterprise of the time. In the Azores, one of the leading figures in production and trade was a naturalized Italian by the name of Bartolomeo Perestrelo for whom Columbus worked for several years and whose daughter he married.

The Azores were not Columbus's only taste of the Atlantic. It is probable that he sailed with the Portuguese south to São Tomé and the Gold Coast. But on any beach in the Azores, he could have seen evidence of the far Atlantic. Flotsam that washed up on the shores could have inspired him to wonder what lay beyond the "Ocean Sea." Still, although flotsam may have been Columbus's inspiration, the winds that carry it eastward across the Atlantic to the Azores are also headwinds impeding ships trying to sail westward. From the Azores, Columbus would probably never have reached America. Atlantic winds dictated that the springboard for a westward quest had to be located farther south in the Canaries.

The Canaries had two important roles in the Europeans' discovery and conquest of the New World. Situated about a third of the way across the Atlantic from the Spanish coast, they afforded Columbus's little ships water, food, and an opportunity to repair damage. From them, the "westerlies" made sailing "downwind" across the Atlantic toward the Caribbean relatively easy and rapid. They would be the staging base for most wind-powered ships for the next two centuries. That was the first and most obvious role of the Canaries.

The second role of the Canaries in American history is no less important. It was there (as mentioned in chapter 1) that the Spaniards first encountered a "colonial" people. The Canaries had been settled in the distant past by peoples from North Africa. These people, known as the Guanche, are thought to have been of Berber origin. Their existence was known to the Romans and to medieval Muslim travelers, but, cut off from the great events that took place in Europe and Africa, they remained essentially frozen in time for thousands of years. When they were rediscovered, they were first noted on a sea chart in 1389. To the Europeans, they appeared utterly alien. Although the Guanche looked Mediterranean, as indeed they were in origin, they were described by the Spaniards as living in caves, wearing only skin and rush clothing, illiterate, without coherent government or civic institutions, and—disastrously for them—armed only with sticks and stones. The Spaniards dealt with them by conquering, enslaving, and ultimately exterminating them—setting the pattern for later relations with the American Indians.

Determined as he was to land on the mainland of Cathay (China) or at least on Cipangu (Japan), Columbus had a shock when he saw the first group of inhabitants of the New World. They were not the silk-robed men of Marco Polo's tale. Like the Guanche of the Canaries, they were a primitive people. As he sailed on and landed on Hispaniola he remarked, "The people of this island and of all the other islands which I have found and seen, or have not seen, all go naked, men and women, as their mothers bore them, except that some women cover one place only with the leaf of a plant or with a net of cotton which they make for that." Columbus noted also that their "skin is the color of the Canary Islanders." That is, they were very like the Guanche, toward whom the Spaniards had already adopted a policy.

However they appeared to him, the natives saved Columbus's life. On Christmas eve the *Santa María* struck a reef. Had the ship been of the Mediterranean shape, curved on the bottom, it could have been rocked backward and forward, by moving men and cargo, until it floated free. But the *Santa María* was not a Mediterranean ship, and it quickly began to break up. Had it not been for the Arawak Indians, everything might have been lost.

The people, equipment, and stores were saved, but Columbus could not crowd forty more people into the tiny *Niña.* Thus the accidental combination of a shipwreck and the Indians' hospitality forced him to establish the first European settlement in the New World.

Gathering up the few gold trinkets the natives had acquired from some distant and unknown source and taking several Indians along as hostages,

Columbus set out for Spain. In his three-month visit, he had already set precedents—hostilities with the natives, establishment of a colony based on exploitation of them, and the single-minded search for gold—that would mark generation after generation of Spanish endeavors in the New World.

The news he brought to Spain was stunning. A new and unexpected source of gold and unlimited numbers of slaves to work the sugar plantations were there for the taking. Electrified, the usually slow-moving Spanish government was quick to act, and ambitious private citizens flocked to Columbus's venture. Within a few months, the first large-scale conquering and colonizing mission had been assembled. About 1,200 men were to be carried by seventeen carracks and smaller ships with provisions for half a year. In addition to soldiers and the usual sprinkling of "gentlemen," there were artisans, a few priests, and some government officials. In the holds of the ships were animals, equipment, and seed to plant the European foods that the Spanish insisted on eating. It was, as the maritime historian John Parry remarked, "a whole society in miniature."

When the new flotilla arrived at Hispaniola, the Spaniards found the first settlement destroyed. The men Columbus had left behind had refused to dirty their hands planting crops to feed themselves and had infuriated the natives by stealing their food and raping their women. By the time of Columbus's return, they were all dead. Conflict between Europeans and Native Americans was to be the central theme of colonialism in the New World.

In 1493, all trade with the New World was made a crown monopoly; by 1503 the monopoly had spawned a bureaucracy, a sort of ministry of colonial affairs, known as the Casa de Contratación. The Casa became purchasing agent, outfitter, and supplier of the fleets sent to the Caribbean. Insofar as the great distances and poor communications allowed, it tried to supervise or control everything that happened in America. But the Spanish state did not stand alone—it always included the Catholic church. Before her death, Isabella had encouraged the church to take on a major role in the new empire. So the policy of her government could be summarized as getting rich and saving souls. Under the supervision of the Casa, merchants and conquistadors took care of looting while, under the Spanish church, Dominicans, Franciscans, and Jesuits were charged with conversion of the

Indians. The two aims of policy were almost immediately in collision. Exploiting the natives was usually easy and sometimes profitable; bringing them into the Christian fold was neither.

Isabella was not deterred. Encouraged by the pope, she sent with Columbus on his second voyage the first group of what would ultimately become thousands of priests. And she sought to prevent others who might contaminate the natives from going there. Insofar as possible, the Spaniards sought to quarantine the New World. Migrants were screened, not for criminal activity—even Columbus's first crew on the *Santa María* were former convicts—but for religious conviction. Muslims, Jews, and later of course Protestants (called *luteranos*) were excluded.

Spain was obsessed not only with religious purity and administrative order but also with judicial justification for its dominion. Only by trying to save their souls, the Spaniards believed, did it acquire the right to exploit the natives. Since, like the people of the Canaries, these natives were primitive barbarians, the state as agent of the church had a sacred obligation to convert them. They were to be issued with the requirement (in Spanish, *requerimiento*) that they accept the authority of the state and the faith and leadership of the church. Those who resisted were to be read a warning by the local military authority (in Spanish, of course, which few natives could have understood): "The resultant deaths and damages [of your rebellion] shall be your fault and not the monarch's or mine or the soldiers'." Against those who remained recalcitrant, war was just, and the legal penalty for those who survived was enslavement. It was on these conditions that the aims of the conquistadors and the priests united. The impact on the Indians was catastrophic.

When Columbus first encountered them, the gentle Arawak people of Hispaniola probably numbered at least 1 million. A contemporary observer described Hispaniola as "perhaps the most densely populated place in the world," and some modern scholars put the number of inhabitants at as many as 8 million. Within thirty years, enslaved to grow sugar, decimated by European diseases, and killed off almost casually, they were nearly extinct. Fray Bartolomé de las Casas's *Brevissima Relación de la Destrución de las Indias* makes sickening reading. He wrote in 1544 that even after half a century of occupation, the Spaniards

> are still acting like ravening beasts, killing, terrorizing, afflicting, torturing, and destroying the native peoples, doing all this with the strangest and most varied new methods of cruelty, never seen or heard of before, and to such a degree that this Island of Hispaniola, once so populous (having a population that I estimated to be more than three millions), has now a population of barely two hundred persons.

As the Spaniards expanded their sphere of control, Indian populations would virtually disappear: in what is now Florida, the Timucua had been estimated at half a million people in the middle of the sixteenth century. By 1728 only a single Timucua Indian was known to be still alive.

As they regularized their new empire, the Spaniards sought to make the Indians work for them. Steps toward the creation of a system of forced labor began as early as 1496, after an Indian revolt. Then when Columbus imposed a tribute or *repartimiento* to be paid, when the Indians had no other means, in labor. Without forced labor, Spain's colonies could not be maintained. Spaniards were told to treat the Indians well, but only as far as was possible. That vague injunction made no impact. What men had won with the sword, they would not give up to the cross.

So bad had oppression of the Indians become that in 1511 the king set out regulations, known as the "Laws of Burgos," to develop a more tolerant policy. To supervise it, Fray Bartolomé de las Casas, who had originally gone to the Caribbean as a conquistador, was appointed "protector of the Indians." He was blocked at every turn by the new sugar planters and, seeing his efforts fail, concluded that the only way to protect the Indians was to import black slaves. The planters eagerly adopted his suggestion, and by the middle of the sixteenth century, sugar plantations on Hispaniola were being worked by about 20,000 blacks. Fray Bartolomé had unintentionally promoted the nightmare with which Americans would live for centuries.

Meanwhile, Catholic priests struggled to convert the Indians. The record of their activities has all the elements of tragedy, comedy, self-sacrifice, and destruction. Some priests were murdered by the people they thought they were saving; and many priests abused their position for personal satisfaction. As one senior Spanish official remarked in the late

1500s, "It is as safe to loose a stallion among a herd of mares as to let a friar out among the Indian women." They were often criticized, but by and large, the Jesuits, Franciscans, and Dominicans were Spain's "best and brightest." Assembled in highly disciplined, centrally directed organizations, with the most intimate connections to the throne, and, after the advent of the Inquisition, in possession of a police and judicial power that no individual or group could withstand, they aimed to create new Indian societies within the old. The way to do this, they decided, was to destroy Indian religions, uproot Indian cultures, create new forms of identity, and forge new ties of community. Learning Indian languages and catering to Indian tastes for ritual and ceremony, they engaged in a highly sophisticated campaign of what we would call brainwashing.

In what is now the United States, the first Spaniards known to the native population were explorers. The southeastern mainland of North America was first visited in 1513 by Ponce de León during Easter, and the name Florida came from the Spanish *Pascua Florida,* "Feast of Flowers." Ponce de León thought that La Florida was another of the Caribbean islands. Unlike the gentle Arawaks, whom Columbus had described as "wonderfully timorous," however, the Florida natives fought back and fatally wounded Ponce. His death brought Spanish exploration to a temporary halt.

Ponce was followed nearly a generation later by Lucas Vázquez de Ayllón, who had made a fortune from sugar in the Caribbean. Not much is known of his expedition, but it may have reached the Chesapeake, and it certainly touched at what he called Santa Elena (roughly halfway between the modern cities of Charleston and Savannah) in a land the Indians called Chicora. Finding nothing of value that was portable, the captain of Ayllón's ship took to kidnapping Indians to sell as slaves.

Learning of his discoveries, the government pushed Ayllón to plant a colony. Without much enthusiasm, he embarked in July 1526 in six small ships with some 500 men, including several Indians who had been taken to Hispaniola and taught Spanish. Along with tools and supplies, they also took eighty-nine of what they had learned was their most fearsome weapon of war, horses. They came ashore near Cape Fear (in what is today North Carolina), where, to their horror, their largest ship ran aground and was lost. The Indians they had carefully prepared to be their guides and translators immediately (and wisely) fled. Without them, the Spaniards were essentially blind on an unknown shore. Using their horses, they searched the hinterland and decided upon an island site just offshore from the modern Georgetown, South Carolina. There they established a settlement they named San Miguel de Guadalupe. The site was not a good choice. Almost immediately the colonists began to die of fever. When Ayllón himself died, the 150 who were still alive gave up and sailed for Hispaniola. Spain's first attempt to create a colony in North America had failed.

An even more ill-fated expedition followed in 1528, when Pánfilo Narváez landed 400 men and about eighty horses at Tampa Bay. Lured inland by the hint of a source of gold, he lost touch with his supply ships. Then he and his men had to live off the Indians. Although initially friendly, the Indians became tired of the Spaniards' thefts and began to harass them. Many of Narváez's men learned painfully that Indian arrows could pierce Spanish steel; other Spaniards succumbed to fevers. When the expedition disintegrated, as quickly happened, it was every man for himself. One of the men was Cabeza de Vaca.

Cabeza de Vaca made off with a few companions and began what would be an incredible eight-year odyssey across America. Not knowing where he was but trying, vaguely, to reach Spanish Mexico, he won Indian hospitality by acting as a medicine man. As word spread that he could heal the sick, he not only survived but gathered about him a party of Indian followers. Traveling with some of them, he finally encountered a Spanish slaving party far to the west. To his disgust, his fellow Spaniards wanted to enslave his companions. He admonished the Spaniards, he later wrote,

> that we healed the sick and they killed the healthy; and that we came naked and barefoot and they well dressed and on horses and with lances; and that we did not covet anything, rather we returned everything that they gave us and were left with nothing, and the only aim of the others was to steal everything they found, and they never gave anything to anyone.

The Spaniards' lack of politeness was soon followed by Spanish genocide.

Hernando de Soto had taken part in the conquest of the Inca empire of Peru; the loot he took there had made him one of the richest Spaniards of his time. He even lent money to the king. Perhaps to get rid of him, the king sent him off to "conquer, pacify, and people" La Florida. For those tasks, de Soto assembled an expedition over twice the size of those of Ayllón and Narváez. Undoubtedly an able man, he was also intensely cruel, reputedly "much given to hunting Indians on horseback." Like other Spanish conquistadors, he also used attack dogs against the Indians, animals "so fierce that in two bites they laid open their victims to the entrails." Dogs and horses terrified the natives, who were not used to dealing with animals. The use of dogs and horses would be given full scope during de Soto's rampage through what became the modern states of South Carolina, Georgia, Florida, Alabama, Mississippi, Louisiana, and Arkansas.

Well prepared in advance by having brought iron chains and collars with him from Spain, de Soto used such Indians as he did not immediately hang or "put to the sword" as pack animals, taking their food, robbing their tombs for trinkets, and raping their women. Leaving behind a five-year-long trail of burned-out villages, rotting corpses, and mutilated Indians but gaining little knowledge and no gold, he reached the Mississippi (which the Spaniards called Río Espírtu Santo) and crossed it into what is now Arkansas. There he died in May 1542, having won a place in history as one of the most brutal and destructive of explorers.

Following on the heels of the explorers, and reaping the hatred they had sown, the Jesuits attempted in 1567 to establish the first of their outposts in La Florida. Within five years, the local Indians had managed to drive them away. The friars learned from this defeat that they could not control Indians who were living "in the state of nature." To render them susceptible to conversion, they had to be "reduced" into something as close as possible to a European pattern. *Reducción* began with relocating Indians in new settlements grouped around churches.

Such church settlements had to become self-sustaining, so the missionaries soon began to import European tools and animals. Since the Indians had had neither plows nor animals to pull them, the ox-drawn plow was the most revolutionary of the Spaniards' initial innovations. Cattle, burros, or horses and sheep or goats followed. Then, the priests taught the natives blacksmithing and weaving. Not much attention was paid to teaching literacy, but much consideration was given to dress, particularly to getting women to cover their breasts, so a local market grew for textiles. Indians, often driven by fear of the murderous human predators left behind by the explorers, began to flock into the settlements. At the peak of their activity, thirty-eight Catholic missions had jurisdiction over 26,000 actual or alleged Christian Indians.

Life in these mission communities was never easy. We now know from archaeological studies that the health of the Indian inhabitants declined drastically: instead of the varied diet they had obtained from hunting, fishing, and farming, they were reduced to corn. Vitamin and mineral deficiencies in this diet were apparent in skeletons excavated in the 1990s at the mission town of San Luis de Apalachee, where tribesmen from the Guale, Apalachee, and Timucua lived. And instead of getting their drinking water from running streams and springs, as they had done in times past, Indians in the mission communities had to draw water from wells that were often contaminated. As a result, bacterial infections were common.

Diet and drinking water were not the only sources of ill health. Skeletal evidence of osteoarthritis indicates that the Indians were worked extremely hard both by the priests and by *caciques,* or Spanish appointed headmen. They had to work hard, since no Spaniard "comes to the Indies to plow or sow," as one of the Spanish viceroys remarked, "but only to eat and loaf." The Indians of La Florida thus became the helots of the New World.

In addition to the score of missions established in the northwestern and northeastern parts of the modern state of Florida, the Spaniards sent out missions to more remote places. In a remarkable coincidence, a group of Jesuits briefly established themselves in 1570 within about five miles of the Chesapeake site where the English colony of Jamestown would be planted thirty-seven years later. The Jesuits had taken along with them an Indian who had been kidnapped from the area. Having been "reduced" during a long residence in Spain and Mexico, he not only wore Spanish clothes and spoke Spanish but was, or pretended to be, a devout Christian. No sooner was he home, however, than he "went native." Having seen Spain and Mexico, he must have had a clear idea of the real meaning of *reducion,* and he hated it. So he assembled a group of Indians and killed the missionaries. The Historian Carl Bridenbaugh speculated that this man, known to the Spaniards as Luis de Velasco, may have been

Opechancanough, who reputedly led the revolt against the later English colonists at Jamestown. That act was far in the future, but whoever killed the Jesuits so disheartened the Spanish missionary movement as to clear the way for the English colonists.

Spanish colonialism in the southeast of North America was not a major success. The early explorers had found the area unsuitable. Hernando de Soto reported that it was "full of bogs and poisonous fruits, barren, and the worst country that is warmed by the sun." Pedro Menéndez de Avilés cautioned King Philip II that "Florida's shoreline was too low and sandy, her countryside too poor in resources, and her harbors too shallow to permit practical settlement." Fevers carried off more would-be colonists than the Indians killed. Hurricanes worked havoc on shipping. And, finally, there was no gold—the one thing that might have drawn a large settlement. As it was, few Spaniards went; those who did were almost all men without Spanish wives. Marriage or concubinage with Indian (and black) women was common and was at least initially favored by the state. In marriage, as in all aspects of life, the Spaniards regarded the Indians as a resource. If not killed or driven away, they had to be refined, reduced, converted—and bred. When the Indians tried to resist, they were flogged or starved, but they were never fully cowed. They rebelled time after time; and finally, between 1680 and 1706, the missions collapsed.

Saving the Indians was not enough to sustain the Spaniards' interest, and colonialism was rarely profitable, but the American Atlantic coast had a strategic value that Spain could not afford to neglect. This coast was along the route on which Spain conducted its main business in the New World: shipping home gold and silver.

Gold and silver from Peru were carried to Nombre de Dios on the eastern coast of Panama and from Mexico to Veracruz; from these two ports, the treasure was placed on ships and then taken to Havana, where, from the 1560s on, one or two convoys assembled each year. These convoys, which the Spaniards called *las flotas*, caught trade winds and the Gulf Stream through the Florida straits and sailed along the Atlantic coast up to the Chesapeake. There they turned east, catching the trade winds to the Azores. Since the *flotas* were Spain's economic lifeblood, they had to be protected at almost any cost. The Spanish government thought of its attempts at colonization of La Florida as part of that cost; it was determined to maintain the security of the route and to prevent others from intruding. That "vital national interest" would bring about a clash with France and Britain; and the attempt of both other powers to intrude on Spain's treasure route set them on the road to colonialism along the eastern coast of North America.

CHAPTER 4

Fish, Fur, and Piracy

Centuries before Columbus's voyage to the Caribbean, Vikings had sailed across the northern ocean toward, or perhaps even to, the North American mainland. By the thirteenth century more than 5,000 of their settlers lived in Greenland. Thereafter, as the "little ice age" began, and the Greenland shores were more frequently and more deeply blocked by ice floes, the colonies lost contact with Scandinavia; at about the time Columbus sailed, they had been abandoned.

In turn, the Vikings were followed by fishermen from the little ports along the Atlantic coast of Portugal, France, England, and Norway. They were not trying to discover or map the distant reaches of the earth and certainly not trying to establish colonies. Their limited aims were purely commercial. As long as they were allowed to trawl and hook close to Europe, they did so; fishing in the North Sea around Britain and Norway was safer and more economical than going far out into the Atlantic. But the powerful confederation of cities known as the Hanseatic League created a monopoly that excluded these fishermen from the North Sea in 1410. Thereafter, they were forced to seek a catch beyond the reach of the League.

The favored catch of that period was the cod. A large fish, it lent itself easily to drying or salting and was extremely rich in protein, which, in Europe, had become desperately scarce. The great famine of 1315–1321 had been followed by a long series of catastrophes that starved the population. Weakened, the survivors were less able to produce food. Even when grain was available, meat often was in short supply. It was also banned by the Catholic church on Fridays and during the profusion of religious holidays. So fishermen had a ready market for their catch.

The challenge fishermen faced in servicing this market was daunting: the little boats they had at the beginning of the fifteenth century were no match for the powerful storms of the North Atlantic. But, responding to the closure of the North Sea, shipwrights along the Atlantic coast adapted the combined lore of the Mediterranean and the North Sea to produce a sturdy, two-masted, deep-hulled ship known as a dogger.

The dogger gave the northern Europeans a vessel in which they could confront the tempestuous northern seas, just as the carrack had given the southern Europeans. Whereas the carrack had evolved from warfare, the dogger was designed purely for fishing and was an early example of private initiative. In a dogger, fishermen were able to chase the cod far out into the Atlantic. It was a desperate chase. The cod nearly eluded them. As great stretches of pack ice chilled the waters around Greenland, the huge schools of cod migrated southwestward all the way down to Cape Cod. At what cost in sunken ships and drowned crewmen we will never know, but the fishermen followed the cod. They had certainly reached Iceland by 1413. By roughly the time that Columbus sailed, thousands of Basques, Portuguese, Frenchmen, Englishmen, and other western Europeans had found their way to the shores of Newfoundland and probably to the North American mainland.

These voyages were financed by merchants in Dieppe, La Rochelle, and Bayonne and across the Channel in Bristol and other towns in the west of England. Unlike the Spanish, the French and English governments did little to help or control their sailors. So these sailors lacked the resources to found colonies and had no interest in founding any. They simply wanted beaches in coves along the coast of Newfoundland and New England where they could clean, dry, or salt their catch. There, while they camped out and were busy processing their catch, they were approached from time to time by curious Indians. We must imagine, since we have no records, what happened then. They must have swapped small items such as nails, fishhooks, twine, or simple tools for the fur capes the Indians were wearing. When they took these capes back to Europe, as we know from portraits painted then, fur became a symbol of status for royalty and rich merchants. The furs the fishermen brought set off a fashion craze that would continue for the next two centuries and would do much to shape America.

As W. J. Eccles has pointed out, the French government initially saw the fur trade as a means to finance the Catholic missions that aimed to convert the Indians; but from about 1700 the trade was used "mainly as a political instrument to further the imperial aims of France."

Inadvertently, the fishermen also introduced smallpox and other European diseases against which the Indians had no immunities. By the late sixteenth century and the early seventeenth century, these new diseases had virtually wiped out at least one tribe, the Patuxet, in the Massachusetts area. In the North, as in the South, disease turned the European encounter with the Native Americans into a silent genocide.

For a century or more, the British, Dutch, and French governments paid little attention to the fishermen. But as the fur trade, fishing, and piracy grew in scale, these governments began to take a greater interest in Atlantic shipping. Unlike the Spanish and Portuguese monarchies, the northern European states had no significant navies. But they soon discovered that they could create a virtual navy from the already armed, sturdy doggers. To enlist the sailors of these little ships, they had to offer an incentive. The easiest (and cheapest) incentive was to let sailors engage in what amounted practically to piracy. From time to time, the governments provided the fishermen with documents known as letters of marque, which turned (private) pirates into (official) privateers.

The French led the way in using privateers against the Spanish. François I spent much of his reign locked in a bitter conflict with Charles V of Spain. On land, François I could do little about the papal bull of 1493 or the Hispano-Portuguese Treaty of Tordesillas of 1494, which effectively cut France off from the New World. Spain's infantry was then unbeatable, as François learned in 1512, but already fearful of the formidable Spanish infantry, he had entered into an alliance with Henry VIII of England (in the grand concourse of the "Cloth of Gold" in 1520), but England added little to French power. If on land he was blocked, François could make at least a symbolic gesture at sea. That gesture was to send a mission to the New World to dispute Spain's monopoly. Since France had no man with the necessary technical skills, François hired another of those restless Italian seamen who went to work for foreign magnates, Giovanni da Verrazzano.

Verrazzano set out in 1524 toward what he, like Columbus, thought was Asia. He hit the North American continent far to the north of where Columbus landed, at about Cape Fear in what is today North Carolina. Although at first he found no place to land, the fires he observed convinced him that the land was filled with "Chinese." Seeking a port or safe inlet, he turned south for about a day's sail and, then, still finding no suitable harbor, he doubled back to the north. Somewhere along the coast, probably in what is now Virginia, he did find a place to land. There, he convinced himself that the people he saw were indeed Chinese.

These "Chinese" were curious rather than hostile. As far as we know, they had had no previous contact with Europeans: Ponce de León, who had discovered Florida in 1513, had not traveled that far north, and the ravages of de Soto would not begin for another fifteen years. Verrazzano wisely did little more than observe. He had been sent only to gather information.

The information he wanted, how to find the supposed passage to the Pacific, the natives could not tell him. So he continued close along the coast to the north, where he found what he took to be the entrance to the grand passage. What he saw was probably Chesapeake Bay. (He does not mention the word, but the Indians later told the Spaniards that they called it Jacán.) Still farther north, he found a waterway later known as the Narrows "in the midst of which flowed to the sea a very great river, which was deep within the mouth." That, we assume, was the Hudson. Again he found the natives to be friendly. Apparently seeing the visit as a supernatural event, they turned out in a grand festival, "clothed with the feathers of birds of various colors . . . to see us."

Quite a sight it must have been, but Verrazzano was not to be delayed. He feared the advent of stormy weather. Pushing on north, he paused at another bay where the Narragansett people—perhaps reacting to earlier visits by fishermen—were less welcoming; continuing, he rounded Cape Cod, where European fishermen almost certainly had been. The farther north he got, the less hospitable the Indians became. Their acquaintance with Europeans had led to aversion: the more the natives knew of Europeans, the less they wanted to see any. On the Massachusetts coast, the Native Americans were actively hostile. While they were willing to trade, they wanted no contact. Gathering on the tops of cliffs, they lowered in woven baskets the goods they were willing to swap. Verrazzano's men, standing in the ship's boat, then placed in the baskets what they were willing to give. As the shocked Verrazzano watched, the Indians concluded their silent trading session with obscene dances and gestures to show that they hated the sailors. Verrazzano made no friends, but he accomplished his mission.

On the basis of Verrazzano's trip, France proclaimed that the Atlantic coast of North America was not La Florida, as the Spaniards had announced, but the rightful possession of France, Terre Francesca. Probably that gesture satisfied François I, since, other than putting the name on maps, the French made no further moves at that time. François's eyes were glued on Spain, and it was the then-powerful Ottoman Empire on which he was basing his strategy. A decade later, however, a new opportunity arose in America. The Vatican, then under the very worldly Pope Paul III, decided that France too should be given a stake in the New World; the pope ruled that the grand division of the world between Spain and Portugal had not covered territories then not yet discovered. What is now the United States and Canada was there for the taking. François decided to commission another voyage, and the French Catholic church offered to help finance it.

This time, the king found a suitable Frenchman, Jacques Cartier, from the little port of Saint-Malo in Brittany, to lead the expedition. Cartier, the king was told, was already an experienced mariner who had visited "Brazil and the New Land." Legal sanction and finance having been arranged, Cartier set out in 1534 with two small ships and sixty-one men to explore the Gulf of the Saint Lawrence. There he found that the Micmac Indians were already quite familiar with European traders, some of whom he also met as he made his way west. Deeper into the gulf, on the Gaspé Peninsula, he encountered a number of Huron Indians from among whom he kidnapped two to take back to France. So far he had not seen any signs of the "great quantities of gold" the king had expected. But he reported that there was hope; on that hope, a new venture was to be mounted the following year, with a larger group of men in three ships. Again in command, Cartier determined to find out if the Saint Lawrence was the route to the "South Sea." Sailing when he could and rowing when he could not sail, and guided by the two Hurons, whom he had brought back from France, he went almost 1,000 miles up the Saint Lawrence until he reached the rapids at the site of modern Montreal.

Where, Cartier asked every native he met, was the way to Cathay, and where could he find gold? The Iroquois who lived in the area of Montreal gave him the answer that had become standard when the southern Indians were questioned by the Spaniards: "Go west or north or anywhere far away from here."

Since the area Cartier reached was south of the latitude of Paris (and no one then understood the effect on Europe of the Gulf Stream), Cartier did not prepare for intense cold. Almost half his crew died during the bitter winter. The hospitable Iroquois tried to help, but living in a subsistence economy, they had little to spare. By seizing what the Indians could not give, Cartier's party turned hospitality into hatred. This early false step was to color Franco-Indian and Franco-British relations for more than a century throughout New England, Canada, and the lands around the Great Lakes.

War in Europe again distracted France, and it was not until six years later, in 1541, that Cartier was able to return, this time to establish a stronghold near modern Quebec, but founding colonies proved far harder for the French in the north than for the Spanish in the Caribbean. The French abandoned their outpost after just two years. Disappointed by what they had experienced in America, and distracted by war with England over Brittany and Normandy and by continued clashes with Spain, France dropped its efforts in the New World for half a century. But there was more to America than frozen Canada. The vast wealth of Spain's treasure fleets had attracted the attention of French pirates almost as soon as the Spaniards found Aztec and Inca gold. In 1559 the Franco-Spanish treaty of Cateau-Cambrésis brought a respite from attacks, but the lure of the treasure *flota* was irresistible. In the shifting politics of Europe, new alliances were replacing old, and, in its hostility with Spain, France occasionally found common cause with England. One Englishman who led the way against Spain was Francis Drake.

A passionate Protestant who hated the Spaniards, Drake became one of the most feared of the English raiders. His raids convinced the Spanish government that its transport of the gold of Mexico and Peru was in grave jeopardy. If any foreign power could establish a base on the Atlantic coast—that is, along the route the treasure fleet had to take to cross the Atlantic—Spain itself would be in mortal danger. As the Spanish general and statesman Pedro Menéndez de Avilés warned his king in 1564, if "French or English or any other nation" were allowed on that coast, "they would be able to establish a site and fortifications that would enable them to have galleys and other ships of war, to take the *flotas.*" Menéndez was right, and his report convinced the Spanish government. Consequently, Spain was determined to deny North American lands to other powers.

The first challenge to the security of the Spanish sea lane came from France. Spain regarded this challenge as particularly dangerous and offensive because France not only was a rival power but was guided in its anti-Spanish policy by Protestant (Huguenot) statesmen whose coreligionists, led by Jacques de Sores, had recently sacked Havana. If the Spanish had caught them, they would have burned them at the stake as heretics. The Spaniards' anger turned to fury when the noted French cartographer Jean Ribault, also a Huguenot, was commissioned to lead the first French expedition in 1562 designed to contest ownership of La Florida.

The French realized what they were getting into; and to avoid rousing the Spaniards, at least until they were in a position to defend themselves, Ribault was ordered to avoid all contact with the Spaniards. An able navigator, he avoided the Canaries and made an unprecedented nonstop passage from Dieppe straight across the Atlantic to a point just south of modern Saint Augustine. Then, sailing up the coast, he found what appeared to be a suitable (and vacant) location roughly halfway between modern Charleston and Savannah. There he established a provisional base, which he named Charlesfort. By that time, the hurricane season was approaching, so he left a small contingent and hurriedly returned to France to get additional supplies.

The France to which Ribault returned was not the one he had left: the first of the terrible wars of religion that pitted the Catholics against the Protestants had begun to tear France apart. Ribault was unable to get the attention of anyone in a position to assemble supplies for the new base in America. Meanwhile the colonists lost heart. No food or supplies were reaching them from France, and they were unwilling to work to feed themselves as farmers. So when they got hungry, as they quickly did, they looted the neighboring Timucuan Indians, who, in return, began to attack them. Dispirited, they decided that they should give up and return to France. But they had no boat. None of their group knew how to build one. But they were desperate and determined. So, cutting down pine trees, they shaped them as best they could into beams and clapboard siding. They caulked the gaping holes left by their poor carpentry and by the green wood with moss; for sails, they stitched together their shirts. The boat had no deck or cabin to protect them from sun and sea. And they had little food to store in it. None of them knew anything about navigation. But twenty-six of them—one boy prudently stayed behind—pushed off into the Atlantic. Drifting more than sailing, they were carried along by the Gulf Stream and the prevailing winds. As the weeks turned into months, the twenty stronger men ate the weaker six. Finally, even the remaining twenty were mere skeletons. Yet almost unbelievably, their nearly swamped boat reached the English Channel. There they were rescued by an astonished English ship. Undecided whether to treat them as survivors of catastrophe or as cannibals, the English put them in jail. Their odyssey is one of the most ghastly tales of American colonialism.

While these events were unfolding to their gruesome end, the departed leader fled from France. As a Huguenot, Ribault feared for his own safety and sought refuge in England. There, the government was unsure what to make of him. Was he some sort of spy or a valuable recruit to England's small group of able navigators? Probably to try to convince the English that he was a prize addition, Ribault published his account of the new colony.

Publishing what he had been doing was not a wise move. It frightened the Spanish ambassador in London into sending a panicky report to King Philip II, who immediately issued orders to establish a new Spanish fort, Santa Elena, near Charlesfort and to scout the entire coastline for hostile invaders. The Spaniards speedily found and burned what remained of Charlesfort.

Meanwhile, knowing nothing of the fate of Charlesfort and despite the terrible conditions of the civil war, the French government managed to put together another expedition. It carried some 300 colonists to establish a new colony, called Fort Caroline, near the mouth of the Saint Johns River. Under the command of Ribault's former deputy, René Goulaine de Laudonnière, the venture was doomed from the start. Looting Spanish treasure ships was what the colonists had signed on to do; they had not come to labor like peasants in the fields. As the English pirate John Hawkins remarked of them, "they being soldiers desired to live by the sweat of other men's brows." So when they got hungry, as they quickly did, they looted the nearby Indians. When Laudonnière tried to stop them, a group of rebels made off with the colony's two small boats to engage in piracy. When they returned, Laudonnière, fearing that his renegades were likely to bring the Spaniards down on him, had them hanged. But he was too late. The Spaniards had learned what was afoot and reacted like bees whose hive had been attacked: for them, Fort Caroline was not a colony but a pirate lair. Their fears seemed confirmed when they learned that John Hawkins, fresh from a raid on Spanish possessions in the Caribbean, had sailed in for a visit. That was the last straw. Philip II ordered an attack.

Acting on Philip's orders, his skilled and ambitious general Pedro Menéndez surprised the French at Fort Caroline and killed all the 132 men he caught; René Goulaine de Laudonnière and about fifty others managed to escape. Menéndez caught up with them and tricked them into surrender. He then "caused their hands to be tied behind them, and put them to the knife." They were killed, Menéndez reported to his government, not as Frenchmen or as pirates but as Protestants. The war of religion had come to ground in America. It would never again be far from men's minds.

The Spanish knife effectively ended French activities in La Florida, but the French would not give America up. Twenty-two years later they prepared a new expedition. During the interval of peace between the two powers, the man who was to lead it had a chance to study the reasons for Spain's success. Sammuel de Champlain spent nearly two and a half years in the Caribbean, where he commanded a French ship chartered to sail with the Spanish *flota.* As the editor of Champlain's *Voyages* wrote, "He alone of all the great leaders in the colonization of North America had the privilege of observing and studying a European colony before he tried to found one."

Champlain was a good observer and writer. Ordered to report in writing on all he saw, he published an account of his 1603 expedition to the Saint Lawrence under the title *Des Sauvages, ou, Voyage de Sammuel Champlain.*

His immediate superior, the Huguenot Sieur de Monts, received a charter from King Henri IV granting him all the lands north of 40° latitude (roughly the site of modern Philadelphia). De Monts was to explore what became the northeastern United States and eastern Canada and settle the territory in the name of France. Under his direction, Champlain began a series of voyages along the North American coast that would take the French sailors south around Cape Cod almost to what is now Rhode Island. There, the Indians received them with the same hostility with which their fathers had received Verrazzano. Perhaps discouraged, he returned to the Saint Lawrence. There in 1608, he helped to found a trading post at what became Quebec.

Unlike the Indians near Cape Cod, those in remote Canada met his party with singing, dancing, and a *tabagie,* or feast. A realist, Champlain was aware that the Indians were delighted to see him because they had decided he could be of use to them. The nature of that utility was immediately evident. The Hurons (Ochateguins) and their Algonquin allies were locked in mortal hostility with the Iroquois. As Champlain witnessed a gruesome scene of torture, killing, and eating of Iroquois prisoners, he concluded that that war was likely to be perpetual and that each side would seek all the help it could get. Since he happened to be with the Hurons and the Algonquians, he chose to side with them. The firepower of his arquebuses would make the difference between life or death for them. He thus established what would become a persistent theme of French (and English) dealings with the Native Americans, intervening in local wars to emerge themselves as the victors.

Alliance with the Algonquians and Hurons against the Iroquois was to be the heart of Champlain's strategy; it was to lay the foundation for French Canada. A similar strategy would later win allies among the Illinois, the Miami, and other Indian peoples who also were terrified of the powerful Iroquois confederation. That policy would shape the America that began to emerge around the Great Lakes. Far more astute and humane than his Spanish predecessors or his English and Dutch successors, Champlain tried to win the "hearts and minds" of the natives—or at least the hearts and minds of one faction of them.

The policies Champlain originated were followed by his successors out of necessity. La Nouvelle France never had a large population of French colonists. In half a century of efforts to build a colony, only about 6,000 Frenchmen managed to establish themselves. They could never displace the Indians, as the Spaniards did and the English were to do, but had to work with one part of the natives against another. Moreover, the French did not take Indians' land; they did not need it. With a vast area available and few people to work it, French colonists were never driven by the hunger for land that permeated "British America." Their major thrust was toward trade, and trade meant fur. Since they could not catch fur-bearing animals themselves, the French had no incentive to plunder the Indians as the Spaniards did in the Caribbean and La Florida. Rather, they distributed what to them were cheap trade goods in exchange for fur. This exchange soon gave rise to a new kind of Frenchmen, the *coureurs de bois.*

The *coureurs de bois* were counterparts of the English and Scots "Indian traders," who later established themselves in Indian villages to swap guns, cloth, and liquor for furs. Dressed in Indian clothing, living in Indian houses, speaking Indian dialects, they often "married" Indian women and begat half-breed children. More than any other group, they would be the interface of Europeans and Native Americans.

CHAPTER 5

Society and Wars in the Old Countries

Apart from the Spaniards who were drawn to the New World by the lure of riches, most colonists were not so much coming to America as going away from Europe. English, French, and German migrants had suffered, often egregiously, in the "old countries." They took with them the emotional baggage of their travail, so that what they created in the New World was in part shaped by their memories of the Old World. For them as for the Spaniards, we can say that American history begins in Europe.

Much about European society and government in the sixteenth and seventeenth centuries would seem unusual to modern Americans. Governments were small and had limited resources. They were made up of the personal servants, retainers, relatives, and friends of the monarch's coterie. When a ruler appointed a man to manage one of the functions of the state, which were originally just the chores of the monarch's household, the appointee brought his own retainers with him. When he was dismissed, he took them away with him. Titles of personnel in the British and French governments betray their original functions as the monarch's personal servants. In the sixteenth century, and for generations thereafter, the entire government of the British empire numbered less than 1,000 men. The government of France was only marginally larger.

With so few officials putting into effect the decisions of the king, his ministers, judges, and the legislature, citizens or subjects, even those living in the capital cities, seldom heard of edicts. A typical French subject, as the

French statesman and philosopher Montaigne remarked, was unlikely to come into contact with the authority of the king more than once or twice in his lifetime; if he was wise, he did all he could to avoid such contact. In Britain, subjects of the king likewise were little affected by government. As one modern historian of England has commented, it was what governments did not do that was impressive.

Having limited aims, government was cheap. In England, the government spent less than 7 percent of the estimated gross national product (GNP) during most of the seventeenth and eighteenth centuries and spent most of that on the army and navy. From the tiny force of just 9,000 in 1685 it still numbered only 27,000 just before the outbreak of the American Revolution in 1775. It usually counted on being able to "rent" foreign armies as needed. For instance, in 1747, during the War of the Austrian Succession, it tried (but failed) to hire 36,000 Russians to fight the French; instead, it hired (as it would do later during the American Revolution) thousands of soldiers from the petty German principalities. However, the British navy was the strongest in the world, with twice as many capital ships as Spain and almost twice as many as France. It was the English government's biggest expense. France's army in the eighteenth century reached 170,000 men and could be doubled in time of war.

What the British and French governments did, other than fighting foreign powers, was "privatized." Government was effected mainly through proxies—contemporary inheritors of feudal fief holders, local authorities, religious institutions, and nongovernmental organizations. Neither France nor England had a national police force until well into the nineteenth century. France relied upon paramilitary vigilantes, known as the *maréchaussée,* who operated as though France were an occupied enemy country, while England employed mercenary "thief takers" to suppress domestic enemies. Thief takers made no attempt to prevent crime; all they did was (for a bounty) catch those who had already committed crimes. Therefore, the more affluent neighborhoods employed their own watchmen. Scores of watchmen walked the darkened streets of London, but they had no official standing and were mostly elderly, often feeble, and always unarmed. To avoid confronting a criminal, they heralded their approach by thumping the cobblestones with their staves, swinging their lanterns, and calling out, hopefully, "All's well." Real protection, or at least retribution, was also private: provided by furious citizens or bored bullies in a "hue and cry" mob.

Britain and France also privatized the collection of taxes. Just as in the "Oriental despotisms"—Ottoman Turkey, Safavid Persia, and Mughal India—England and France auctioned to syndicates of merchants or granted to court favorites the right to collect taxes in a given area for a specified length of time. The "tax farmers" paid the government lump sums for the right to squeeze whatever they could get from the peasant farmers, artisans, and merchants. Thus governments got the money they needed to make wars while their most important supporters were enabled to live in grand luxury. This system of privilege and exploitation was later to disgust the frugal, hard-working "middle sort" of Americans such as Benjamin Franklin and convince them of the ultimate corruption of European states. In their eyes, the sordid reality of Europe justified the American Revolution.

The French state was much more centralized, and it was centralized earlier than the English state. Whereas in seventeenth-century England the great landowners tended to live on their country estates, Louis XIV compelled the French nobles to reside at court. In England, usually under the leadership of the aristocrats, local councils took charge of social issues, including provisions for the indigent and sick; great and small landowners organized themselves to be represented in Parliament. Contrariwise, in France the Estates General was "simply forgotten," and local institutions hardly existed. Rather, the central government appointed officials (*intendants*) who acted as its agents to execute the policies it handed down.

Church in the European countries of the sixteenth, seventeenth, and eighteenth centuries also meant something quite different to contemporaries than it does to most of us. It resembled more closely what today's Christian, Jewish, Muslim, and Hindu "fundamentalists" attempt to institute than what the "mainstream" actually experiences in Asia, Europe, and the Americas. Church took up some of the slack of government inaction. It not only ordered "religion" in the narrow sense of that word, but it also defined community; provided or stimulated a number of commercial and cultural activities; and, when given a chance, defined and controlled the economy and organized and ran government.

Under the surface of organized religion was a substratum of custom amounting at times virtually to nationhood. It was this combination of custom and belief that expressed itself as "church." Sixteenth-century monarchs sought to define and dominate religious institutions and to suppress those that did not fit in. Henry VIII broke with the Catholic church, confiscated most of its property, and established the Church of England. In France, a hundred years later, Louis XIV nearly followed suit. In his clash with Pope Innocent XI, he cut off contacts between the French clergy and the Vatican; and in the archbishop of Paris, he found a Catholic churchman willing to become "patriarch of the Gauls." The split stopped short of the English schism only because Innocent died and Louis was distracted by his European wars. Heterodoxy was everywhere regarded as tantamount to treason. Each ruler sought to enforce orthodoxy as he defined it; no ruler was willing to tolerate diversity of faith. Henry VIII's break with the papacy was not intended to create religious freedom, although that was, in part, its effect. Henry's successors all sought to impose their own orthodoxy on the whole country. Charles I cracked down on "dissenting" Protestants; Cromwell brought the Puritans to power; the restored monarchy under Charles II and James II then attempted to reimpose a quasi-Catholic orthodoxy. Repressive acts under Charles II—collectively known as the Clarendon Code—amounted to a purge. By the Act of Uniformity in 1662, members of the clergy who did not conform lost their state-granted benefices; dissenters from the Church of England were excluded from the universities and schools; were forbidden to assemble in groups of more than five persons; and upon suspicion or denunciation, were imprisoned, deported, or hanged. Under James II, an act of attainder declared thousands of Protestants guilty of treason and made possible the confiscation of their lands. Terrified, many would leave for America.

When Calvinism took hold in France, the French Reformed communion—the religion of the Huguenots—was eventually adopted by about 10 percent of the country. The Catholic church then used the state to fight it. The massacre of Huguenots on Saint Bartholomew's Day in 1572 set off a century of turmoil. As an English newsletter reported in 1682, "The King is resolved to make his Huguenot Subjects grow weary either of their lives or of their Religion." Three years later, Louis XIV decided to drive the Huguenots into exile. Almost half a million fled France for Holland, the

German principalities, and England; like the English dissenters, many eventually found their way to the American colonies.

These actions by the French and English governments, in addition to the general insecurity of life in the German states as a result of wars between the powers, explain why so many members of the dissident communities of Lutherans, Calvinists, Quakers, Anglicans, Puritans, and Moravians came to America. Because they were to play so important a role in America, their experiences are pertinent.

In England by 1629, the attempt by the Church of England, actively supported by King Charles I, to destroy Puritanism had reached a level of intensity that the Puritan leaders found intolerable. John Winthrop, later governor of Massachusetts Bay Colony, wrote that England had become "weary of her Inhabitants, so as man which is most precious of all the Creatures, is here more vile and base than the earth they tread upon." From their leaders the call went out all over England for the Puritans to gather their families, their animals, and other possessions; sell their property; and take ship for the New World. Whole communities answered the call. In the next year, seventeen ships sailed for Massachusetts in the vanguard of what would become a flood of some 21,000 people.

Quite different from the exclusive and intolerant Puritans, the Quakers were in practice what they called themselves, "Friends." They were multinational in origin; whereas most Puritans came only from southern England, many Quakers were Welsh and Irish. But like the Puritans, the Quakers migrated because of religious persecution in the Old World. Their meetings documented their travails in the *Books of Sufferings.* In addition to physical abuse by their neighbors, they were prosecuted, were jailed, and had their property seized by the government for refusing to pay taxes to the government-recognized church from which they dissented. Marching to their own drummers, the Quakers were sometimes treated as "sturdy beggars and rogues," arrested, and sent off as indentured servants to the New World. The courts in England had no sympathy for them, and they got little in America either. Both the Anglicans of Virginia and the Puritans of Massachusetts drove the Quakers out, had them whipped, or even executed them. Treated as pariahs in the established colonies, the Quakers too felt the need to form a self-governing community. Their first significant settlement was in West Jersey, where in 1675 they founded Salem (a name adapted from the Hebrew *shalom*). Then in 1682 some 2,000 sailed across the Atlantic with William Penn to found Pennsylvania.

The Moravians (Brüdergemeine) had a different background and different reasons for going to the New World. Their church grew out of a fifteenth-century central European offshoot of the Catholic church; after a long period of persecution and eclipse, this offshoot spread to Germany. In 1732, Moravian missionaries reached the West Indies, where they worked among the black slaves; eight years later, a group of Moravians established their first settlements, Nazareth and Bethlehem, in Pennsylvania. Thus, although the Moravians began their movements from central Europe as a consequence of persecution, they went to the New World to spread their interpretation of the Gospel.

Socially, as well as religiously, seventeenth- and eighteenth-century England and France were deeply divided. The division was not so evident as the division by color in present-day American society; but culturally, economically, and politically it was nonetheless sharp and painful. I will focus on England, but much of what I say is also applicable to France.

The English middle and upper classes, who have bequeathed to us elegant furniture, exquisite silver, serene paintings, and graceful architecture, floated above a swamp of misery. "Swamp" is not just a metaphor: the narrow, nearly airless streets of English cities, particularly London, were also open sewers. The refuse of butchers, tanners, and pliers of other noxious trades lay rotting in the streets, along with the bodies of animals and people. The bulk of London's inhabitants endured poverty, filth, disease, and ignorance that today could probably be matched only in the worst slums of Calcutta. Hordes of rats, fighting with the living for food and gnawing their way through the dead and dying, give an English actuality to the German legend of the Pied Piper. Crowded into that slough, the poor lived precariously on the brink of starvation. We see the English "poore folkes" today mainly through the eyes of novelists such as Charles Dickens, Daniel Defoe, and Henry Fielding; the playwright John Gay; and the artist William Hogarth.

The London poor lived segregated in slums where unemployment reached monumental proportions, and hunger was a constant companion.

The great famine of 1315–1321 was followed by a long series of "hungers" including one in 1594–1597 just on the eve of the English move toward the New World. Driven over the edge, the poor died in the streets of many towns and cities. An entry in the Newcastle town accounts is eloquent in its simplicity: "October 1597. Paid for the charge of buringe 16 poore folkes who died for wante in the strettes 6s.8d."

"Poore folkes" were not just the unemployed. Even the employed suffered. Occupations we now think of as menial were then regarded as privileged; for others we can find no comprehensible parallel in our experience. Probably not even the most desperate immigrants from the Third World would work as "pure-finders"—collecting dog excrement to sell by the bucket to tanneries.

We do not know much about how the seventeenth-century poor escaped from their daily misery and hopelessness; but by the beginning of the eighteenth century those who could afford it escaped as often as possible into gin-induced oblivion. Marx was wrong—it was not religion that was the opiate of the people; it was gin. Gin, which became popular after 1720, was the crack or heroin of that age. By the middle of the eighteenth century, London was said to have had one gin shop for each 120 inhabitants. As far as they were able, the English poor stayed drunk, consuming an average of 8 gallons of gin a year. Henry Fielding wrote that gin was "the principal sustenance of an hundred thousand people in this metropolis." William Hogarth immortalized the scene in "Gin Lane." Daniel Defoe thought that Parliament, dominated by the gentry, favored cheap gin because the members' estates produced the grain from which gin was made. Probably a more persuasive consideration, although not so openly admitted, was that gin immobilized the poor, the "criminal classes," who might otherwise revolt. The criminal classes appeared to the upper class an alien race. Even their language was virtually unintelligible to the wellborn. Writers and legislators agreed that there was no practical way to integrate these lower classes into "English" society. It followed that they had to be rigidly confined, severely repressed, and constantly reduced in number. Gin was a tool of repression.

Neither in England nor anywhere else was it the absolutely poor, the group Marx called the *Lumpenproletariat,* who migrated. Migration, at least voluntary migration, required a minimum amount of cunning, craft, and capital, which the poor did not have. But the appalling poverty all around them must have deeply influenced the marginally better-off who made up the bulk of the nearly half million eighteenth-century emigrants to the Western Hemisphere: even relatively affluent artisans must daily have peered over the edge of the abyss into which they knew they could easily tumble as a result of sickness, accident, or bad luck.

One of the arguments for colonialism Richard Hakluyt made to Queen Elizabeth was that enforced migration would rid the country of the "great number of men which do now live idely at home, and are burthenous, chargeable, & unprofitable to this realme." Parliament opened the abyss, sweeping into it a wide variety of the poor: beggars, out-of-work sailors and laborers, street entertainers, peddlers, tinkerers, crippled army veterans, and "counterfayte Egipcians." All these were "to be stripped to the waist" and whipped until bloody, and then sent to their birthplace or place of last residence or to a house of correction for a year. They could then be banished. The mouth of the abyss was wide, and slipping into it was very easy.

For abandoned children, falling into the abyss was inevitable. The streets of London swarmed with bands of such children; they were the survivors. Most others died. Some were captured; if guilty of a crime, they were hanged; if not known criminals, they were placed in workhouses or hired out to factories. Some were no more than four years old. Still, they proliferated. In 1617, to try to get rid of as many as possible, the lord mayor of London collected charitable contributions to send 100 poor children to Virginia. They were followed by other groups: about 1,400 were rounded up and "transported" that same year. Other groups followed in 1620, 1622, and 1627, when "fourteen or fifteen hundred children gathered up from divers places were being sent to Virginia."

For London's poor each new day could bring starvation, sickness, or death. Gin was one release. Another escape was crime. Crime was so broadly defined that it was, literally, a way of life. When the numbers got too high, pressure on the "criminal class" was sometimes slightly lifted. For example, town officials in Bristol reported that "the poor of our city were all relieved and kept from starving or rising." "Or rising" are the key words. In 1671, the duke of Albemarle wrote a memorandum on how to prevent civil war in England. He set out four recommendations, one of which was "the diversion, occupation, and control of the hopeless poor." But if bread and circuses did not work, the lower class had to be held in place by the awful certainty of inexorable and brutal punishment.

The punishments inflicted in seventeenth- and eighteenth-century England were garish, painstaking, and grisly. Convicted political offenders who were not lucky enough to go directly to the block were usually disemboweled while still alive. Then they were quartered. Some criminals were burned at the stake. Accused but not yet tried prisoners who refused to plead were often slowly "pressed" to death under heavy weights. However, the most common form of execution was hanging.

Hanging became London's main entertainment. If the poor had little bread, the government at least gave them circuses. One of the more celebrated hangings, in 1767, is thought to have been attended by one in ten inhabitants of London. As many as 25,000 routinely watched lesser hangings. Age was not a mitigating consideration. Children were hanged alongside adults. But over time the ruling class lost its stomach for cruelty: juries, judges, and even the king began to fudge the requirements of the law. If a prisoner could read, he could plead "of clergy," since by medieval custom it was assumed that anyone who could read was a clergyman. A pregnant woman could plead her "belly" as being "great with child." They thus avoided the noose.

The man, woman, or child who was not hanged had to be imprisoned. Consequently, the prisons were soon overflowing. Building prisons was expensive, so lower-class prisoners were bundled down into dark cellars with scores or hundreds of others. There they sat or lay, often in chains and in their own excrement until they rotted, starved, or died. But if a prisoner did not die quickly, despite being given every opportunity by malnutrition, unsanitary conditions, and utter hopelessness, what could be done with him? The answer set forth at about the time the American colonies were being established, in the "Acte for Punishment of Rogues, Vagabonds, and Sturdy Beggars," was being sent to the colonies as a virtual slave. In the eighteenth century this practice resulted in a flood of emigrants: about 6 in each 1,000 inhabitants were being shipped out of England.

Since exiles were often sold into slavery, kidnapping became a profitable business. In 1665, a merchant in Edinburgh—one of many who made such a request—petitioned to be allowed to abduct "strong and idle beggars, Egyptians [Gypsies], common and notorious whores and thieves and other dissolute and louse persons" for shipment to America. His petition was granted. To earn bounties, press gangs, "spirits," or crimps often kidnapped men and women who could not defend themselves. Women became particularly attractive, since "by their breeding they should replenish the white population" in the colonies, as a Venetian ambassador wrote in 1655.

Government after British government, royal and republican, Catholic and Protestant, was insensitive to the humane, moral, and legal condition of poorer Englishmen. And not only England was affected. From late medieval times, the English government sporadically pushed outward toward the "natural" boundaries of Great Britain; to do so, it had to overcome the neighboring Irish and Scots. Centuries of English attempts to subdue, exterminate, or exile them had a major impact upon the flow of migrants to America.

Border wars with Scotland, which had been fought since the time of William the Conqueror, became genocidal under Henry VIII. He instructed his generals to "put all to fire and sword." Unsubdued, the fiercely independent Scots time after time fought English armies. The last great rebellion, the "Forty-Five," which was ended by the battle and subsequent massacre at Culloden in 1745, released a flood of Scottish migrants to the New World.

Ireland had been suppressed and partly colonized in the twelfth century, but Henry VIII was the first English king to make a serious effort to extend the "Englysshe Pale" (the already subdued area enclosed in a palisade) over the whole of the "Great Irishry." When Henry's daughter Elizabeth became queen, she was advised that Ireland was too dangerous to be left to the Irish. Its very geography invited intervention of fellow Catholic Spain. Like many strategic assessments, this one led to self-fulfilling policies: when the English drove the Irish out of their more productive lands, the Irish tried to resist. Their resistance justified a campaign of brutal search-and-destroy operations. These operations were harbingers of campaigns the British army and the colonial militias later mounted against Native Americans: razing houses, burning fields of grain, killing livestock, spoiling caches of food, and enslaving or murdering men, women, and children. Prisoners of war were often shot, hanged, or drowned; and their women and children were used for target practice. To justify the slaughter of children, the English worked on the principle, as one general put it, that "nits make lice." Despite such ferocity on the part of the English, or perhaps because of it, the Irish fought on.

Observing England's Irish "problem," the Spaniards saw an opportunity to turn Ireland into England's unwinnable war, its "quagmire," and so to prevent England from harassing Spain in the New World. To understand Spain's role in the British colonization of America, it is necessary to consider a process that began in Ireland, but first we must objectively compare the Spain and England of that time.

During the sixteenth century Spain was the western superpower. It deployed the largest fleet of merchantmen in the world. Its soldiers were regarded as the finest infantry of the time. The looting of the treasures of the Aztecs and Incas gave it riches beyond the dreams of the other European powers. Its vast and relatively efficient bureaucracy controlled an empire stretching over five continents. The embodiment of medieval Christianity, Spain regarded itself as the protector of law and order: it was the conservative, "establishment" power of its day.

In contrast, the England that Henry VIII had brought into being and that was then led by Elizabeth I was what today would be called a rogue state. Henry destroyed the economic base and shattered the structure of England's medieval church; in the already complex affairs of Europe, he fostered dissident movements and used his espionage service and small army to dabble in revolution. Being weak, England used the weapon of the weak, terrorism. Men like Hawkins and Drake were the terrorists of their time. Drake's savage raid of 1586 on the Spanish Caribbean involved the massacre of populations and the looting of treasuries on a scale and with a ferocity that would have been almost unimaginable even a few years before. While some European states managed to acquire Spanish gold by selling their products, the British got theirs through piracy. Piracy, as John Maynard Keynes wrote, enabled Britain to pay off its entire foreign debt and establish its great trading ventures.

The Spanish government was well informed of English plans and actions by its diplomatic and intelligence-gathering networks (which included disaffected English Catholics). On the basis of this information, Philip II had to answer two questions: what exactly was the danger, and what could he afford to do about it? The danger was clear; it was losing the flow of gold and silver from Spanish-controlled America. What to do to prevent it was less clear. However spectacular and costly, raids like Drake's in 1586 were so far only sporadic. What England appeared to be planning, however, was a concerted interdiction of the treasure fleets and thus the destruction of the Spanish economy. The point of danger, what modern military strategists call a choke point, was Virginia.

Virginia was a choke point because it was just below Chesapeake Bay, where the fleets that carried gold and silver back to Spain veered east with the prevailing winds to cross the Atlantic. The French had already tried, twice, to set up bases there. Now, the Spaniards discovered that the English were planning the same move, precisely at the choke point.

As the Spanish evaluated the intelligence they were getting, the only logic they could find behind England's interest in establishing some sort of base at the Chesapeake was piracy. Spaniards had explored the Bahía de Santa María, as they called the Chesapeake, in 1570; and, as described in chapter 4, they found it unsuitable for colonists or even for missionaries. There were no riches, no great societies like the Incas and Aztecs to plunder, and no possibilities for profitable trade. But from a base like Roanoke Island, small, fast pirate ships could sally out to intercept Spanish galleons and then retreat rapidly behind the sandbars, shoals, and banks of the Virginia coast where large, heavy, deep-drafted Spanish warships could not follow them. It seemed obvious to the Spanish authorities that England was creating what would now be called a nest for terrorists. This was not their first such attempt. In 1563 an English privateer by the name of Thomas Stukeley (or Stucley) had attempted to establish just such a base in what is now South Carolina near where the French had built their forts. The Spaniards rightly saw that a base for privateers was their major danger.

Warnings of this danger came from many quarters. Foremost among them was the Spanish ambassador in London, Pedro de Zúñiga, just two years after the planting of Jamestown: "All the pirates who are out of this kingdom will be pardoned by the King if they resort there [Virginia], and the place is so perfect (as they say) for piractical excursions that they will ruin the trade of Your Majesty's vessels, for that is the purpose of their going [there]." After reviewing the government papers, the historian David Beers Quinn concluded that the Spanish ambassador was right. As Quinn wrote, "There is little doubt that a primary consideration in the plans of Ralegh and Grenville was the creation of a strong military base at a reasonable distance from Spanish Florida, but capable of protecting vessels assembling for raids on the Indies and refitting after their return."

Attacks were certainly expected. What to do about them was debated by Spanish officials for years. Attempts were made to create or improve fortifications in the Caribbean and along the Florida coast: "Much money and many men are required to keep that territory safe," lamented the president of the Council of the Indies in 1585. But Drake's attack had shown that a static land defense could never prevail against highly mobile seaborne forces. The Caribbean was not the place to organize protection.

If not the Caribbean, where? First, the Spanish government issued orders to find and destroy the "pirate base" at Roanoke, but the Spaniards soon concluded that nothing they did against the English in North America would ensure the security of their treasure fleet. Philip decided that he must attack the sponsor of the terrorists, England itself. Thus the first English landing on Roanoke was one of the actions that led Philip II to launch the "invincible Armada" of 132 ships—mounting 3,165 cannon and carrying 19,000 soldiers—which he sent to "shock and awe" England in 1588.

After the first armada failed to subdue England, the Spaniards' next move came in Ireland. Even if the Irish could not defeat the English, the Spanish government judged, they could distract or weaken England. The Spaniards did not start the war, which had been going on for centuries, but they tried to make it more equal and, for the English, less winnable. So Spain sent yet another fleet to aid the Irish in 1597 and a third in 1601. Like the first armada of 1588, they failed in fire and storm, but the survivors who made it ashore served as "advisers," to use the modern euphemism, teaching the Irish the use of cannon. Both sides fought with a ferocity seldom equaled until the twentieth century. The English took no prisoners among the Spaniards and determined upon a policy of genocide against the Irish in which, as the English general put it, "We spare none of what quality or sex soever." But still the Irish fought on.

When military campaigns proved ineffective, the English hit upon two other policies. The first was to replace the natives with English and Scots colonists. This was the policy adopted by James I: by 1603, with help from the Scots, the English had almost completely cleared Ulster of Irish farmers. Then in 1607, the same year as the first successful move to Virginia, the British sent commissions to Ireland to survey the country and organize it into "plantations" for incoming English and Scots colonists. They also began to remove the Irish to reservations, as they would later do to the Indians.

Even isolated on reservations, the Irish remained a potential danger, so over the following years, the British also engaged in what today we would call "sanctions"—economic restrictions, particularly in the Navigation Acts of the 1660s, that weakened and starved Ireland. Irish beef could not be exported to England, and the Irish woolen industry was crippled. As the Spaniards would do to the Indians in America, the English tried to suppress the religion of the Irish to break their cultural tradition and morale. As a result, as Jonathan Swift observed, the Irish were "dying and rotting by cold and famine and filth and vermin." Thus began the process that would devastate Ireland in the eighteenth century, when about 10,000 gave up and left for America.

Similarly, Scotland was reduced. In addition to commercial restrictions, its civil liberties vanished. If juries failed to return the convictions officials demanded of them, they were imprisoned; when the accused did not voluntarily confess, they could be tortured with a new device imported from Muscovy, the thumbscrew—it was found to be better than the torture "boot" then in common use in Scotland because, as an English historian commented, the boot "was unsuitable for [malnourished] people with thin legs."

Embattled as it was, Britain needed an ally in Europe; Holland was the logical candidate. Like England, Holland was a dynamic commercial society, and like England it was deeply divided religiously. In the middle of the seventeenth century, nearly 1 million of the 2 million Dutch people were Catholic, but the government was Protestant. The Protestants were themselves split into congeries of sects which forced upon them a degree of religious tolerance unique in that intolerant century. (This tolerance made Holland a logical refuge for the English Puritan group that became the American Pilgrims.) Religious diversity also made Holland a target for Catholic Spain and France, and its lack of major natural barriers made it vulnerable to attack by their armies.

England was at least theoretically a potential counterweight. Realizing this, and looking for allies, the lord protector of England during the Puritan-dominated Commonwealth, Oliver Cromwell, sent a mission to Holland in 1650 with a truly revolutionary proposal: instead of competing against one another for trade, the two countries should form a union. Fearing English domination, the Dutch imprudently declined his offer. Rebuffed, Cromwell enacted the first of the so-called Navigation Acts in 1651. This act quickly led to the first of three Anglo-Dutch Wars and would cost Holland its first colonial venture in North America, the area that became New York. The seizure of Dutch ships also gave the British colonists their means to get to the New World.

Even more than Holland, the scores of petty German states were targets of aggression. French troops devastated their farmlands in 1688, and other British and French armies trampled their fields and burned their houses during the first decade of the eighteenth century. In fear for the present and despair for the future, 13,000 people fled during the single year 1708–1709. Some of these eventually made their way to America and especially to Pennsylvania, where they became known as the Pennsylvania Dutch (German, *Deutsch*). Even distant North Carolina received some 650 Rhinelanders in 1710. They were the advance parties of many tens of thousands who followed during the eighteenth century.

They certainly were driven by fear, but most of these people had at least the theoretical ability to refuse to go. Those who had no choice were the millions of Africans who went as slaves. To them, I now turn.

CHAPTER 6

The African Roots of American Blacks

On the West African coast, the causes of migration to the New World were as multifarious as, and even more compelling than, in western Europe, but we know far less about the societies from which the people came. Indeed, our ancestors knew almost nothing about Africa. As Jonathan Swift scoffed in the early eighteenth century,

> So geographers, in Afric-maps,
> With savage-pictures fill their gaps;
> And o'er unhabitable downs
> Place elephants for want of towns.

In fact, Swift (and everyone else in Europe) was as wrong as the geographers he mocked. In much of the area from which most of the migrants to America came, what has been called Atlantic Africa, the population was denser than in Europe and inhabited thousands of towns. But Swift was right in saying that for Europeans the map of central Africa was a blank. To fill in the void, they invented what they could not see: not just "savage-pictures," they created an empire. Louis XIV's geographer, Guillaume Delisle, produced a map of Africa in 1700 with a vast mythical realm almost the size of western Europe that he called "Nigritie."

Despite the slave trade, which necessarily taught those engaged in it much about the Atlantic fringe of West Africa, ignorance of the interior remained profound. For Europeans and Americans, Atlantic Africa was "darkest Africa." So, in the nineteenth century, they were fascinated by the brave forays of men like David Livingston. In fact, however, as one of the world's major sources of gold, Atlantic Africa had been linked with the Mediterranean world for centuries through well-established trade routes radiating out from such great markets as Timbuktu.

Even more profound was ignorance about society, culture, economy, and politics. Until recently Africa was generally neglected by universities and research institutions; and few Afro-Americans knew anything about their own past. In their accommodation to white American culture, they had lost touch with their African roots. Memories, legends, and folkways were blurred or even forgotten. What had happened was not only that they had become Afro-American but that the process of enslavement had "homogenized" what had been scores of quite distinct African cultures. In this way, peoples whose ancestors had been Ibos, Mandingos, Tebous, Bantus, Ayois, Aqueras, Aradas, Nagos, Fons, and scores of others—who spoke different languages and had different customs—were blended into a single category, Negro. So to get back to origins, even from the perspective of eighteenth-century America when less homogenization had taken place, something akin to a social archaeology would have been required.

Until recently, few scholars even knew what questions to ask about African society and the experience of slavery. The answers that the pioneer scholars began bringing forward in the early twentieth century sometimes turned out to be unproductive or at best simplistic. The reality proved far more complex than they imagined. Now, thanks to a generation of research, the discovery of new materials, and the meticulous compilation of information, the view is becoming clearer, more exact, and more detailed. Still, what we know about America's African roots is meager in comparison with what we know about European roots. That being admitted, the effort is as necessary for Africa as for Europe in seeking to understand the birth of America. Getting as close as is now possible to an understanding is the purpose of this chapter. I begin with an overview of the geography and peoples.

Atlantic Africa is an enormous area beginning in the northeast at the Senegal River (which forms the frontier of the modern state of Senegal) and extending south down to roughly the middle of Angola. No precise eastern frontier can be established, but a reasonable approximation is at the highlands of Rwanda and Burundi. Sandwiched between two huge deserts, the area fades gradually into three zones. South of the great Sahara and north of the Namib Desert, the land becomes progressively wetter toward the middle. First come the semiarid steppe (Arabic, *sahel*) in the north and the Bié plateau of Angola in the south. These grasslands gradually give way to a wooded savanna and then, as rainfall increases, to a true forest on both sides of the equator. Plunging through the heart of the area are a number of great rivers including the Senegal, Gambia, Volta, Niger, and Congo (or Zaire).

This geographical description was not how Westerners of the sixteenth, seventeenth, and eighteenth centuries viewed Atlantic Africa. They knew practically nothing about the interior, but, drawing upon the accounts of sailors and slavers who frequented the shoreline, they saw it as a sequence of "coasts." It was the Portuguese who had led the way in learning about them, in the fifteenth century.

The Portuguese were driven by the lust for gold. European commerce was then hobbled by a lack of specie; so in addition to being a source of personal riches, gold was as strategic an objective as oil would become in our time. Africa was the main source then known; at that time, gold was available to Europeans only through Berber and Tuareg middlemen who brought it on camelback across the Sahara to markets on the North African coast. The Portuguese wanted to get to the source, where they expected to obtain gold more cheaply. Since the Berbers controlled the desert roads, the only way to go was by sea. That was the objective for which Prince Henry the Navigator mobilized the Portuguese voyages of exploration. Step by step, from about 1434, Portuguese sailors dropped down the African shore.

Sailing along the seafront of the vast Sahara, the Portuguese arrived in 1450 at the mouth of the Gambia River, which was deep enough to give their small ships access to the interior. Rowing and sailing upriver brought them closer than they suspected to Prince Henry's dream, but they did not find the mines. So, like the explorers in the New World, they constantly asked where gold came from and how to get there. They initially thought that the then still mythical city of Timbuktu was the source. Timbuktu was Africa's El Dorado. No one from Europe had yet been there. Unlike El

Dorado, it was real, but another twenty years would pass before a lone Florentine merchant was thought to have reached it, and nearly a century would pass before "Leo Africanus" described the city in his *Descrittione dell'Africa.* Even then Timbuktu remained mythical. It was off limits to Westerners, like Mecca or Tibet. The Portuguese stuck to the coast.

This, the northernmost coast, would be called by later sailors Senegambia—the name was a combination of the Senegal and Gambia rivers. There, on the *sahel,* the Portuguese first met black African societies. Just as Columbus found the inhabitants of the Caribbean different from what he expected of "Cathay," black Africans were different from what the Portuguese had expected of Timbuktu. The Wolof people were settled villagers with little to sell the Portuguese except slaves. Slaves were already of value to the Portuguese, although the Portuguese sugar plantations, which later would be the great consumers of slaves, were then only in their infancy; so while the Portuguese entered into trading relationships with the Wolof, they kept moving south.

From Senegambia, they passed along what later sailors knew as the Grain Coast. It was given that name because ships were able to buy foodstuffs there and because the area produced the Malaguetta pepper which is the source of a spice known as "seeds of paradise." One English captain commented that although the climate was oppressive, its agriculture was so bountiful that "by this fruitfulness the Sunne seemeth partly to recompence such griefes and molestations as they otherwise receive by the fervent heate thereof." The Grain Coast was made up of the Atlantic fringe of modern Sierra Leone, Liberia, and the Ivory Coast.

Later sailors regarded it as a particularly dangerous area because the Portuguese had often kidnapped the inhabitants when they found no gold and no slaves for sale. So bad were relations that in the eighteenth century trade was usually conducted on shipboard, as the sailors feared to go ashore and the natives did not want them to land. This too was a parallel to the American experience: along the New England coast, European fishermen had created such hostility that when Giovanni da Verrazzano visited in the sixteenth century, his men were not allowed ashore. What mischief the English, Dutch, and French must have done in North America, the Portuguese certainly did in Africa.

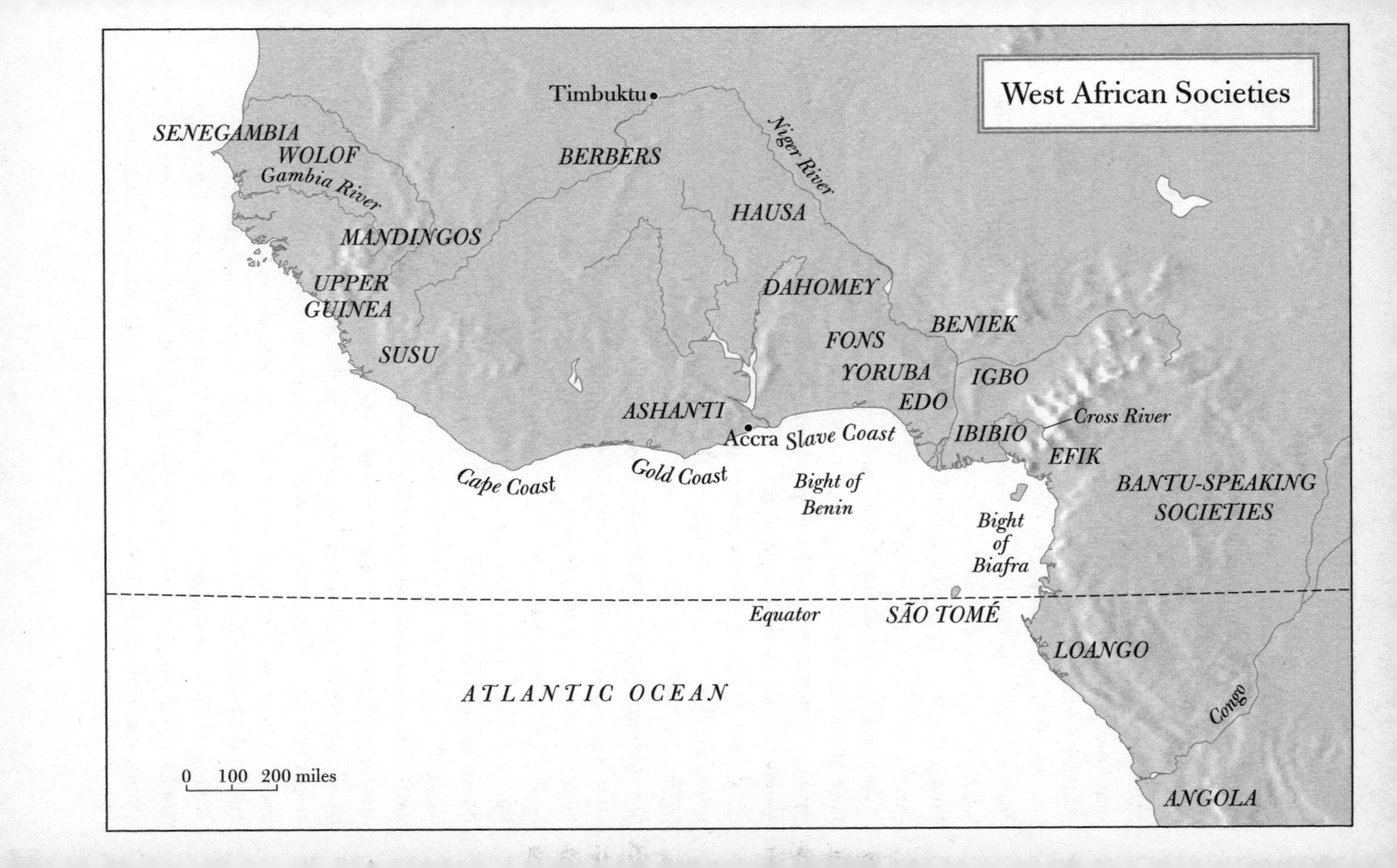
West African Societies
Timbuktu
SENEGAMBIA
WOLOF
Gambia River
BERBERS
Niger River
HAUSA
MANDINGOS
UPPER GUINEA
DAHOMEY
BENIEK
SUSU
FONS
YORUBA
IGBO
EDO
ASHANTI
Cross River
Accra
Slave Coast
IBIBIO
EFIK
BANTU-SPEAKING SOCIETIES
Cape Coast
Gold Coast
Bight of Benin
Bight of Biafra
Equator
SÃO TOMÉ
LOANGO
ATLANTIC OCEAN
Congo
0 100 200 miles
ANGOLA

So the Portuguese plunged farther down the coast, around the great western bulge of Africa, until in 1471 they reached the area that became known as the Gold Coast (roughly modern Ghana). There, they encountered the prosperous Akan people. The Akan welcomed the Portuguese, whose trading goods they found attractive; and from 1471, eleven years after the death of Prince Henry the Navigator and ten years before Columbus sailed for America, the Akan began to supply the Portuguese with gold.

Gold the Akan had, but unlike the Wolof they had no slaves for sale; in fact, what they wanted from the Portuguese as much as goods from Europe were African slaves to work their gold mines. The Portuguese discovered that if they could get slaves elsewhere in Africa, they could trade them to the Akan for gold. They ultimately sold the Akan about 12,000 blacks from other states. So they settled down in a convenient anchorage and built the first of the European trading factories, São Jorge da Mina, later known as Elmina. Columbus himself visited it before his trip to the New World. So profitable was the trade, and so difficult was the return voyage to Portugal, that the Portuguese soon built two more forts. These became the prototypes for the later stations built by the Dutch, English, and Danish slave traders.

Although they were noted for their brutal imperialism, the Portuguese soon learned to respect the native rulers, to pay rent for the lands on which they built their forts, and (usually) to honor the agreed terms of trade. This policy was hardly altruistic; rather, it was a response to African military power. As John Thornton has pointed out, some of the little city-states on the West African coast "had a well-developed specialized maritime culture that was fully capable of protecting its own waters." The first Portuguese visitors learned this the hard way. When they engaged in assault rather than trade, as they initially did on the Grain Coast in 1444, the natives struck back: black soldiers armed with bows and arrows and javelins and carried into battle in dugout canoes wiped out the next Portuguese raiding expedition. Although these canoes would not have seemed impressive to sailors riding high above the water in a carrack, they were large enough to carry up to 100 armed men, were agile enough to dodge among the slow-moving European sailing ships, could cross sandbars and plow through breakers where none of the Portuguese craft could follow, and were so low in the water as to be almost impossible for Portuguese gunners to target. Occasionally, they were able even to board and seize the Portuguese (and later, other European) ships.

That was the beginning of African resistance. It would prevent the Portuguese from achieving the second of their objectives, establishing large-scale sugar plantations, and so it encouraged the growth of European colonization of the Americas when that became possible. Resistance spread and grew. It was so effective that in 1456 the Portuguese began negotiating treaties with the coastal city-states to get by trade what they could not win by force of arms. But the Portuguese, like other nations, rarely learned from experience, so a century later, in 1590, when they were trying to conquer Angola, they suffered another stunning defeat by the Ndongo state. From that debacle, the Portuguese were able to recover in 1615 only by recruiting African mercenary soldiers. But by then, all along the coast, African states, such as Calabar on the Bight of Bonny, were restricting the Portuguese and other foreigners, refusing to allow them to build forts or even to reside ashore.

Initially, as I have said, the trade with Europeans was not for slaves but for gold, ivory, and pepper. The Portuguese were the leaders, but they were soon followed by rival European states. The Dutch were the first rival. After building a fort for themselves, they attacked the Portuguese and drove them away from the Gold Coast. Then, being careful not to anger the natives, they took over the partnerships the Portuguese had established. In turn, the Dutch were followed by the British, the Swedes, and the Danes. By the late seventeenth century, the Prussians and the French were also trying to gain foothold in the gold country; and by the early years of the eighteenth century, the 300-mile stretch of coast would be crowded with twenty-five European forts.

These forts had originally been strongholds in which gold could be safely stored. By the middle of the seventeenth century, the gold trade was being overtaken by the slave trade; from being importers of slaves, the Akan people of the Gold Coast had become exporters. So the buildings that had originally been lockboxes were converted into prisons for captives. These prisons, or concentration camps, were called barracoons. It was through them that most of the 10 million to 12 million Africans began the journey across the Atlantic. Roughly half a million of them would become the ancestors of today's Afro-Americans.

Shipment of slaves to the New World would become massive only in the eighteenth century, but meanwhile the lure of gold was irresistible not only to the Portuguese but to sailors and merchants of several other lands. For them the lure was often deadly. Whereas Europeans brought diseases to the American Indians, who had no immunities against them, the Europeans contracted from the Africans diseases against which they themselves had no defense. It made no difference how well and strong the venturers were when they began; they were often struck down when they entered the infamous Bight of Benin.

Richard Hakluyt recounts one voyage that makes this somberly evident. In 1553 when "two goodly ships . . . being all well furnished aswell [sic] with men of the lustiest [that is, the healthiest] sort" set out from Portsmouth, they thought they were on the way to riches. They were. When they reached the Gold Coast, they bartered their goods for 150 pounds of gold. That would be a fortune if they could get it back safely to England, but, lured on by greed, their commander insisted that they sail farther east into the Bight of Benin, to what was later called the Slave Coast, to buy pepper and "elephants' teeth." Arriving at the Benin River, they made use of their small pinnace to row up the river to a village whose ruler spoke Portuguese. There, they made a deal to buy Malaguetta pepper, whose seed was enormously valuable in England. Getting 80 tons to fill up the ship took weeks. During that time, the crew began to die, "sometimes three & sometimes 4 or 5 in a day." Finally the captain too died. Of the original 140 men on the ships, fewer than 40 made it back to England. Waxing philosophical after the ordeal, Hakluyt's informant wrote a commentary that turned into a prediction: "as fortune in maner never favoureth but flattereth, never promiseth but deceiveth, never raiseth but casteth downe again." That was to be the European experience in Atlantic Africa for the next 300 years.

Long before they were pushed out of the Gold Coast by the Dutch, the Portuguese had sailed east beyond the Slave Coast. Passing the Bight of Bonny, where the little port of Calabar was to become one of the major sources of slaves in the eighteenth century, the Portuguese explorer Diogo Cão in 1485 turned south. After more weeks of sailing, he observed a huge slick of muddy waters. Curious, he turned into the coast and so came to the mouth of the great Zaire (or Congo) River. There Cão made contact with people he realized must come from a kingdom even grander than Timbuktu. When he sailed back to Portugal, he brought a small group of Kongolese. As the Spanish were to do with some Indians half a century later, the Portuguese baptized the natives, instructed them in Portuguese, and, in 1491, returned them to Africa. They did not just return the Kongolese; rather, the Portuguese sent them back with a full-scale expedition designed to create a colony.

Warmly greeted in the Kongo, the new arrivals immediately "converted" King Nzinga Kkuwu and his court to Catholicism. Why they were so warmly received soon became evident. Just as Champlain would later find in the New World, when he was welcomed by the Hurons, the kingdom of the Kongo needed the Europeans. Just as the Hurons were locked in war with the more powerful Iroquois, Kongo was being raided by the neighboring Teke people. With their matchlocks, the Portuguese routed the Teke. Astonished by the effects of firearms, the Kongolese saw the Portuguese as virtually a divine visitation. If this was Christianity, they wanted more of it. The heir to the throne, Nzinga Mbemba, took the Christian name Afonso, agreed to send young men to Portugal to study, dressed his court in Portuguese clothing, and applied for (and soon gained) the pope's recognition of his country as a Christian kingdom. The Portuguese took their prisoners of war back as slaves and sent to the Kongo as colonists Portuguese subjects, many of whom were felons (like many of the Englishmen, Scots and Irishmen who came to the New World). Just before Columbus sailed to America, Kongo was becoming the first European colony in Africa.

These African coasts—Senegambia, Grain, Gold, Slave, Benin, and Kongo—were the transit stations on the way to the New World from inner Africa. There, two features stand out: the first is the diversity of African cultures; the second is their complexity and sophistication. I begin with the diversity.

The inhabitants of the areas just inland from the ocean shores were divided by culture, language, religion, and political organization. It is important to understand these divisions, as they not only facilitated the flow of slaves but also affected how enslaved peoples were integrated into the lands of their destinations. First consider race. Although there is a greater variation of physical forms in Africa than in almost any other part of the Old World—at the extremes are the Pygmies in southwestern Africa and tall Nilotic peoples such as the Masai and Dinka in the northeastern part—modern Americans are taught to consider all Africans as a single black or Negro race. Slave traders knew better. They did not need a sophisticated understanding of biology to identify different ethnic backgrounds: the people for whom they bargained had distinctive facial scars that slave traders called "country marks," which were signs of coming-of-age ceremonies. Since each ethnic group had its own patterns, at a glance the traders could tell one group from another. Some were regarded as more fierce and others as more docile; some as better workers and others as less trainable. Slave traders and those Americans who ultimately purchased the slaves became connoisseurs.

Linguistically, Africa was similarly divided. Its 2,000 to 3,000 dialects are said to be grouped into roughly 150 distinct languages which belong to four "families," each comparable to our Indo-European family. On the northern fringe of Atlantic Africa are Berber and Hausa speakers of a language in the same family as Arabic, Hebrew, Amharic, and various ancient Semitic languages of Egypt and the Levant. Linguists call that family Afroasiatic. On the eastern fringe of the great central rain forest are speakers of the Nilo-Saharan family. To the south are the Khoisan languages, also known as "click languages," used by the Bushmen and Hottentots. In the center, Atlantic Africa, is the Niger-Congo (or Niger-Kordofanian) family, which is composed of dozens of languages and hundreds of dialects. So varied were these that people even in neighboring villages spoke sufficiently differently to appear alien to one another.

Often, being unable to understand one another must have contributed to a sense of hostility among the villages and small states in the interior. People felt no compunction about raiding and raping their neighbors and razing these neighbors' villages. "The archaeological evidence," Oliver and Atmore wrote, "leaves no doubt that warfare between neighboring towns must have been a regular occurrence, for defensive walling was already widespread by the second half of the first millennium A.D." In the one contemporary account we have by an African, Olaudah Equiano remembered that when, as a boy about 1750, he was kidnapped and carried away from his village in the little state of Essaka, he could understand only some of what he heard even in nearby villages.

Neighbors to one another but also strangers, the little states were perpetually locked in hostilities. Each house in his village, Equiano wrote, was surrounded by a wall "made of red earth tempered: which, when dry, is as hard as brick. . . . And when they [the townsmen of Essaka] apprehend an invasion, they guard the avenues to their dwellings by striking sticks into the ground, which are so sharp at one end as to pierce the foot, and are generally dipped in poison." When the inhabitants go "out to till their land, they not only go in a body but generally take their arms with them for fear of a surprise."

These little communities—we might think of them as something like the tiny (and also warring) city-states of archaic Greece—numbered in the hundreds. Each "village-state" would be composed of several hamlets focused on a market and would have a population of only a few thousand people. Although the tendency was for the larger states to absorb the smaller, particularly after the introduction of firearms, the difficulties and costs of travel gave even those that were parts of empires or kingdoms a large degree of autonomy. Equiano comments that "our subjection to the king of Benin [the larger neighboring inland state] was little more than nominal. Every transaction of the government, as far as my slender observation extended, was conducted by the chiefs or elders [*embrenche*] of the place." These men held court, tried accused prisoners, and carried out sentences; they led their people in warfare with other societies; and they presumably ensured that local customs were taught to the young so as to preserve the moral authority of their rule.

Although Africa was thought of as an isolated jungle inhabited by primitive savages, many of the political structures of the village, town, kingdom, and empire evince a high degree of political organization. Equiano's description is borne out by the seventeenth-century Dutch merchant Willem Bosman. He went so far as to suggest that in at least some little states there were rather elaborate village councils. But, being small and unable to form protective alliances with neighbors who seemed alien, the village-states were apt to be gobbled up by more powerful neighbors. Thus from at least the thirteenth century, Africa produced true empires; and from diversity there emerged a high degree of sophistication in at least some African states.

The first of these was the Mali empire, which from the thirteenth to the fifteenth centuries covered most of the modern states of Senegal, Guinea, Mali, Upper Volta, and Nigeria, an area about the size of modern France. Mali's ruling elite came from a single language group, speakers of Mande, but it absorbed or dominated states speaking a number of languages. The empire was divided into semiautonomous townships, from each of which it drew taxes. Royal slaves, whose status was roughly comparable to that of European serfs, farmed the great basin of the Niger River as intensely as the Nile, Indus, and Yangtze basins have been farmed. The bounty of the river gave Mali an agricultural surplus, which was supplemented by long-range trade in ivory, gold, and slaves with faraway Egypt and North Africa. Together agriculture and commerce produced riches that encouraged the growth of schools, libraries, and other civic institutions manned by an educated elite. Although he visited it only in the sixteenth century, long after its great period, Hassan az-Zaiyati ("Leo Africanus") found the capital, Timbuktu, to be a city some 12 miles in circumference, replete with stores and artisans' workshops. The government was well organized, with a large and effective bureaucracy, a stable system that made good use of local autonomy and a professional army composed, as Zaiyati tells us, of "3,000 or so cavalry and uncountable foot soldiers."

Although contemporary Europeans regarded Timbuktu as almost unimaginably remote, its contacts with the Islamic world were impressive. Its well-financed, experienced, and venturesome merchants spread trade networks over much of Africa. Since it was an Islamic state—rich, devout, and well-organized—the emperor Musa was able in 1325 to lead some 10,000 Malians thousands of miles across Africa through Cairo to perform the *hajj* in Mecca. This unprecedented and quite remarkable journey leads me to discuss the eastern influence on Atlantic Africa, religion.

In Musa's time, large parts of Atlantic Africa had become Muslim, as Islam was spread among their peoples by merchants. This is well known. However, there is an even older thread of influence from north and east that shows the depth of the cultural network of Africa. Since the ancient societies were almost all illiterate, we cannot trace it with any precision, but vestiges remain. One such vestige on the Gold Coast is suggestive. According to the English ethnographer A. B. Ellis, the Tshi-speaking inhabitants of the Gold Coast believed in something they called *kra. Kra* has been interpreted as "a second individuality, an indwelling spirit residing in the body. . . . *Kra* in some respects resembles a guardian spirit." This notion, allowing for the long distance and the great passage of time, echoes the ancient Egyptian notion of *ka,* which the Egyptologist John Wilson called "guiding and protective forces in life and death."

The Egyptians believed that the gods permeated all things. So too the West Africans believed that all things—bushes, trees, and rocks as well as living things—had a *kra* or spirit capable of causing events. This enabled them to explain the otherwise inexplicable. For the Tshi, if a man drowned, water alone could not explain his death. After all, as Ellis reported, a Tshi would say, "Water, alone, is harmless: he drinks it daily, washes in it, uses it for a variety of purposes. He decides, therefore, that water did not cause the death of the man." Rather, the cause must be the river's *kra,* its indwelling spirit. And just as the Egyptians believed that upon death the *ka* of a person goes to an afterworld, so the West Africans held that *kra* goes to what Ellis translated as the "dead world." Both cultures believed that after death the desires of a person could be conjured by those gifted with oracular powers.

Other concepts are shared among the ancient Egyptians and various African peoples. And it is not just in religion that we can see the pervasive currents of custom. One noteworthy custom is in diplomacy. Given the hostility among the village-states, some means of safe passage had to be created for messages, for trade, and (occasionally) for making truces. Throughout Africa, there is evidence of a figure known along the Nile as the "chief of the path." His ritual status—what we would call his diplomatic immunity—was symbolized by a spear or wand. Carrying it, he was inviolate. So too among the people of the Gold Coast, ambassadors were known as *mosi* or "bearers of the stick of office." The stick or wand, called in Yoruba an *edan,* had "an almost sacred character, and it is an unheard of crime for an ambassador, furnished with this emblem, to be molested."

Influences did not, of course, come only from the East. As I have mentioned, the king and aristocracy of the Kongo kingdom not only converted to Christianity but sent a number of their younger men to Portugal to study. When these students returned, others learned from them, so that local beliefs came to be as interwoven with Catholicism there as they would later be in the New World. One result of this impact led to an event in some ways comparable to the story of Joan of Arc.

Africans also were strikingly inventive on their own in the practical affairs of daily life. In their use of fertilizers, crop rotation, and irrigation, they were so advanced that some of their techniques, carried to the New World by slaves, were adopted by eighteenth-century American colonists. The reverse process was also evident, with New World crops like corn (maize), millet, and sweet potatoes being introduced in Africa. In industry, as John Thornton has pointed out, "Africa possessed a manufacturing sector that supplied the population's needs for tools and clothing as well as luxury goods. As with agriculture, African manufacturing was done with fairly simple tools and techniques, yet the quality of output was as high as that from any other part of the world." Africans were expert metalworkers, turning out iron and copper weapons and tools at least 1,000 years ago; they had also discovered how to make high-quality steel which, as I will point out in chapter 14, was beyond the capacity of eighteenth-century American colonists. The Kongolese "had become a nation of miners with an aristocracy of smiths and traders in metal goods, which gave them an economic and political significance which spread far beyond their own ethnic homeland."

In textiles, the Kongolese are thought to have been one of the largest producers in the world. So esteemed was their cloth that it was exported to Europe and America. And in at least one field of medicine, the people of the Gold Coast had made a literally vital breakthrough: at a time when smallpox was a scourge in Europe and in the American colonies, with as many as one person in ten or twelve dying of it before age twenty, the Akan and perhaps other societies were successfully practicing inoculation. We know this because Reverend Cotton Mather of Boston wrote to the Royal Society in London in July 1716 to say that he had learned "from a Servant of my own, an Account of its being practiced in *Africa*." Mather's Akan slave, known as Onesimus, inspired Mather, when the next smallpox epidemic broke out in Boston in 1721, to begin the experiment that saved thousands of lives.

Impressive though these aspects of African societies were, it is in the structure of states and the mobilization of armies that they particularly demand attention. After the decline of Mali, other African states built themselves into empires by absorbing neighbors, organizing production and commerce, and creating bureaucracies and armies. In each of these areas, they were more impressive than American Indian and many European societies.

The most remarkable state in late seventeenth-century Africa was Dahomey. It might be considered Africa's Prussia, but it was even more coherently organized. Military service was universal, with young boys assigned to mature soldiers to be trained "up in Hardships from their Youth." The standing army numbered about 10,000 and was composed "entirely of women of remarkable physique and fierceness in combat." Western observers referred to them as "Amazons." Given the enlistment of women, Dahomey's army was capable of extraordinary expansion. In the season of war, about one in each four inhabitants was available for military service, that is about 50,000 of the 200,000 or so men, women, and children. And it was not, as some other African (and many European) armies were, just a rabble. It was described by a French observer as "elite troops, brave and well-disciplined, led by a prince full of valor and prudence, supported by a staff of experienced officers." Perhaps even more remarkable than the size and efficiency of the army was the fact that it was under civilian control except during combat. As Karl Polanyi observed, Dahomey's administration and law were remarkably honest and reliable. The monarchy was despotic, but the way in which religion was woven into the economy and administration made "superfluous the governmental apparatus of constraint with the masses of the people." Even taxation, according to Polanyi, was "linked to an efficient system of collection, accounting, and control." So deep was the penetration of government into every aspect of economic and social life that it regulated gifts, the "bride price," and even prostitution. Nor was Dahomey an isolated case. At least half a dozen other empires were nearly as impressive.

Warfare, of course, existed long before the arrival of the Portuguese, but it increased in scale and frequency under the impact of European firearms. Whereas the Portuguese had sought to keep their matchlocks and cannon to themselves, other Europeans used arms to trade for slaves. A Dutch memorandum of 1730 commented that the "great quantity of guns and powder that the Europeans have brought here . . . has caused terrible wars." Firearms changed the balance of power among the African states, as would later happen among the American Indian nations. Those who acquired them were able to overwhelm their neighbors. Minor states that could buy arms—like Denkyira, Akyim, and Akwamu—now rose to new power. Societies lacking firearms, like the Ewe, Whydah, and Aja, lost tens of thousands as prisoners of war and then slaves. At war with one another, the little African states could no more than the American Indians resist trading for firearms. In their struggles with one another, muskets were the key to survival. That absolute requirement accounts, at least in part, for their willingness to supply the slave trade.

But why, some writers have asked, did slavery exist at all? The short answer is that in Africa, as virtually everywhere else in the world from before the earliest records were kept, societies used human beings as domesticated animals. African villages, kingdoms, and empires followed the same practices, staffing their administrations, conscripting their armies, and running their economies with slaves. Probably most slaves fell into slavery by birth, but thousands of others were enslaved as prisoners of war. Still others were condemned to slavery for crimes, as judged by the application of traditional law by town councils. A person might also be forced into a limited period of slavery for debt; this might be compared to the practice in England in the seventeenth and eighteenth centuries whereby a debtor might be "sold" into exile. Finally, a person might be sold into slavery, as Olaudah Equiano was, after being kidnapped. Only kidnapping, particularly if members of the upper class were the victims, was thought morally wrong or legally actionable.

No one paid attention to Olaudah Equiano; but when two princes of the Efik people in the little slave-dealing state of Calabar on the Bight of Bonny were enslaved, slave traders attempted to find and return the victims lest African authorities close ports, seize European goods, or take hostages. These actions were sufficiently successful that on more than one occasion,

European or colonial American governments sought to locate the offenders and restore the lost people or property. Two famous cases involved Americans.

In 1645, the *Rainbow,* the first American ship known to have taken part in the slave trade, engaged in a bloody raid one Sunday on a village in Gambia, in modern Senegal. In the division of the captives with the British participants, the captain of the *Rainbow* received two blacks as his share. When he returned to Massachusetts, the captain was charged with "murder, man-stealing and [worst of all] Sabbath-breaking." The court decided that it lacked jurisdiction, since the event had happened in Africa, but it confiscated the slaves and returned them, at the expense of the Massachusetts legislature, to Africa.

Almost a century later, in 1731, the first known Muslims—probably of the Fulani people, many of whom were literate in Arabic—were brought to America as slaves. One of them was from the little kingdom of Jalot and was known to his masters as Job Ben Solomon or, as he would have said, Aiyub ibn Sulaiman. He had been kidnapped by men of the Mandingo tribe, eventually sold to an English trader, and shipped to Maryland. From there he wrote (in Arabic) a letter to his father. How he managed to do this or how he thought the letter would ever reach Africa would tell us a great deal about both Africa and America but, unfortunately, we do not know. In any case, his letter was intercepted and passed to James Oglethorpe, one of the founders of Georgia, who somehow managed to get it translated—there cannot have been many Americans or Englishmen who were literate in Arabic. For reasons that are not specified, but perhaps because of Aiyub's family's importance in what is today Mali and along the West African coast, he was sent to England in 1733. From there, he was escorted by an employee of the Royal African Company to his home on the Gambia River.

Conditions of slavery varied from place to place and time to time, from awful to tolerable. In Atlantic Africa, according to contemporary observers, slaves may not have been notably worse off than serfs in France or Russia at that time. Indeed, some slaves in the military forces and the bureaucracy of the various states were probably better off. Generally, all Africans regarded slavery as normal and proper. John Thornton argues that slavery was the only form in which Africans could accumulate wealth, since "African law did not recognize the right to own land. . . . Private wealth derived instead

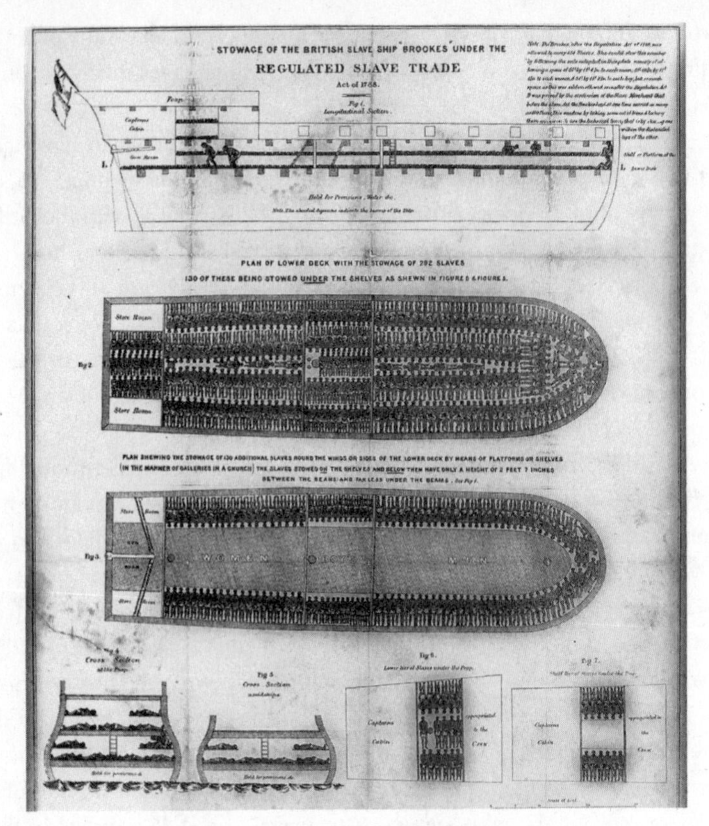

Diagram of stowage on a British slave ship

from rights over dependents. . . . By far the most important form of dependent labor was slavery. It was common for travelers to find whole villages of slaves producing for a master along major trade routes."

Slave raids were always vicious, but they seem to have developed a particularly cruel pattern in Calabar. There slave traders mounted expeditions in war canoes, armed even with cannon, up the Cross River to fall on hapless Igbo and Ibibio villages. Overpowering those they wanted, the Efik raiders bound or shackled them, and according to an English sailor who witnessed one event, threw them into the canoes and rushed away before any sort of defense or pursuit could be organized. Between 1725 and 1775, Calabar alone sold about 80,000 men, women, and children into slavery.

Slaves on Bark Wildfire

As soon as possible, the captives were sold to the captains of waiting slave ships, as Robert Harms learned from the log of the French ship *The Diligent,* at the port of Jakin. Often, ships could not collect a cargo quickly enough to prevent epidemics. *The Diligent* took three months to collect its cargo, so both the prisoners and the crew began to die of fevers and dysentery, and the prisoners of malnutrition and despair. Prisoners commonly believed that they were to be eaten by white cannibals. During this time, each prisoner was branded with a red-hot iron to indicate the ship on which he was being placed and his owner. I find no authority for this, but I suspect that branding was done not only for identification but also to terrorize the captives, who were always thought to be planning a revolt.

Meanwhile, the ship's carpenter was building a barricade to separate the forward deck on which the males, fettered in pairs, were occasionally allowed to exercise; and the rear section or quarterdeck, where the females were given relatively more freedom. When that task was finished, the carpenter hammered together a rack or platform on which the male slaves would be forced to lie. On *The Diligent* each captive had about 2 feet of headroom above the rack on which he slept and a horizontal space of only about 1 foot by 5 feet so that he had to "spoon in" on his side with another captive.

Conditions on any one of the 36,000 or so voyages could hardly have been exceeded by conditions on Devil's Island or in a Nazi or Soviet concentration camp. Slaves were fed only a gruel that left them vulnerable to scurvy, were so jammed together that the sickness of any one quickly spread to all, and at the first sign of resistance received a ghastly and usually lethal punishment. The younger women were almost always raped by the crew. Not surprisingly, many slaves attempted suicide by jumping overboard to the sharks or by attempting to starve themselves to death. The death rate was appalling. Perhaps 15 percent, or upwards of 2 million, died on the journeys.

PART II

The Establishment of British America

CHAPTER 7

Early Days in the Colonies

As the Spaniards had guessed, the English were initially more interested in creating a base to raid Spanish ships than in creating a colony on the Virginia coast. When Sir Humphrey Gilbert and his half brother Walter Ralegh sponsored a series of expeditions to the New World from 1584 to 1590, they chose Roanoke Island off the coast of Virginia, where the Spanish treasure fleets were most vulnerable. Although the English did not advertise this objective, and indeed doctored the record of their adventures to disguise it, the Spaniards were well aware of their plans.

When Gilbert vanished in a mighty storm on the Atlantic, Walter Ralegh picked up the challenge. To encourage him, Queen Elizabeth in 1584 did what the English monarchs had been doing for years in Ireland: she gave him a "patent." His was "to discover, search, finde out, and view such remote, heathen and barbarous lands, countreis, and territories, not actually possessed of any Christian prince, not inhabited by Christian people . . . to have, hold, occupy & enjoy." (It was understood that the same choke point where the treasure fleet was vulnerable was to be the focus.) Ralegh was empowered to take "so many of our subjects as shall willingly accompany him," and the queen, who was always conscious of money, specified that the crown was to get one-fifth of the proceeds. Thus encouraged, Ralegh organized a mission at his own expense to gather intelligence; and, forewarned by Gilbert's disaster in the stormy northern sea, he sent it south along the route Columbus had pioneered.

After a rapid and safe trip, the mission came back to report splendid prospects. To be sure of continued royal support, Ralegh decided to rename the land the resident Indians called Tsenacomacoh after Elizabeth, the "Virgin Queen," Virginia. Meanwhile, he commissioned the younger of the two Richard Hakluyts to write *Divers Voyages Touching the Discovery of America and the Islands Adjacent,* to assure the queen and other potential participants of the multiple benefits to be derived from creating a colony in Virginia. Ralegh hoped that these efforts would open the queen's purse—which was notoriously hard to open—but he and Hakluyt failed. Although the idea of a base of operations for piracy strongly appealed to her, Elizabeth was skeptical about most of Hakluyt's arguments. If the colony was to be established, she decided, Ralegh would have to find the means himself.

Despite his disappointment, Ralegh moved to the next stage. In 1585, he assembled a team of about 100 men to probe deeper into the possibilities of Virginia. To the acute annoyance of the Spaniards, this group first raided Spanish territory in the Caribbean to collect animals and plants to be tested in Virginia's soil. The Spaniards were impressed that "the members of the expedition include men skilled in all trades . . . accompanied by two tall Indians, whom they treated well, and who spoke English." Ralegh had prepared for the expedition long in advance by bringing perhaps as many as twenty Indians to London to be taught English. He was well aware, as Alden Vaughan has written, that knowledge of Indian languages "was an essential instrument of empire." Ralegh's protégé Thomas Hariot even compiled a dictionary of "the Virginian speache, 1585." Ralegh set a pattern: during his lifetime, perhaps fifty Amerindians were in England for training, and later many others would work with the colonists in their settlements.

But however skilled they were and whatever help they got from their Indians, the early Englishmen spent a hungry winter on Roanoke Island and were so discouraged that their chief, Ralph Lane, concluded that they must discover either a passage to "Southsea" (the Far East) or a gold mine; "nothing els can bring this country in request to be inhabited by our nation." Luckily, just at that point, the two aims of English policy came temporarily into a single focus: colonist and privateer joined hands when Sir Francis Drake sailed up from plundering Spanish cities in the Caribbean. It was obvious, at least to Drake and Ralegh, how valuable a base in Virginia could be, and they determined to support it. They received support in London when Thomas Hariot's *Brief and True Report of the New Found Land of Virginia,* illustrated with engravings of John White's vivid watercolors, was published. It painted a glowing picture of the colonial prospect. Still, the public was skeptical. Ralegh found only a small group of 117 people—among them seventeen women, two of whom were pregnant; and nine children—who rose to the bait. It was not much of a beginning, and worse was to come.

Today, it is difficult for us to appreciate how tenuous communication was in the Elizabethans' time. Their America was more "distant" than the moon: a round trip was at best a yearlong venture; messages might never arrive or might arrive in circumstances that made acting on them impossible. When no relief ships came, the little group at Roanoke grew desperate; so John White, the artist of the previous expedition who had become its governor, returned to England to try to expedite aid. He found all eyes fixed on the imminent danger of attack by the Spanish "Invincible Armada." There must have been some people in England who blamed the threat on exactly what Ralegh and Drake were doing; in any event White received no assistance. The privy council banned the departure of all ships that might help defend England. It would be four years before White managed to return to Roanoke.

Meanwhile, in 1588 Spain sent Captain Vicente González to search the Virginia coast, and he discovered the little base at Roanoke. There he found "signs of a slipway for small vessels, and on land a number of wells made with English casks, and other debris indicting that a considerable number of people had been there." But none still were. The Spaniards were relieved; they thought the English had given up. The worried king, however, anticipated correctly that the English would return and find a better location.

It would be nearly twenty years before they did. When White got back to Roanoke in 1591, he too found only a scene of devastation: the little camp had been looted and all that remained of the inhabitants was an undecipherable message carved on a tree trunk. Forced by bad weather to sail home, White found England desperately preparing to defend itself and Ralegh bankrupt.

Virginia had been the ruin of Ralegh, but the subsequent ruin of the

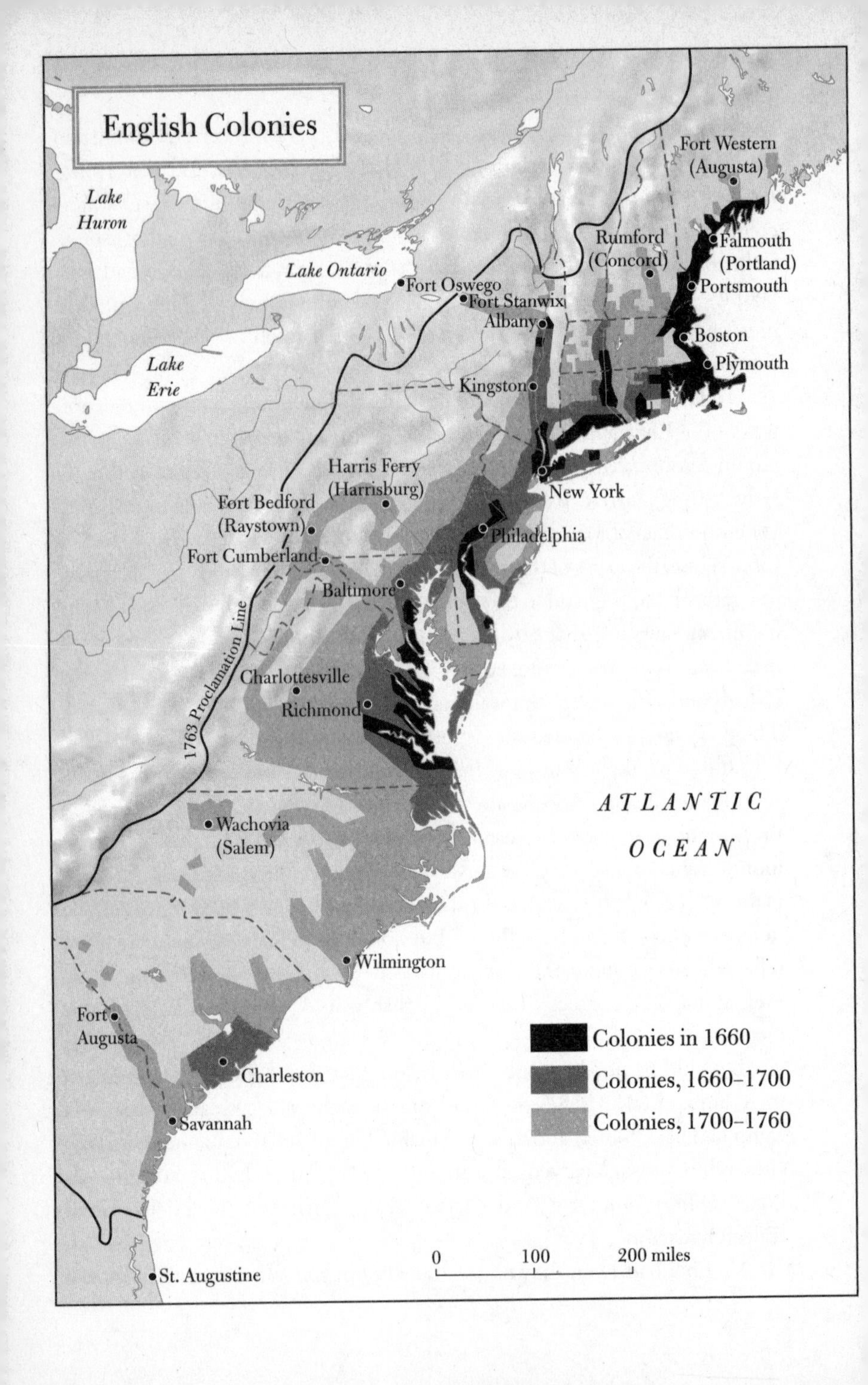
English Colonies
Lake Huron
Lake Ontario
Lake Erie
Fort Oswego
Fort Stanwix
Albany
Kingston
Fort Western (Augusta)
Rumford (Concord)
Falmouth (Portland)
Portsmouth
Boston
Plymouth
New York
Harris Ferry (Harrisburg)
Fort Bedford (Raystown)
Fort Cumberland
Philadelphia
Baltimore
1763 Proclamation Line
Charlottesville
Richmond
Wachovia (Salem)
Wilmington
Fort Augusta
Charleston
Savannah
St. Augustine
ATLANTIC OCEAN
Colonies in 1660
Colonies, 1660–1700
Colonies, 1700–1760
0
100
200 miles

Armada made colonialism in North America feasible. With the mortal threat to England ended, a new venture could be mounted. What the government would not do and what an individual could not do might be accomplished by a coalition of wealthy and determined men. Ralegh turned over his patent to a group of merchants who in 1606 organized the Virginia Company of London.

The Virginia Company had its roots in associations that Muslims and Jews began to create centuries before to facilitate Mediterranean commerce, banking, and insurance. Their ways of spreading risk were adopted later by Italian entrepreneurs whose joint-stock companies made possible the commercial revolution that "reawakened" Europe. Reaching England in the sixteenth century, the Muslim-Jewish-Italian experience gave birth to what became the powerful arm of English imperialism in Asia: the East India Company. The charter of that company, in turn, was the model for the London Company of Virginia. And it was not only the model: the key figure in both companies was the greatest English merchant of the time, Sir Thomas Smyth. Smyth would pick up where Ralegh had faltered.

Almost immediately, merchants in the port towns in western England, who had been the pioneers in the Atlantic trade and who felt left out of the new project, agitated to get a piece of North America for themselves. Their activities produced a sequence of ventures of which the Plymouth Company, also chartered in 1606, was the first. Attempting to set up a trading station at the mouth of the Kennebec River in 1607, it aimed to do in "New England"—as Captain John Smith would later name the place—what the London Company was starting to do in Virginia.

Like many ventures in our own times, the London Company began with a tax break: it would pay no customs until it became profitable; further, it was constituted as virtually an autonomous kingdom under the English crown. Immediately, the directors organized a campaign to mobilize public support by revolutionary advertising; even the great poet John Donne, then dean of St. Paul's cathedral, was hired to trumpet the prospects. Advertising paid off: in 1612, the company organized a "great fleet" loaded with 600 new settlers, supplies, and a small herd of cattle, pigs, and poultry. To help it raise money, the company was given permission to run a national lottery. The lottery was the only endeavor in which the company was profitable: over the decade and a half of its life, its investment of the then huge sum of £50,000 yielded no return.

The fate of the company was less interesting to colonists than their own immediate prospects. As they set out on their voyage into the unknown, they must have pondered long and hard over what they should take with them. Some had little to choose from, but those with at least a few resources needed guidance. In 1624, in *The Generall Historie of Virginia, New-England and the Summer Isles,* Captain John Smith summed up advice they had been given: in essence, everything they would need to survive for a year. His "particular of such necessaries" included three suits; three "paire of Irish stockings"; four pairs of shoes; 10 yards of canvas "to make a bed and boulster, to be filled in Virginia, serving for two men"; and "Victuall for a whole yeare for a man." The estimated cost was seven pounds, three shillings—or about the wages of a laborer for four months. Arms and ammunition were separate items.

Arms would not do the settlers much good against the most feared of possible enemies, the Spaniards. The last armada had sailed against England in 1602, and no peace was patched together until two years before the colonists sailed; the Spaniards had not written off their claim to the whole Atlantic coast as the English ambassador in Madrid reported. So the colonists prudently placed their settlement out of sight of the sea, some 30 miles up the James River. It was well that they did, because the Spaniards had guessed they would choose the Chesapeake. In 1611 Spain confirmed that guess by sending a caravel into the bay. Three men landed from the vessel, asking "to lett them have A pylott to bringe their shipp into the harbour the wch was grawnted. Butt haveinge the pylott noe soener a board hoysed upp their sayles and caryed the pylott quyte away wth them Leaveinge the thre wch were surprysed in his steade behynd them." The colonists, who were surprisingly well informed of Spanish (and French) colonizing ventures in Florida, accused the captives of being spies. They sent the "principall" to England, presumably to be interrogated. The second Spaniard died in Virginia. The third man, a "hispanyolated Inglishe man," who was an experienced pilot and therefore a particularly dangerous person, was sent back to England; but just before he arrived, he was "hanged upp att the yardes Arme."

The earlier explorers had written glowingly about Virginia. It was touted as a re-creation of the garden of Eden where fruits and vegetables flourished, wild animals virtually walked into the cooking pot, the natives were friendly, and with little effort every man would live like a king. But a Spanish observer, Alonso Suarez de Toledo, who was experienced in the New World, was closer to the reality the colonists would find: "The land itself would wage war against them!"

It did. As they stepped ashore, the colonists faced the massive defensive wall of that land—forest, stretching as far as the eye could see. Brambles and thickets grew out of the swampy ground; and for the first days the settlers must have just huddled, uncomfortably, on the muddy riverbank, plagued by mosquitoes and occasional downpours of rain. Once they got themselves and their supplies ashore and made tents out of sailcloth or cobbled together temporary huts, their troubles really began. The location was poorly chosen, and the colonists did everything they could to worsen their situation. They argued among themselves, infuriated the originally friendly natives by stealing corn, and refused to dirty their hands in American soil. Their toehold on the American continent would have slipped from under their feet had it not been for the flamboyant soldier of fortune, Captain John Smith.

Whether or not Smith exaggerated his own role—his determined enemy George Percy certainly thought so, calling him "an Ambityous, unworthy, and vayneglorious fellowe Attempteinge to take all Mens Authoreties from them"—Smith certainly played a constructive part in the new colony. He had come to America under particularly unhappy circumstances. Accused of conspiring to mutiny, he was clapped in irons and locked belowdecks. When the ship's captain opened his sealed instructions, he was horrified to find that they named Smith as one of seven members of the colony's ruling council. The other members promptly ejected Smith; so when the ship landed, Smith went off on his own to explore the hinterland.

Also exploring was George Percy, who shared Smith's romantic view of the wilderness. For Percy, it was

> flowing over with fair flowers of sundry colours and kinds, as though it has been in any garden or orchard in England. There be many strawber-

> ries and other fruits unknown. We saw the woods full of cedar and cypress trees, with other trees which issues out sweet gums like to balsam. We kept on our way in this Paradise.

When Percy and Smith returned from their explorations, they found the colony in disarray. As Percy wrote:

> There were never Englishmen left in a foreign country in such misery as we were in this new-discovered Virginia . . . lying on the bare, cold ground. . . . Our food was but a small can of barley sod in water to five men a day; our drink cold water taken out of the river, which was at flood very salt, at low tide full of slime and filth, which was the destruction of many of our men.

What was almost the final blow came when a fire destroyed most of the settlers' remaining provisions and possessions. Fifty of the small group had died between May and September, and the rest were starving.

Horrified, Smith took off again to offer the leader of the nearby Indians, the Powhatan, a deal: in return for food, Smith would use the colony's marvelous firearms to help him conquer two neighboring Indian societies. The delighted chief responded that if the colonists did this, "all his subjects should so esteeme us, and no man accout us strangers . . . and that the Corne, weomen and Country, should be as to his own people." "Weomen" aside, Smith returned with a boatload of corn.

The hungry settlers soon threw aside the remaining members of their administrative committee and elected Smith de facto governor. He immediately began whipping them into shape. He got them to dig a well for safe water, hoe the ground to plant crops, cut timber to build or repair their huts, and attend to their few surviving animals. But his efforts were insufficient. The 100 who were still alive were preparing to abandon the colony when they sighted the first ship of an expedition bringing supplies and more colonists. It was a close call.

In fact, it was a closer call than they knew, since the first ship was part of a fleet that had been scattered and partly sunk by a "West Indian horacano." Thanks to the hurricane, this first ship brought no supplies. Instead, ravenously

John Smith

hungry, the passengers tore into the colony's small store of corn "and in three days, at the most, wholly devoured it." Their own leaders had been either drowned or delayed, and they would not accept Smith's tyrannical rule.

They did not have to accept it, since Smith was partly incapacitated when gunpowder he was carrying in his pocket caught fire and badly burned him. Without his determined, hands-on leadership during the winter of 1609–10, the colonists collapsed into despair, sunk in the "the extreme beastly idleness of our nation . . . [who] will rather die and starve than be brought to any labor or industry to maintain themselves." That winter became known as the "starving time."

As George Percy remembered it:

> That sharpe pricke of hunger $w^{c}h$ noe man trewly descrybe butt he $w^{c}h$ hath Tasted the bitternesse thereof . . . having fedd uponn horses and other beastes as long as they Lasted we weare gladd to make shifte $w^{t}h$

> vermine as doggs Catts Ratts and myce. . . . And now famin begineinge to Looke gastely and pale in every face that notheinge was spared to mainteyne Lyfe and to doe those things w^{c}h seem incredible As to digge up dead corpses outt of graves and to eate them. . . . And amongste the reste this was moste Lamentable Thatt one of our Colline murdered his wyfe Ripped the childe outt of her woambe and threw itt into the River and after chopped the Mother in pieces and salted her for his foode.

Of the original 220, the 100 who survived found that even their dreams had turned into nightmares. Their suffering was for naught. Whatever profits there might be were to be held in a lockup, and even the food, clothing, and shelter promised were in default. All they had received was their passage to America. In short, whereas the allurement held out in London was a modern garden of Eden, the colonists found themselves hungry, sick, and abandoned.

When finally a new governor—a "high marshall"—arrived, he threw himself into a desperate effort to save the colony. A veteran of England's savage wars in Europe, Sir Thomas Dale instituted draconian punishments for shirkers and deserters. He was almost too late, since

> dyvrs of his men beinge idile and not willeinge to take paynes did Runne Away unto the Indyans many of them beinge taken againe Sr Thomas in A moste severe mannor cawsed to be executed. Some he apointed to be hanged Some burned Some to be broken upon wheles, others to be staked and some to be shott to deathe all theis extreme and crewell tortures he used and inflicted upon them To terrefy the reste for Attempteinge the Lyke and some w^{c}h Robbed the store he cawsed them to be bownd faste unto Trees and so starved them to deathe.

Miserable though they were, they reacted against other intruders as the Spanish had done. When a group sixty men from Jamestown sailed north to fish, they surprised a French group who were setting up a base on the Maine coast. The Jamestowners attacked, captured the Frenchmen, and took some of them and two Jesuit priests back to Jamestown as prisoners. Perhaps in part to divert discontent from his harsh measures at home,

Governor Dale mounted an expedition to destroy all signs of French occupation along the northern coast. Luckily for them, the French had all departed, or their fate would probably have been the same as they met at the hands of the Spanish in La Florida.

Draconian measures and foreign expeditions were unlikely to pacify people made desperate by starvation or infuriated by vicious punishments; moreover, such means were also unlikely to increase production. What the colonists needed was something to fill their stomachs and give them hope. By 1613 it was clear that even under harsh discipline the collective farming then practiced in Jamestown was not efficient. Yields were low because no one wanted to work harder than his neighbor, and no one had an incentive to cultivate carefully. Governor Dale decided to assign older settlers plots of 3 acres each to cultivate privately (as the contemporary phrase put it, "in particular"), just as would be done in Soviet collective farms in the 1960s. Each recipient also had to agree to work one month each year for the general community and to pay 2½ barrels of grain. The experiment worked so well that, a few years later, the governor decided to award plots of up to 100 acres. Some of these lands began, for the first time, to be farmed by imported "indentured" or contract workers from England, Scotland, and Wales.

Recognizing that man does not live by bread alone, the company also arranged, in 1619, the first of many importations of women. In the land of the blind the one-eyed man is king, and certainly in the land of bachelors any two-legged woman was a queen. About women, at least, the settlers were open-minded: any female, regardless of health, previous means of earning a living, or beauty, was welcome. The home government and the company viewed women simply as engines of colonialism.

No sooner had the colonists stepped ashore than they prudently set about constructing some sort of stronghold. The first such strongholds were so flimsy that nothing remains to show how they were built. Probably, they were not much more than piles of brush and brambles gathered around the tents and campfires on the beach. Many of the immigrants "were for some Time glad to lodge in empty Casks to shelter them from the Weather, for want of Housing" or in "Tents of Cloath" and "Canvis Boothes" as colonists did in Massachusetts Bay. But weather was not their only problem: soon the settlers began building a "pale" such as the English had built in Ireland. Getting behind a wall was almost the only protection they had, since few knew how to use a gun.

Housing came next. The log cabin that is so firmly fixed in American mythology was not those settlers' choice—they tried to build houses like those they had known in England. But they lacked many of the necessities. At first, and for many years, they had no paint, brick, or plaster. Timber they had in plenty, but it had to be shaped by hand with inefficient tools made of soft iron that dulled or cracked easily. Moreover, the American climate was more severe than the English. Out of necessity, the colonists quickly improvised a partial solution: they dug a cellar, 6 or 7 feet deep, into which they sunk their houses. Almost the only part that was visible aboveground was the roof, covered with thatch, sod, or bark. The surrounding earth acted as insulation to keep the house warmer in winter than it would have been aboveground. Such "sod houses" would last three or four years before rotting.

But sod houses had a serious disadvantage: in the marshy ground of the first settlement at Jamestown, they were wet. As soon as possible, therefore, the settlers began to build aboveground. They did so by planting upright poles, just slightly dressed logs, at the corners of an area usually no more than 15 by 12 feet. Into these uprights they drove wooden pegs—iron nails were too precious to use—that would hold roughly trimmed boards. Over these boards they fixed thatch on which they smeared mud or clay. These wattle-and-daub buildings resembled the huts of Irish and Scottish peasants. They were not elegant, but as long as the inhabitants continued smearing any cracks and fissures with mud, the houses kept out most of the cold winds. Covering the bare earth floor with reeds kept it dry.

Houses were not only cold and dank but also dark. The entrance was a low door, and there were no windows. As windows began to be introduced, they were just gaps in the wall, which could be closed by shutters or covered by oiled paper to admit light but not too much wind. Glass did not become available until much later in the seventeenth century; even then it was still so rare and valuable that in his raid on Maryland in 1645, the English pirate Richard Ingle removed every pane he could find.

The main source of warmth was human bodies. In a space measuring no more than about five steps in either direction lived not only a family but often also one or more servants. In one house in Jamestown's satellite settlement of Martin's Hundred, about 1622, the Emersons—a man, his wife, and their eleven-year-old son—lived with two servants. As blacks began to be imported after 1619, they too lived with colonists' families, along with indentured white servants. In Virginia, blacks did not move into separate quarters until nearly the end of the seventeenth century. Then they too attempted to reproduce what they remembered of the styles of the "old country." For them, of course, that was not England but Africa.

Given the severity of the winters all along the eastern seaboard and need for means to cook, fireplaces were a necessity even in the tiny, crowded houses. In the absence of brick and mortar, they were made of logs insulated as well as possible by clay to prevent fire from "breaking through" and lighting the highly flammable roofs.

Inside the houses there was not much room for furniture. That was just as well, since the settlers had little. As Captain John Smith told immigrants, they should take canvas to make sacks, which, stuffed with straw and laid on the floor, served as beds. Bedsteads did not become common until later in the century. In inventories dating from around 1630, chairs were still rare, even in the more settled areas; and they would remain rare for more than a century on the frontier. Tables, benches, and stools were made of split logs held up by sticks inserted as legs. Even necessary equipment, such as spinning wheels and churns, was rare.

The first colonists were naturally governed by the habits they brought with them from England. Their basic diet was limited initially to the dried food that Smith had urged them to bring. Colonists were not adventurous gourmets. They rejected the potato and most other American vegetables, though they did eat corn, probably because it lent itself to the mush and stews that were their common recipes. They ate shellfish because it was readily available at low tide, but they were not fond of the salmon or shad that swarmed in the sea and rivers. As soon as they could get enough pigs, pork became a staple: smoked, salted, or pickled, it was easier to preserve than other meat. As their houses became more sophisticated, they also learned to cure venison in "smoke-holes" abutting their chimneys. Adopting most native foods took years of hunger and experiment.

As was done in England at the time, colonists cooked in iron pots or skillets and then poured the food onto wooden "trenchers." A trencher resembled a chopping block but had a declivity in the middle to keep stew from running out. Ceramic mugs and cups did not become common until after the middle of the seventeenth century, and plates came even later. Porcelain was uncommon until the eighteenth century.

People usually ate with their fingers, but they scooped up mush with a shell or spoon—in 1635 in Maryland immigrants were advised to bring "Platters, Dishes, spoones of wood"—and cut chunks of meat with sheath knives. When the table knife appeared, it had a sharp point on which meat was speared and carried to the mouth. Forks were not used in early America. Governor Winthrop of Massachusetts famously owned one fork in the middle of the seventeenth century; but no forks have turned up at any archaeological site dating before the eighteenth century. The first mention of a fork in Plymouth Colony dates from 1721.

The amount and the quality of food and drink were, of course, factors in well-being. The first English explorers were impressed by the healthiness of Virginia. Ralph Lane found "the climate so wholesome, that wee had not one sicke since we touched the land here." He did not stay long enough to find out that throughout the low-lying lands surrounding the Chesapeake, "agues and fevers" were prevalent. The later colonists found that "seasoning"—living through the onset of sicknesses against which they had no immunities—was a hurdle many did not cross. Later in the seventeenth century malaria became much more virulent when a new strain, falciparum, came from Africa. Malaria was the most obvious danger to health because it weakened the body's resistance to dysentery (known as "Grypes of the Gutts") and influenza. In Maryland, most immigrant men did not live beyond age forty-three, and about seven out of ten died before age fifty; women usually had even shorter lives. Few marriages lasted more than ten years before one spouse died. Among babies the death rate was about one in four before the first birthday.

Succeeding in the New World was not just a matter of staying alive, hard as that was. Investors in the Virginia Company were businessmen who were disinclined to provide charity. To get what they could not make themselves, settlers had to produce something that they could barter or sell. The first move was obvious. All around them were forests, and timber was in desperately short supply in England which had been deforested by shipbuilding, heating, and iron smelting. So even before Jamestown was settled, the colony sent a shipment of timber to the mother country. We know about it only because Sir Walter Ralegh's little 70-ton *Job* was forced by bad weather into a French port, where it was found to contain 16 tons of "Cedar wodde." That probably was the first timber exported from America to England. It was followed shortly thereafter by shipments of clapboard, logs for masts and spars, and staves for barrels. In 1621, the Pilgrims at Plymouth freighted the *Fortune* with "good clapboard as full as she could stow." (That cargo plus two hogsheads of beaver and otter skins was thought to be worth the princely sum of 500 pounds but was looted by French privateers.) Just a few years later, settlers in Maine were harnessing waterpower to run sawmills. And by mid-century, ships were being purposefully built to carry long white-pine logs to be made into masts for the Royal Navy.

The Indians in Virginia also cultivated and used tobacco (Algonquian, *apooke*). In its original form, it was too bitter and strong for the English market, but it was catching on, as we know because King James I "blasted" the "filtyie novelitie" of smoking, "so vile and stinking a custome." It was, he said, "loathsome to the eye, hateful to the nose, harmful to the brain [and] dangerous to the lungs." His attack did not deter his subjects, and after 1612, when John Rolfe adapted West Indian or "Spanish" methods of curing Virginia tobacco, it gained popularity. Almost overnight, it became the salvation of the community. As a colonist later wrote, "Tobacco is our meat, drinke, cloathing and monies." So frenzied were the colonists that they planted it even in the lanes between their houses.

Another local product also caught the attention of the settlers in Virginia, as it had attracted the French farther north—furs. Colonists could not get furs themselves. They were generally unable to use firearms, did not know how to hunt, and were afraid to venture into the interior. So the colonists had to work out a relationship with the 13,000 or so nearby Algonquian-speaking Indians. That should be easy, Sir George Peckham had predicted: just treat them as childlike barbarians:

> Considering that all creatures, by constitution of nature, are rendered more tractable and easier wonne for all assayes, by courtesie and mild-

> ness, then by crueltie or roughness . . . there must bee presented unto them gratis, some kindes of our pettie marchandizes and trifles: As looking glasses, Belles, Beades, Bracelets, Chaines, or collers of Bewgle, Chrystall, Amber, Jet, or Glasse &c. For such be the things, though to us of small value, yet accounted by them of high price and estimation: soonest will induce their Barbarous natures to a liking and a mutuall societie with us. But if after these good and fayre meanes used, the Savages neverthelesse will not bee herewithall satisfied, but barbarously will goe about to practise violence eyther in repelling the Christians from their Ports & safelandings, or in withstanding them afterwards to enjoy the rights for which both painfully and lawfully they have adventured themselves thither: Then in such a case I holde it no breach of equitie for the Christians to defend themselves, to pursue revenge with force, and to doe whatsoever is necessarie for the attaining of their safetie.

Settlers in Virginia soon followed that advice, but they got the sequence wrong: although they obviously needed the Indians' help, since they did not know how to plant or to hunt, they antagonized the Indians before having fully learned from them. Without their assistance, the colonists died like flies. During the first years after 1607, when they were finally established, about one in each two died each year. By the end of the first decade, 1,700 settlers had arrived but only 351 were still alive.

From the start, Virginians were more bellicose and less cooperative than settlers in Maryland and New England. Having begun to take the land of the Indians and finding that virgin lands required massive amounts of labor—either hoeing around trees to plant crops; girdling them so that they would die; or, more laboriously, cutting them down and digging up the stumps—the early colonists could manage only small plots of land. No matter how much land might be granted by the colonial authorities, little was actually cultivated. Clearing sufficient land was beyond the settlers' capacity. That is what made hostility with the Indians inevitable. As Captain John Smith had observed during his trip around the Chesapeake basin in 1607, "many plain marshes, containing some twenty, some a hundred, some two hundred acres, some more, some less," had already been cleared by the Indians. Since clearing land was such an exhausting and time-consuming task, it was obviously cheaper to take land from the Indians. But Indian agricultural lands could not satisfy the increasing number of colonists; Indian hunting grounds were needed as well.

From 1618 on, newcomers were offered a "head right" if they would clear and work plots; that program would remain a feature of colonial America throughout the century. Larger areas were also granted or sold. As expected revenues failed to materialize, the Virginia Company began in 1616 to sell "particular plantations" to individuals or companies who contracted to bring over and settle colonists at their own expense. Organized like the subdivision of an English county or shire known as a "hundred," each settlement became legally semiautonomous. By 1620, nearly fifty such grants had been made.

Except for passing out land, government in Virginia and Maryland was restricted in its interventions in society and economy. One reason may have been that farmers settled so far apart as to be out of reach of any authority, but the restriction was also an echo of what was familiar in England. In the original patent given to Sir Walter Ralegh, Queen Elizabeth had specified that whatever arrangements Ralegh or his successors made to govern the territories they colonized must be "as nere as conveniently may bee, agreeable to the forme of the lawes, statutes, governement, or pollicie of England . . . nor in any wise to withdrawe any of the subjects or people of those lands or places from the alleagance of us, our heires and successours." That is, Ralegh could do whatever was necessary as long as he did not violate English law or wean his new state away from the crown. Walking that fine line would be accomplished initially, but the attempt was always precarious and ultimately failed completely, eventually resulting in the American Revolution.

Accommodation with the Indians was never seriously considered. The "chiefe adventurer" of the voyage to Roanoke in 1583 had written that by international law, the Indians had no right to resist the colonists: "I say that the Christians may lawfully travel into those Countries and abide there: whom the Savages may not justly impugne and forbidde in respect of the mutuall societie and fellowshippe betweene man and man prescribed by the Law of Nations." In practice, the Virginia colonists took this as license to raid Indians' food caches to steal corn. And, as their descendants so often did, they suspected the Indians of plotting retaliation and so moved peremptorily to attack them. In June 1586, they killed the chief of the Roanoke Indians. When they suspected the Indians of stealing a silver cup, they conducted, as Ralph Lane reported, the first of countless search-and-destroy missions (which they called "marches" and the French knew as *chevauchées*): they "burnt, and spoyled their corne, and Towne, all the people being fled."

The experience of those Englishmen who went to Maryland differed in important respects from what happened in Virginia. In Virginia, Jamestown was set up and controlled until 1625 by a company of merchants; but Maryland was granted to one man, Lord Baltimore. Except for having converted to Catholicism, Baltimore was fairly typical of the English aristocracy. As a young man, he became the private secretary of a leading public figure, then clerk of the privy council, and a member of Parliament from 1609 to 1624. During that period, he was also an investor in and a member of the Virginia Company of London. The grant King Charles gave him in 1632 was for about 10 million acres of land uninhabited except by "Barbarians, Heathen and Savages." The original intent had been to locate the grant south of Virginia, but to block the southward move of the Dutch, who had already set up a small colony called New Amsterdam in the Hudson valley, Baltimore's grant was shifted to what became Maryland. The king treated him royally: he was to get the rights of an absolute lord with complete authority over every aspect of life in the colony; all writs were to be in his name rather than the king's; and in the event of a dispute, the outcome was to be "beneficial, profitable and favorable" to him. In return for all this, Baltimore was to pay a symbolic yearly "rent" of just two Indian arrows.

Remarkable as these terms were, even more remarkable was the fact that in an age of intense interfaith hostility, Protestant England made Maryland a haven for Catholics. From the beginning, however, Protestants outnumbered Catholics, and the disproportion between the Protestants' status and their aspirations would later cause havoc in Maryland's civic life.

After receiving his charter, Lord Baltimore soon died. His son, Cecilius, as "lord proprietor," then opened an office in London to register would-be colonists. Merchants who had invested in the Virginia Company of London feared competition and tried to prevent the initial group of about 200 colonists from sailing. When his ships *Ark* and *Dove* finally got under way in the winter of 1633, Cecilius warned his captains not to stray into Protestant Virginia—they actually did, though, and were well treated there—and ordered his passengers not to argue over religion. Peace was probably kept less by his order than by the hardships of the stormy passage.

Landing at Blakiston Island on March 3, 1634, after three months, Father Andrew White, S.J., was thrilled by the sight of Chesapeake Bay:

> the most delightfull water I ever saw, between two sweet landes [and the] Patomecke . . . the sweetest and greatest river I have seene, so that the Thames is but a little finger to it. There are noe marshes or swampes about it, but solid firme ground, with great variety of woode, not choaked up with undershrubs, but commonly so farre distant from each other as a coach and fower horses may travale without molestation.

He found the local Indians "(as they all generally be) of a very loveing and kinde nature." Unlike the Virginians, the Marylanders sought "to avoid all occasion of dislike and the Colour of wrong." Conferring with them, helped by "one Captaine Henry Fleete an English-man, who had lived many yeeres among the Indians, and by that meanes spake the Countrey language very well," the governor-designate of Maryland colony gave the chief man and his "court" presents of axes, hatchets, hoes, knives, and cloth, in return for which the chief

> freely gave consent that hee and his company should dwell in one part of their Towne [which the Indians called Yoacomaco], and reserved the other for themselves; and those Indians that dwelt in that part of the Towne, which was allotted for the English, freely left them their houses [Algonquian, *wigwang*], and some corne that they had begun to plant: It was also agreed between them, that at the end of harvest they should leave the whole towne; which they did accordingly: And they made mutuall promises to each other, to live friendly and peaceably together, and if any injury should happen to be done on any part, that satisfaction

> should be made for the same, and thus upon the 27. day of March, Anno Domini, 1634, the Governour tooke possession of the place, and named the Towne Saint Maries.

Even more important, the Indians taught the Maryland colonists how to plant corn and bartered enough of it that they were able to swap, in perhaps the earliest example of intercolonial trade, 1,000 bushels of corn for New England salted fish. "Experience hath taught us," White continued, "that by kind and faire usage, the Natives are not onely become peaceable, but also friendly, and have upon all occasions performed as many friendly Offices to the English in Maryland, and New-England, as any neighbour or friend uses to doe in the most Civill parts of Christendome." With the help of the Indians and the fact that their group was made up "mostly [of] handicraftsmen, laborers, and servants, men and women, with a few of the yeoman-farmer class," the Marylanders avoided the "starving time" that had nearly wiped out the colony at Jamestown. Lord Baltimore also profited from the experience in Virginia by immediately awarding "headrights" of land, 100 acres each for each settler and the same amount for all servants up to a maximum of five, with only the obligation to pay a yearly "quitrent" of 2 shillings (payable in kind) for each 100 acres.

In England, life had meanwhile become much harder for the radical Protestants. A small group from the town of Scrooby were harassed into migrating in 1607. They went to Holland, which was known to be tolerant toward dissident religions; but they soon found it too tolerant—their children began to marry Dutch men and women. To keep the community together, they decided to move to the New World. Despite feeling that "we are well weaned from the delicate milk of our mother country," they petitioned King James I for permission to settle in the territory of the London Company of Virginia; he agreed, and in 1620 the company granted them a patent for land and a small subsidy. A London moneylender (whose sharp practice was to become the bane of their existence for the next twenty years) underwrote the rest of their initial expenses. Finally, these "Puritans," thirty-five of the original group from Holland and twice that many "strangers," as those who had not been in Holland were called, boarded the *Speedwell* and the *Mayflower* to sail in September 1620.

From the start, the trip was a disaster waiting to happen. The voyage was delayed by the leaky *Speedwell* (which turned back), so that Robert Cushman wrote before they left, "Our victuals will be half eaten up, I think, before we go from the coast of England, and if our voyage last long, we shall not have a month's victuals when we come in the country." The passage was beset by storms in which "the winds were so fierce and the seas so high, as they could not bear a knot of sail, but were forced to hull [heave to] for divers days together." When they finally sighted land, "they fell amongst dangerous shoals and roaring breakers." Trying but failing to reach their intended destination, the Hudson River, they turned into Cape Cod. Fearing worse weather and the onslaught of winter, they decided to go ashore.

What they saw terrified them. "For summer being done, all things stand upon them, with a weather-beaten face," wrote their leader, William Bradford, "and the whole country, full of woods and thickets, represented a wild and savage hue. If they looked behind them, there was the mighty ocean which they had passed and was now as a main bar and gulf to separate them from all the civil parts of the world." They had arrived at a place where, under British law, they had no right to be, and in which there was no fixed community authority. Worries about these two problems led them, on the last day of their voyage, to draw up one of the most famous documents in American history, the Mayflower Compact.

A remarkable document, the Mayflower Compact effected the transition postulated by the great French and English philosophers of the seventeenth and eighteenth centuries from "the state of nature" to organized society. After confirming their loyalty to "our dread Sovereign Lord King *James,* by the Grace of God, of *Great Britain, France,* and *Ireland,* King, *Defender of the Faith*"—but not, as their move had made clear, their faith—the Pilgrims entered into a social contract with one another, to "Covenant and Combine ourselves together into a Civil Body Politic."

As sophisticated as their beginning was politically, their physical prospects were dim. Since their first landfall was not acceptable, they spent precious days at the beginning of winter exploring in their tiny shallop until they found Plymouth harbor on December 11. There, like the settlers at Jamestown, they soon fell out with one another. As William Bradford wrote,

> In these hard and difficult beginnings they found some discontents and murmurings arise amongst some, and mutinous speeches and carriages [deportment] in other; but they were soon quelled and overcome by the wisdom, patience, and just and equal carriage of things, by the Governor and better part, which clave faithfully together in the main.

Cleaving faithfully together did not protect them from the bitter cold; within three months half of them were dead. At first they could make no contact with the local Indians, who had suffered grievously from earlier contacts with whites; but at the end of winter one Indian came to their little settlement. To their astonishment, since they did not know about the fishermen who had frequented those shores, the man spoke understandable English. He offered to bring them another man, named Tisquantum (Squanto, as the Pilgrims pronounced it), who had been kidnapped and taken to England and so could speak better English. Tisquantum became their mentor. First, he introduced the local chief, with whom a peace agreement was made that would last twenty-four years. Then he "directed them how to set their corn, where to take fish, and to procure other commodities, and was also their pilot to bring them to unknown places for their profit, and never left them till he died." Most important of all, Tisquantum acted as their agent in what became their economic salvation, the trade in beaver and otter furs.

Despite their good relations with the nearby Indians, these settlers received a declaration of war—a bundle of arrows bound by snakeskin—from the more distant, numerous, and warlike Narragansett Indians, who had learned to hate the visiting whites. So, as quickly as the colonists could muster the manpower to do so, they constructed a "good strong pale, and . . . flankers in convenient places with gates to shut, which were every night locked, and a watch kept."

It was not only the Indians the Pilgrims learned to fear: it was also one another. Despite the high moral and religious principles that had driven them to abandon their native land and venture into the "wilderness," they found disturbing evidence of sin. Much of their first harvest was stolen "both by night and day before it became scarce eatable, and much more afterward. And though many were well whipped, when they were taken for a few ears of corn; yet hunger made others, whom conscience did not restrain, to venture."

The guiding ideal of the Pilgrims was to live as a band of brothers, a community of like-minded people who shared what they had in a sort of primitive communism; but desperation soon drove them, as it had driven the settlers at Jamestown, to "set corn every man for his own particular." They found, as the people of Jamestown had, that this policy "made all hands very industrious, so as much more corn was planted . . . [even] the women now went willingly into the field, and took their little ones with them to set corn."

Having to cater to greed was dispiriting to the community leaders; worse, they perceived a deeply disturbing moral rot in their community. The most dangerous manifestation was among those who sold arms to the Indians. One man, Edward Ashley "had for some time lived among the Indians as a savage and went naked amongst them and used their manners, in which time he got their language . . . [and may have] committed uncleanness with Indian women." Going over to the Indians was bad, but far worse was what happened in the colony itself. "Marvelous it may be to see and consider how some kind of wickedness did grow and break forth here," William Bradford wrote. The Pilgrims themselves were caught engaging in "drunkenness and uncleanness. Not only incontinency between persons unmarried, for which many both men and women have been punished sharply enough, but some married persons also. But that which is worse, even sodomy and buggery (things fearful to name) have broke forth in this land oftener than once." Given the Puritans' need to plunge honestly to the depths of the transgression, Governor Bradford then related what must stand as the most remarkable set of sins any religious community ever witnessed: the case of young Thomas Granger, who was "this year detected of buggery, and indicted for the same, with a mare, a cow, two goats, five sheep, two calves and a turkey. Horrible it is to mention, but the truth of the history requires it."

How, Bradford sadly wondered, could such a thing happen in a community committed to "preserve holiness and purity"? "I say it may justly be marvelled at and cause us to fear and tremble at the consideration of our corrupt natures, which are so hardly bridled, subdued and mortified."

Then, more soberly, as an experienced leader of his community, Bradford speculated on the very nature of the Puritan philosophy:

> that it may be in this case as it is with waters when their streams are stopped or dammed up. When they get passage they flow with more violence and make more noise and disturbance than when they are suffered to run quietly in their own channels; so wickedness being here more stopped by strict laws, and the same more nearly looked unto so as it cannot run in a common road of liberty as it would and is inclined, it searches everywhere and at last breaks out where it gets vent."

The same pressures manifested themselves politically. From the small, compact, unworldly community they had set out to become, the Pilgrims flew apart from one another. Ironically, even their move toward representative government was an indication that they could no longer assemble in one place. And soon they scattered throughout New England.

Toward the end of the first generation of Pilgrims in the New World, another group of Puritans decided to come. They felt they had little choice because under Charles I, Archbishop Laud began in 1628 to purge the churches and schools of England. As tension mounted, virtually the whole Puritan community determined to emigrate. Unable to protect themselves, a group of merchants, led by John White, calling themselves "the New England Company" sent an initial group of forty settlers under John Endecott to establish a foothold at the Algonquian settlement of Naumkeag, which they renamed Salem. Meanwhile, they lobbied, as we would say, to obtain a royal charter as the Massachusetts Bay Company. When they successfully concluded their negotiations, they sent a well-equipped party of an additional 400 of their members across the ocean in April 1629. At the next stage, in 1630, they sent a whole flotilla of eleven ships with some 700 passengers, a storehouse of equipment, and livestock.

Themselves owning the company, rather than being its lessees as the Virginians were to the London Company; being its serfs as the Marylanders were to Lord Baltimore; or having no clear legal standing, as happened to the Pilgrims at Plymouth, they felt themselves to be in control of their destiny. That sense of destiny was underpinned by religious unity and sense of mission. As John Winthrop put it when he was leaving England and was soon to be governor of the colony, "We shall be as a city upon a hill; the eyes of all people are upon us."

The physical embodiment of the "city upon a hill" was to be Boston, which the settlers began to lay out in 1630. The choice of the name Massachusetts was apposite, since in the Natick Indian dialect it meant "at the great hill." From the beginning, the Puritans would congregate in towns and so emphasized the creation of governments based on town meetings. Within a few years, they had established more than a score of settlements along the coastline. And they quickly began to fill these settlements with immigrants. Although about one in each four of those who went with Winthrop died in the first winter, and others were discouraged and turned back, by the summer of 1631 they numbered over 2,000. Numbers counted, but also important was the fact that the English Puritans were rich and disciplined. They were in word and in deed establishing themselves as not only a city on a hill but also the kingdom that owned the hill.

CHAPTER 8

"Mother England" Loses Touch

As Queen Elizabeth had made clear to Sir Walter Ralegh, the English bridgehead in the New World was of minor interest to the English government apart from its possible use as a base for pirate raids on the Spanish treasure fleet or Caribbean cities. The English government was far more concerned with matters closer to England and of more commercial value than the Chesapeake. From the time they stepped ashore, the early colonists were cut off from Europe by months of dangerous and unpredictable sailing over 3,000 miles of ocean. Neither their sponsors nor their government knew much about the New World. The home government intervened only if private sponsors failed (as they did in Virginia in the 1620s), if the settlers and "proprietaries" were caught in an irresolvable conflict (as they were in Maryland in the 1650s), or if their leaders and merchants were transgressing imperial interests and laws (as they did in Massachusetts in the 1660s). Looking back on this period from much later, the great English parliamentarian Edmund Burke summed up the administrative mess that had been created by saying that the establishment of the colonies was "never pursued upon any regular plan; but they were formed, grew, and flourished, as accidents, the nature of climate, or the disposition of private men happened to operate." This was, perhaps, an exaggeration, but the colonists certainly were unsure of their relationship to England. As they gazed wistfully out to sea, they thought of themselves as loving daughters, but soon they suspected that their "mother" regarded them as an "orphant Plantation."

Just eleven years after the first colonists began to create Jamestown, the Virginia Company of London, acting without royal sanction, decided that it would be cheaper, more productive, and less troublesome to allow them some discretion. It had its governor announce that henceforward they should choose from among themselves their own leaders to formulate laws and regulations based on English common law. The "burgesses" were to be selected from "from every Town, Hundred or other particular Plantation" and, as the "General Assembly," were to meet once a year with the governor's "Council of State." The council of state would gradually evolve into an upper council, regarded as something like the House of Lords; and the "House of Burgesses" would copy the pattern of the House of Commons.

The newly elected burgesses held their first meeting on July 30, 1619, on the ground floor of the Jamestown church while the governor and his advisory council met in the choir loft. After a few days of sweltering in the July heat as swarms of mosquitoes buzzed in from the dank surrounding woods and marshes, the delegates fled. But they had managed in their short session to pass a number of laws, including the first tax levied in America.

Although historians may find this assembly the first step on the road to independence, the colonists do not seem to have found it so. They were much more concerned with clearing land, bartering for food and furs with the Indians, and above all growing tobacco. Tobacco was their lifeline to the good things they wanted from England. Tampering with the tobacco market was much more sensitive than being allowed to levy their own taxes. So, when in April 1623 the British government declared the tobacco sales contract void and sent a mission to reassess Virginia, the burgesses boycotted it. More pointedly, when one of their clerks agreed to assist the mission, they had him put into the pillory and sliced off his ears. That, together with an expensive and disastrous war in 1622, when the neighboring Indian tribes nearly destroyed the colony, caused the government in 1625 to get a court order declaring the company in default, to order it disbanded, and to proclaim the tiny triangle between the James and York rivers, which was all there was to "Virginia," a royal colony.

During the 1620s and 1630s, people were flooding into North America. Virginia was being subdivided as plots of land were sold off to separate groups; and new colonies were being established in Maryland and Massachusetts. By 1641, the population of the "British" colonies had reached 50,000; this critical mass gave a reassuring sense of capacity, stimulated the ambitions of politicians and merchants, created local interests, and made possible thoughts of autonomy. The course these events might have taken under other circumstances is, of course, speculative; what did happen was that as the colonies grew, England was virtually cut off from America for a whole generation, because in the late 1620s, England drifted toward chaos. Then, from roughly 1640 to 1660, it was the scene of a religious revolution, which was compounded by a clash between the king and Parliament that erupted in civil war; this was followed by wars with Ireland, Scotland, and the Netherlands; and the country was then reconstituted under a new form of quasi-republican government. No one in England had time for the distant and as yet not very important American colonies.

Meanwhile, the colonists embarked upon programs, acquired habits, and developed capacities that would shape events far beyond the restoration of the monarchy in 1660 and, in sum, would constitute a virtual seventeenth-century revolution. The purpose of this chapter is to show how the revolution came about, what happened in the several colonies, and what the effects were to be for the future. I turn first to Massachusetts where it began, where it reached its high-water mark, and where its legacy was most evident.

In 1629 the newly incorporated Massachusetts Bay Company had acquired a patent for lands stretching "from the Atlantick and Westerne Sea and Ocean on the Easte Parte, to the South Sea [Pacific] on the West Parte"—that is, far beyond anything even claimed by Britain and across territories which were being claimed by France and Spain. This territory was nearly the size of western Europe. The company also was awarded an extremely permissive royal charter constituting it "one Body corporate and politique [for] the Government of the People there." In Massachusetts, unlike the other colonies, the "proprietary" company was owned by the colonists, who as "Dissenters" were motivated by a sense of alienation from England and as "Puritans" were motivated by a mission to create an entirely new kind of society. In their scheme, the king and Parliament had no role—that is, except for the first act: they had granted Massachusetts virtual independence.

The leaders of the new colony moved immediately to effect that independence. In October 1630 eight senior Puritans named themselves "magistrates" and assembled together as the colony's first "Generall Court" or legislature. That body then appointed itself a ruling council as the "Governor's Court of Assistants" in which all executive, legislative, and judicial powers were to be lodged. Sure of their mission and driven by the urgency of getting the community organized, the governor, John Winthrop, and the "assistants" set to work. They made no pretence of election, entertained no concept of the division of powers, and certainly planned no tolerance for opposition. They acted as they thought "meete," that is, in line with their interpretation of biblical commandments and current needs. They moved fast and comprehensively. With control of both the proprietary company and the charter, they granted land to incoming Puritans, incorporated towns, assessed taxes, and appointed officers of government. Since they always intended Massachusetts to be a self-governing theocracy as defined by their church, they made no pretence of allowing freedom of religion to others; "non-Puritans" were outsiders, living on sufferance so long as they conformed to strictly enforced codes of dress, drink, and deportment. If they did not conform, they were to be punished, driven away, mutilated, or even executed. As though to prove their growing alienation from England, the new government required all inhabitants to pledge allegiance not to the king but to Massachusetts.

Foreshadowing the great issue of the eighteenth century, the first significant dissent from the government of Massachusetts came over the issue of "taxation without representation." The Puritan minister at Watertown, Deacon George Phillips, advised his parishioners in 1632 that it would be wrong to pay taxes they had not been represented in levying. Stung by this criticism, Governor Winthrop took the first small step toward representative government by allowing each township to elect two delegates "to advise with the governor and assistants." But they were not to enact laws. Laws were to be created only by the Puritan elders. In fact, Massachusetts moved away from popular participation in government and law by mandating, in 1641, a comprehensive and intrusive list, not of rights but of crimes and punishments.

Meanwhile, in England, Archbishop William Laud, the chief ideologist in Charles I's authoritarian government, undertook a vigorous program to revitalize the Church of England. Watching the departure of a growing number of parishioners to New England, he concluded that he could combat dissenters effectively only if he followed them to their new refuge. In 1634, Laud got the king to appoint a "Commission for Regulating Plantations" under his leadership to subordinate Massachusetts. Massachusetts's vulnerability, he found, was the doubtful legality of its original grant. If he could overturn the Puritans' claim to their grant, Laud hoped he could humble them. So he went to court to sue for repeal of the company's patent. The court took its time, but two years later it complied. The king then appointed a governor to effect Massachusetts's transformation into a crown colony. Sensing the drift of affairs and fearing for his life, this man wisely declined to go. There the dispute rested without resolution, as the British became increasingly distracted by the events leading toward civil war. Meanwhile, thousands more Puritans moved to Massachusetts.

The civil war; the collapse of the royal government, in which both the king and Laud lost their heads; and the Puritans' assumption of power under Oliver Cromwell seemed literally a godsend to the Puritans in New England. They took the victory of the Parliamentary forces not only as a confirmation of the rightness of their cause but also as the completion of their move toward independence. Marking this, in 1652 they declared Massachusetts a replica of Cromwell's new English regime, a "commonwealth," and assumed the attributes of a sovereign state. While favoring the Puritans, Cromwell was not willing to go that far. So the matter rested—the Massachusetts colony claiming sovereignty and Cromwell's government not sanctioning it—until the Restoration.

During the twenty years before the Restoration, New England underwent a reversal of sentiment. Some of the Puritans had become less willing to enforce the authoritarian way of life of the colony's first generation. Many, indeed, began to evince nostalgia for the "old country." The zeal of the first generation abated. Merchants were making money; towns were prospering; farming, fishing, and fur trading occupied people's thoughts; and many were tired of the rigidity and tyranny of the Puritan leadership. Those developments had already caused what became Connecticut and New Hampshire to break away from Massachusetts and set themselves up, without government sanction, as quasi-independent colonies. The Massachusetts regime reacted angrily and took advantage of "those sad distractions in England" to begin a sort of imperial expansion, annexing scattered settlements all the way into New Hampshire and Maine. Rhode Island, ostracized by Massachusetts, became a sanctuary for non-Puritans and the freest, most tolerant of any of the colonies; not surprisingly, in 1642 Massachusetts considered invading it.

By 1643, feeling secure in virtual independence, Massachusetts set up a confederation of the New England settlements including Plymouth, Connecticut, and New Haven—"being all in Church-fellowship with us"—known as the United Colonies of New England. The purposes they agreed on were to protect one another against the Indians (who "have formerly committed sundry insolence and outrages") and to prevent indentured servants from escaping. Although each member retained its "peculiar jurisdiction," the confederation acted like a sovereign state rather than a colony when it negotiated the Treaty of Hartford with New Netherland in 1650. It was the first American experiment in federalism.

When the monarchy was restored, Massachusetts was unsure what to do. Having enjoyed virtual independence during the adult lifetimes of most of its citizens, and feeling no affinity with English monarchy, it was tempted to sever the tie with Britain completely. On June 10, 1661, The General Court opted for autonomy within the empire: Massachusetts was, the court stated, "a body politicke, in fact & name [with] full power and authoritie, both legislative & executive, for the gouvernment of all the people heere . . . without appeal [to English courts] excepting lawe or lawes repugnant to the lawes of England." The proclamation was not immediately challenged, because the restored monarch, Charles II, had his hands full asserting royal authority in England. To curtail the powers of Massachusetts, he granted charters to Connecticut and Rhode Island; but not until 1664, after receiving a deluge of complaints, did he made a serious effort to come to grips with Massachusetts's challenge to royal authority.

What triggered Charles's sudden assertion of authority was Britain's takeover in 1667 of New Netherland, which the Dutch had claimed since 1626. During the troubled period of the civil war and the restoration of the monarchy, England fought three wars with the Dutch. These wars had two significant results in American history. The first was that the British acquired approximately 1,000 Dutch ships, the *fluyts,* thus creating an "instant" merchant marine that would carry passengers and cargo to and from the New World. The second result was that in the Treaty of Breda of 1667, the Dutch ceded the territory of what became New York in return for Britain's recognition of their claim to Surinam. New Netherland was a vast tract of land, stretching along the coast from Massachusetts down to the Delaware River. Only partially developed by the Dutch and a small number of Swedes (who had established a trading post on the Delaware), it was to become the private preserve of the king's brother, James, duke of York.

Just what they were getting and how to deal with the new colony were obviously complex and difficult questions. To answer them, settle boundary disputes, reform local administrations, and prepare for the granting of lands to new settlers, Charles II sent a commission to the New World. In due course, the commissioners arrived in Massachusetts. There, they were met with defiance that stopped just short of armed resistance. As Bernard Bailyn has written, the commissioners appeared as "devil figures, incarnations of evil to the inflamed Puritans. The General Court declared their commission invalid on the ground that the authority it conveyed conflicted with that of the Massachusetts charter, and refused to authorize their activities within its jurisdiction." Massachusetts, the governor wrote to the king on October 19, 1664, had for "above thirty yeares enjoyed the aforesaid power & privilege of government within themselves, as their undoubted right in the sight of God & man." The Puritan community did not intend to live in any other way.

When the king received the report of his representatives, he ordered Massachusetts to send agents at once to explain this response. Cleverly, Massachusetts evaded the summons but sent the king a present of twenty-five tall New England white pine logs to be made into masts for the Royal Navy—a most welcome gift, since England no longer had trees tall enough to serve as single-pole masts. Governor John Endecott also sent an overtly obsequious but substantively firm letter describing the history of the relationship between Britain and Massachusetts. This letter ended on the practical note that if the king attempted to impose direct rule, his officers would have to "expend more then [sic] can be raised here."

Having thus made his case, the governor "neglected" to send representatives to London to discuss the charges. In turn, the king equivocated; he did nothing but urge compliance. Unconcerned by mere words not backed by force, Massachusetts continued as before, running its own affairs.

Also during this period, New England had become a "great mart," increasingly engaged in overseas trade in fish, lumber, and other items, in which it paid little or no attention to the regulations, the Navigation Acts, by which the imperial government sought to control commerce. Massachusetts's evasion of duties may have cost the royal treasury 100,000 pounds yearly in lost revenues. That fact got London's attention, and another mission was sent in 1676.

This mission, like its predecessor, met with evasion, noncooperation, and threats. Massachusetts had learned that it could simply "stonewall" when ordered to cooperate. When the head of the mission confronted the governor, he was told "that the Laws made by Our King and Parliament obligeth them in nothing but what consists with the Interest of New England." In his report, Edward Randolph sought to shame the king into acting by alleging that Massachusetts had given sanctuary to some of the people who beheaded Charles I and that, in violation of royal prerogatives, it had set up a mint to coin its own money. Again, the government issued statements but did little.

Convinced that the government would continue to do little, Massachusetts imprisoned and threatened to try the chief royal representative, under its own laws, for the capital offense of "attempted subversion." Britain's lack of a consistent policy—its tendency to waver between action and inaction, condemnation and sufferance, repression and laxity—was apparent again in the 1770s, during the confused period before the Revolution. As it was to do then, the British government in the seventeenth century allowed, indeed caused, a spirit of rebellion to build up; then in 1684 it attempted to smother the spirit of independence. In that year, an English court declared the Massachusetts charter null and void. The following year, when the duke of York became James II, he decided to cut to the heart of the dissidence: to abolish the existing colonial structure and its alleged "rights" by bringing the northern colonies together in the Dominion of New England.

The Dominion of New England was to begin afresh without troublesome promises and contentious legislatures; it would be governed by a royal official: the "captain generall and governor in chief." Sir Edmund Andros took the strongest possible view of his powers and very soon managed to antagonize even those who had initially welcomed him. Insensitive to local opinion, he flaunted his power even in Puritan Boston by commandeering a Congregational church for Anglican services, thus resurrecting in America the very issue that had caused the Puritans to quit England. Even more than most royal officials, he quickly earned a reputation for corrupt practices.

Corruption might ultimately have brought him down, but what actually did so was the arrival of news from London: James II had been overthrown and had fled the country; Prince William of Orange had invaded England; the "Glorious Revolution" had begun. The news became known in Boston on April 18, 1689, and before the end of the day a mob seized the governor and most of the royal officials. The townsmen were spurred on by Robert Small, a carpenter from the Royal Navy frigate *Rose.* Small organized a gang of armed men and managed to seize the ship's commander; thus incited, mobs of Bostonians surrounded the fort, and soon companies of militia were assembling at the town center. The old Puritan leadership put itself at the head of the insurgents and by the end of the following day had resumed command of the entire colony. The governor's final humiliation came when he was apprehended as he attempted "to make an Escape in Woman's Apparel, and pass'd two Guards, and was stopped at the third, being discovered by his Shoes, not having changed them."

Having overthrown James II's appointee, the Bostonians set up a "Council of Safety" composed of twenty-two prominent citizens but for months were unable to subdue the mobs who roamed the streets. Thus it was that beginning in the 1660s Massachusetts began to carry out activities which would be echoed with eerie precision there in the 1760s, on the eve of the Revolution.

In the 1660s, with Massachusetts in turmoil, the aggrieved and dispossessed Indians, inspired by the French, struck back at—the settlers who had been taking their lands. All along the frontier, they burned houses, killed livestock, and scalped settlers. In terror, large numbers of settlers fled the frontier. In Boston itself, government virtually ceased to function; there it was not Indians but foreign merchantmen and even pirates who took advantage of official incapacity, sailing into the harbor to swap the goods they had looted at sea.

Meanwhile in England, some members of the new government perceived what had long been evident: that New England, now grown to about 300,000 people, was rapidly drifting away from Britain. By then, since commerce had flourished, there was much to be lost. Responsible officials were worried. But they were not alone in seeking the ear of the new king and queen. Men from America arrived to proclaim that they had acted against the agents of James II in support of principles for which William and Mary stood. Foremost among these men was the Bostonian firebrand, prelate, and later president of Harvard, Increase Mather, who was sent by the General Court to seek a new charter to restore "their auncient priviledges."

In the new charter of 1691, Massachusetts received rather less from the king and queen than its "auncient priviledges." William and Mary decided that the governor was to be a royal (not a church-designated) appointee; the "Great and Generall Court of Assembly" was to consist of the governor, his council, and men elected by freeholders (who need not be members of the Congregational church); oaths were to be taken to the king (not to the government of Massachusetts); and all inhabitants were to have the same "Libertyes and Immunities of Free and naturall Subjects within any of the [British] Dominions." In the most sensitive area, for all Christians except Catholics, "liberty of Conscience [was to be] allowed in the Worshipp of God."

Finally, however, came one concession to Massachusetts that the British would have much cause to regret in the days before the Revolution of 1776: the Massachusetts General Court was empowered to appoint judges and constitute courts. Those courts would become a painful thorn in the heel of British imperialism.

I have dwelled at length on Massachusetts because it was in many ways the most interesting of the rebellious "daughters" of the English mother, but it was by no means unique. Virginia, the oldest of the colonies, was also the one that most easily meshed into the imperial economic system: it produced agricultural products that the English either consumed or could resell, and it was a major customer for British goods. It had developed a society based on plantations that paid the English landed gentry the supreme compliment of imitation. The Virginia elite read the English press, wore English clothing, affected English manners, worshipped in the Anglican church, exchanged visits with English friends, and sent their sons to study in English institutions. The popular royal governor, Sir William Berkeley, was an appointee of Charles I; not surprisingly, even after the execution of the king, Governor Berkeley and his Virginia Anglicans remained loyal to the house of Stuart.

When Berkeley finally had to step down, Virginia functioned essentially as a self-governing republic under its House of Burgesses. Not until 1652 did a parliamentary mission arrive to accept Virginia's "surrender" to Cromwell's government. Surrender came easily. Form rather than substance was its hallmark: the House of Burgesses did elect a Puritan governor, but his appointment scarcely affected the colony's life. Scattered widely in their plantations—rather than being grouped together in a city as the Puritans were in Boston—Virginians went about their daily lives more or less oblivious of the great events in London. Then, having suffered not at all, the burgesses received news of the death of Cromwell with mild relief, and news of the Restoration with joy. The transition was easy: they simply reelected the former royalist governor, Sir William Berkeley, and in short order the new king confirmed their choice. Berkeley was to enjoy sixteen more years of watching his colony grow and prosper.

Then, in 1676, he faced another challenge, a revolt led by his young cousin by marriage, Nathaniel Bacon. The issue on which Bacon based his movement was the hostility felt by frontiersmen to the crown's and the coastal Virginians' relationships with the Indians. The more settled peoples along the coast were profiting from trade with the Indians and were supplying them with arms that, although intended for hunting, were occasionally used against whites trespassing on Indian lands. This relatively simple issue was complicated by the fact that "the" Indians known to Virginians were not a single group but several mutually hostile societies. Those with whom the Virginians had earlier been in contact had all but disappeared. Their numbers had declined from about 10,000 to only 3,000 or 4,000 men, women, and children. They had lost most of their lands, and the little land on which they still lived was theirs only at the sufferance of the colony. As Governor Berkeley wrote in 1671, "The Indians, our neighbours, are absolutely subjected."

The nearer societies, particularly the Susquehannocks, had long worked with whites as intermediaries to the more distant Indians in the fur trade; some had been encouraged to think of themselves as allies of the Virginians. Farther inland were tribes like the Senecas and other branches of the Iroquois who were fiercely independent and were as apt to strike against Indians as against whites. They remained a formidable military force; even as late as 1750, they and the Atchatchakangouen (whom the whites called the Miami) and their immediate allies, the Shawnees, Delawares, and Wyndots, could muster 1,500 to 2,000 warriors.

Since the frontiersmen wanted the land and did not care about furs, they blamed the coastal Virginians and specifically Governor Berkeley for giving the Indians the means to defend themselves. Bacon put himself at their head and led them successfully against the Indians. His courage and decisiveness immediately made him a popular hero. On that base, he gathered support by advocating lower taxes and a more open electoral franchise. Governor Berkeley tried to resist but was defeated; he then fled across the Chesapeake to the Eastern Shore of Virginia to gather forces for a counterattack. Meanwhile, showing his and his followers' hostility to the coastal Virginians, Bacon captured and burned Jamestown. He had hardly begun to organize his government when, apparently after a heart attack, he suddenly died. Berkeley quickly took his revenge. From London, Charles II remarked that the "old fool has hanged more men in that naked country than I have done for the murder of my father." In an effort to make peace, the king recalled him. But the issue that Bacon illuminated would persist down the years until, on the eve of the Revolution, it burst into a similar rebellion in North Carolina known as the War of the Regulation.

Maryland differed from both Massachusetts and Virginia in two significant and interlocking ways. First, it was a proprietary colony in which political discord arose not between king and colony but between the colonists and successive lords Baltimore. Second, while the population was overwhelmingly Protestant, the government was Catholic and favored the Catholic minority. The already volatile mix of religions was further complicated when a group of Puritans "and other well-affected people in Virginia, being debarred from the free exercise of Religion" moved over to Maryland. When told that they must take an oath of fealty to Lord Baltimore, they refused.

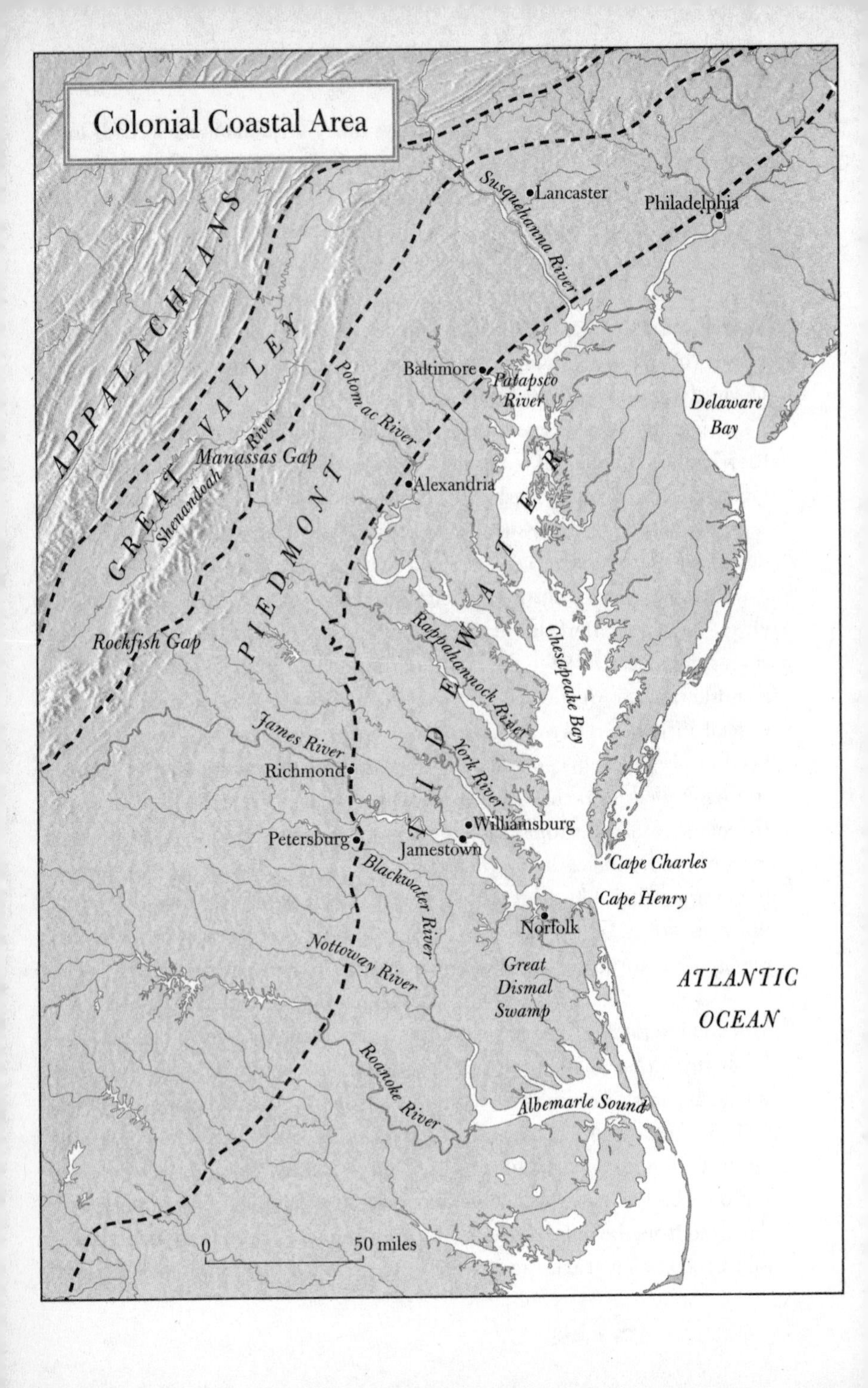
Colonial Coastal Area
APPALACHIANS
GREAT VALLEY
PIEDMONT
TIDEWATER
Susquehanna River
Lancaster
Philadelphia
Baltimore
Patapsco River
Delaware Bay
Potom ac River
Shenandoah River
Manassas Gap
Alexandria
Rockfish Gap
Rappahannock River
Chesapeake Bay
James River
York River
Richmond
Williamsburg
Petersburg
Jamestown
Blackwater River
Cape Charles
Cape Henry
Norfolk
Nottoway River
Great Dismal Swamp
ATLANTIC OCEAN
Roanoke River
Albemarle Sound
0
50 miles

Meanwhile, in London, to the astonishment of the local Protestants, Lord Baltimore won over Cromwell. Cromwell empowered Baltimore to reinstate his officials; he did, and they threatened to charge the Protestants with sedition. So each side mustered its forces: the Protestant leader called his opponents "Papists and other desperate and bloody fellows," and Baltimore's agent derided the Protestants' "barbarous and bloudy actions."

Angry shouts quickly turned to musket shots. About 250 of Baltimore's men marched on the capital, which was still held by the Protestants. The Protestants rallied their forces, which were somewhat smaller than Baltimore's, and enlisted help from the *Golden Lion*, an armed merchantman that happened to be anchored nearby. The guns and crew of the *Golden Lion* swung the balance. Baltimore's force was routed with a loss of about fifty wounded or killed. The civil war that convulsed Maryland was a small-scale version of the one that had racked England. Over the next few years, the Protestant proportion of the community grew larger, reaching a ratio of about twenty to one.

In 1685, the Catholic duke of York became King James II. Protestants feared the advent of a new Counter-Reformation with the English monarchy allied to Louis XIV of Catholic France. In a widely discussed local affair, the governor, a Catholic nephew of Lord Baltimore, stabbed to death a Protestant customs collector. In the already poisoned atmosphere of religious strife, Baltimore sent over in 1688 a governor who astonished the assembly not only by proclaiming the antiquated doctrine of the divine right of kings but also by charging that Maryland "is full of Adulterers . . . as I hear some do not only abroad but even at home under their wives' noses, where strumpets rule and the wives obey to the scandal of all honest and good men." It was not, perhaps, the most tactful maiden speech for an incoming governor.

With these and other goads driving them, and upon hearing of the fall of James II and the arrival of his Protestant successors, William and Mary, the Protestant community of Maryland formed an "Association in arms for the defence of the Protestant religion and for asserting the right of King William and Queen Mary to the province of Maryland." Since Baltimore issued no instructions to recognize the new monarchs, Protestants feared a backstairs deal like the one Cromwell had made with an earlier Lord

Baltimore. Deciding to act first, 250 poorly armed men pulled off what amounted to a coup d'état on August 1, 1689. They marched on Saint Mary's and seized Baltimore's officials, displacing the proprietary government, and petitioned the king to make Maryland a crown colony. The king accepted on June 27, 1691, but generously allowed Baltimore to retain "his rights to the soil."

Not all incentives to revolt were religious. In Carolina (not yet divided between North and South), the planters felt squeezed by the tightening of customs control after 1673. Fearing a rise in prices, Carolina farmers staged a revolt and formed a revolutionary government. In 1680, the leader of the revolt, John Culpeper, sailed for England to plead the planters' case before the proprietors. He was arrested and tried for treason but was allowed to return to Carolina because the group he represented, the government decided, was "the only one that could be trusted with the colony's well-being."

Meanwhile, on the Hudson River, the Dutch had formed settlements the same year the English began to establish Jamestown. In that year, the explorer Henry Hudson had sailed up the Hudson to the future site of Albany, where his *Half Moon* touched bottom. It was there, in 1624, that the Dutch established Fort Orange. Two years later, they purchased the island we know, from the name of the Indians in those parts, as Manhattan for the celebrated price of 60 guilders. In their New Netherland, the Dutch set about organizing a colony patterned on the English model. Settlers, under the *patroon* system (comparable to the "hundreds" in Virginia), functioned as a semi-autonomous lords. A war that the Dutch provoked with the Indians in 1643 had the same result as the two wars between the settlers at Jamestown and their Indians: devastation of their newly built properties. However bad their conduct toward the Indians, the Dutch were (relative to their neighbors) enlightened toward two other groups: they provided black slaves with land to enable them to buy their freedom, and they established religious freedom. The colony prospered and by 1664 was populated by nearly 10,000 settlers.

In that year, having seized it from the Dutch, Charles II gave New Netherlands to his brother, then the duke of York, on terms even more generous than those granted earlier to Lord Baltimore. Although the duke thought representative assemblies are "of dangerous consequence [and] prove destructive to, or very oft disturbe, the peace of the government," he agreed to let New York have one when the settlers on Long Island refused to pay taxes unless he did. The first assembly of eighteen deputies met in Manhattan on October 17, 1683, some sixty years after the first assembly in Virginia, and passed a "constitution," the Charter of Liberties and Privileges, which remained in effect until 1686 when, as king, James II formed the Dominion of New England.

The overthrow of James II and the accession of William and Mary did not bring happiness to New York. In fact the people of New York did not hear of the events in England for nearly three months after most of the other colonies; and then they learned about them only from the insurrection in Massachusetts. Already agitated by local issues, the various communities along the Hudson and on Long Island projected their anxiety, anger, and fear onto the acting governor. In New York, as in the other colonies, one man took the leadership of discord.

Jacob Leisler was one of the oldest residents in the colony, having come there as a soldier for the Dutch company in 1660. A German by birth, he had become a rich merchant with a large following outside the established Dutch elite. Never an easy man, he thrived on conflict and focused his anger on the acting governor. With some 500 like-minded men, he seized the fort to preempt a supposed plot by the governor to burn the town. (Actually, there was no such plot.) Deserted by the militia, the governor wisely if not bravely departed for England. Leisler then convoked a popular assembly that duly proclaimed him "captain of the fort and commander-in-chief." With this authority, he declared the colony loyal to the monarchy of William and Mary and established, as other colonies had, a "Committee of Public Safety." The danger, he said, came not so much from within the colony as from the Indians.

Shortly thereafter, in seeming confirmation of Leisler's prediction, an Iroquois band led by Frenchmen wiped out the new village of Schenectady. So shocking was this event to the settlers that even those who had opposed Leisler rallied to his side. Unlike Bacon in Virginia, Leisler used his popularity and power to rule New York more wisely and more humanely than anyone had yet done, solving in his one year in office many of the problems that had plagued it for decades. But it was not in New York that the ultimate decisions were being made: Leisler's enemies had taken their case to London, where the new government saw Leisler as a troublemaker.

To end the presumed disturbance, London appointed a new governor and sent before him a small military force whose commander quickly allied himself with Leisler's enemies. When Leisler refused to give up the power he had accumulated, the commander arrested him and tried him for treason. Leisler and his son-in-law were convicted and sentenced to be "hanged by the Neck and, being Alive, their bodies be Cutt Downe to the Earth that their Bowells be taken out and they being Alive burnt before their faces that their heads shall be struck off and their Bodys Cutt cut in four parts and which has be Desposed of as their Majesties Shall Assigne."

That "assignment" took care of Leisler, but not of the cause he had embraced: a few years later, in 1704, the governor wrote to the Board of Trade in London that the colonial legislature of New York aimed "to make themselves an independent people, and . . . to divest the administration . . . of all the Queen's power and authority and to lodge it in the Assembly."

Before these events were played out, the western part of what had become New York had been given by the duke of York to two of his friends. What became New Jersey (known then as Nova Caesaria) went to Lord John Berkeley and Sir George Carteret. Then the southeastern territory along the Delaware River was acquired by William Penn.

Meanwhile, in 1681, William Penn acquired the patent to what became Pennsylvania. Not only did the Society of Friends gain a new homeland, but their emigration, Penn told the king, rid England of the troublesome Quakers. Apparently, the king found the argument persuasive, since, as with Baltimore, the monarchy charged Penn only a nominal rent: two beaver skins to be delivered to Windsor Castle each year.

To settle his grant, Penn mobilized an unprecedented real estate marketing venture. Agents toured Europe and broadcast pamphlets and broadsheets in French, Dutch, and German as well as English. Each settler was expected to pay his way: 6 pounds for a husband and wife; 5 pounds for a servant. One chest of their belongings was to be shipped free. Land would be free except for a small quitrent. Although Penn was active in promoting its interests, he did not visit his colony until 1682—and then he almost did not make it, because smallpox devastated the passengers on his ship. By the time he left the colony, two years later, the population had reached almost 7,000.

They soon became 7,000 disaffected, ungrateful, avaricious people, engaged in what was virtually a revolt against the government Penn had devised. By about 1720, so many European immigrants had poured in—about 13,500 in 1709 alone—that the colonial administration was swamped. As the Penn family's agent and land commissioner, James Logan, wrote in 1727, the horde of Germans who poured in were "a surly people . . . generally well-arm'd." Penn's dream of a vast new feudal but beneficent, tolerant, and "friendly" state evaporated.

CHAPTER 9

The Growth of the Colonies

For us it takes a leap of imagination to see and feel what it was like to live in the American colonies in the half century before the Revolution. The past truly is another country, hard to visit and harder to understand. Consider first the vast cities we know today: New York with its millions was a village of 11,000 people in 1743 and did not reach 25,000 until 1775. During the same years, Boston actually fell from a recorded 16,382 inhabitants to 16,000 while Philadelphia, the American giant, rose from 13,000 to 40,000.

The very landscape would be shockingly alien to a modern American. Much of today's downtown Boston was under water. New York City hung like an overripe fig at the southern tip of the leafy branch that was Manhattan Island. The island was mostly a wild tangle of brambles and forest sliced by gullies and blocked by imposing, now mainly leveled, heights. Washington, D.C., did not exist. Huge areas of America were undrained swamp or virgin forest. The villages and towns that had begun to dot the map were far apart and isolated from one another by vast spaces.

As the colonists grew in number, they began to imprint the land with styles they remembered from the mother country in language, dress, diet, mores, legal systems, and religious organizations. The ways are many and subtle, but one that I personally have found fascinating was how the early settlers built their houses. I learned this as I worked to restore a "center chimney" house, built in Harvard, Massachusetts, about 1692 by a shipwright. The builder had no cement, plaster, or paint—none was then available in colonial America—so, to make the rooms of his house look as much as possible like the rooms he remembered in the England of his childhood, where ceilings were customarily coated with white plaster and partitioned by supporting wooden beams, he whitewashed the floorboards of the second story between the beams. Many of the adaptations of early America were, like that ceiling, almost English.

As David Freeman Hawke wrote, each itinerant master builder

> arrived on the scene with a blueprint in his head, a two-foot folding rule in his pocket, and only an experienced eye to judge whether beams were plumb and level. . . . The tools he carried had changed little since the days of the Roman empire—crosscut and rip saws, hammers, chisels, mallets, augers, gimlets, planes, hatchets and axes. He favored huge timbers for the frame—a ton or more of wood went into even a small house.

In fact, so much oak, chestnut, and pine was used that the Massachusetts colonists began as early as 1636 the first serious attempt at conservation. Having been in England when the depletion of forests was already recognized as a serious problem, the colonists appointed overseers to prevent unnecessary felling of large trees, particularly on common land. As each town was incorporated, it took on the responsibility of regulating the use of nearby timber.

In setting forth his thesis on the role of the frontier in American history, Frederick Jackson Turner mentions that as early as 1645, the General Court of Massachusetts forbade settlers to quit towns on the frontier—those frontier towns are now suburbs of Boston. A few years later, between 1669 and 1675, fearing attack by the Indians, Bostonians considered building a wall of wooden poles or stones to enclose the entire hinterland of the city, "which meanes that whole tract will [be] environed, for the security & safty (vnder God) of the people, their houses, goods & cattel; from the rage & fury of the enimy." As the frontier was pushed outward, additional towns were added to the protected area.

The house I renovated was one of the results: it was made proof against arrows and musket balls by being sheathed in oak planks, each about 2 inches thick. Thus it became a "garrison house" for the little settlement of Harvard. The walls outside the sheathing were overlapping "clapboard," just as the shipwright-builder made ships waterproof. Climbing up to the attic, one could see that the house indeed resembled an inverted ship: the supporting beams were pegged to a "keel" that served as a roof beam. What was different from a ship was that the whole structure hung on a massive central chimney, which occupied about a quarter of the total interior space of the house. The main fireplace, large enough to walk in, was used both for heating and for cooking. Each of the original rooms had a fireplace, and opposite the kitchen fireplace was a smoke room, large enough for hanging and preserving several carcasses of deer or wild boar. The house was built around 1692, but not until at least half a century later was the outside of the house painted. Paint was a luxury; bulletproof sheathing, heating, and cooking were necessities.

My house in Harvard was relatively sophisticated. Roughly half a century earlier, the thirty houses that Manhattan then boasted were mainly made of tree bark on Indian patterns, since there were as yet no sawmills or brick kilns; as late as 1685 many families in Pennsylvania were living in caves or lean-tos called "half-faced camps." As the settlers prospered during the eighteenth century, particularly along the Atlantic coast, they began to turn to England for guidance on building their houses. The works of Christopher Wren and Inigo Jones were published in what amounted to early versions of coffee-table books; the neoclassical works of the great sixteenth-century Italian mason and architect Andrea Palladio, especially, inspired the buildings being constructed in the colonial South. When, shortly after the Revolution, a French traveler, the marquis de Chastellux, visited Monticello, which Thomas Jefferson had begun in 1768, he complimented Jefferson on being "the first American who has consulted the Fine Arts to know how he should shelter himself from the weather."

Houses in the new cities were jammed together. Benjamin Franklin complained in 1753 that "This Town [Philadelphia] is a mere Oven. . . . I languish for the Country, for Air and Shade and Leisure, but Fate has doom'd me to be stifled and roasted and teased to death in a City."

To complement their houses, the colonists imported English furniture. Thomas Chippendale not only made furniture but also published designs; his *Gentleman and Cabinet-Maker's Director* (1754) was much admired in the colonies, where it stimulated the growth of a new furniture-making industry. Soon, American adaptations of English styles were announced by furniture makers in Boston, New York, and Philadelphia, although the Shakers and the Amish continued to fashion more austere, less English styles to express their religious faiths. In poorer houses, particularly on the frontier, furniture was made of split logs; only gradually, as sawed lumber became available, chairs and tables resembling ours came into common use. Beds, however, long remained a luxury. Most people slept on straw-filled bags or on the rushes that covered the tamped-down earth floor.

Houses were without insulation, and walls and roofs let in the cold and damp. Such heat as could be provided came from open fireplaces, which used prodigious amounts of wood. Even a small cabin would consume each winter at least twenty cords or a pile 160 feet long, 4 feet high, and 4 feet wide. A medium-size house would use more than double that amount. But no matter how much wood was burned, fireplaces threw out little warmth. Without dampers, flues not only carried up the chimney most of the heat the fire produced but also sucked streams of cold air through cracks or windows and doors. Water would freeze just a few feet away from a roaring fire, as I learned in my house in Harvard when the newly installed furnace broke down one winter day.

That man of all talents, Benjamin Franklin, rushed into the American breach in 1740, when he came out with a new stove. Inspired by European models, he tinkered with them until he had a stove that was more efficient, was cheaper, and could be manufactured in the colonies. He had a prototype made, tested it, and then published an advertisement listing all its advantages. As revolutionary in its time as air-conditioning in ours, Franklin's "New-Invented Pennsylvania Fire-Places" quickly spread from city to city.

Housing was more rudimentary along the frontier than along the coast. Out of necessity, rooms were built small. An average ground floor, usually a single room which served as parlor, bedroom, kitchen, and dining room, was determined by the length of locally available tree trunks; it was seldom more than about 24 by 18 feet. When families grew too large for the one room, as they quickly did, a duplicate of the first house would be built nearby and connected by what was called a "dog run." Only the richer people built a second story. As late as the first half of the nineteenth century, even prosperous families lived in these "double log cabins." The Harding family, who within a few years would build the great Belle Meade plantation house, were still in their log cabin as late as the 1840s.

No houses had bathrooms. Water was brought in by bucket. Bathing was infrequent even among the rich and fastidious. Thomas Jefferson famously washed his feet almost every day, but he was probably as unusual in that as in his more intellectual activities. And toilets, known as privies (from the French word for private, *privê*) or outhouses, were usually simple lean-tos over a pit. This arrangement was not particularly inconvenient, except in the winter chill and in crowded cities. The answer for the well-to-do was the chamber pot, known as a "necessary."

The well-to-do also announced their status by their clothing. For men, a ruffled shirt offset by a scarlet or yellow vest atop knee breeches joining silk stockings fitted into elaborate shoes with silver buckles was the mark of the "better sort." And, having given up the long, flowing wigs of King Charles's era, they began to affect the shorter periwig, which is still worn by English judges and barristers. In such an outfit, a man could never be accused of having to stoop to sordid physical labor. For a woman, a billowing skirt atop awkward high-heeled shoes and an occasionally fantastic hairdo would prevent her from being taken for a scullery maid.

Clothes made in America were considered less desirable than those made in England. On June 24, 1765, George Washington wrote in his diary that he had ordered a suit of clothes of "fashionable coloured cloth . . . in the best taste to sit easy and loose as Cloaths that are tight always look awkward and are uneasy to ye wearer." Benjamin Franklin wrote on May 9, 1759, that the better-off American colonists usually

> wear the manufactures of Britain, and follow its fashions perhaps too closely, every remarkable change in the mode making its appearance there within a few months after its invention here [in London]; a natural effect of their constant intercourse with *England*, by ships arriving almost every week from the capital, their respect for the mother country, and admiration of every thing that is *British*.

As they took each step forward, Americans looked back over their shoulder toward England.

Although English elegance was much sought after, even the rich had few clothes. In the new cities, the "middling" class had only a "Sunday best" outfit and work clothes. Along the frontier, clothing was even more simple. For shoes, when they wore any at all, frontiersmen and their families usually had moccasins. Men wore buckskin "hunting shirts" and leggings, suitable for work in newly cleared fields; women spun wool, flax, or cotton cloth from which they cut and sewed their own dresses. Women were used to nursing and tending numerous children and to continuous heavy labor not only at the cooking hearth but alongside men in the fields; such style as a frontier woman affected was strictly utilitarian. Whatever they had to wear, Americans usually wore it long after we (or our near neighbors) would have insisted it be washed.

Along the Atlantic coast, more effort was made at cleanliness. A recently arrived Scots tutor in the house of a Virginia grandee was astonished at the ostentatious show of cleanliness. "They wash here the whitest that ever I seed," wrote John Harrower in his journal in 1774, "for they first Boyle all the Cloaths with soap, and then wash them, and I may put on clean linen every day if I please." If he did, he must not have worn it next to his skin. Soap was so harsh that the rags used for babies' diapers were usually just dried in front of the fireplace rather than washed. Perfume had not come into fashion in America, so the body odors that we have been advertised into believing objectionable were accepted as normal. Along the frontier, the use of bear grease as a hair "conditioner" did not help, particularly as it became rancid; and a diet of heavy, fatty meats with few vegetables other than corn contributed to bad breath.

Bad breath also arose from the almost total lack of dental hygiene: few people, even among the rich, had good teeth or, in later adulthood, any teeth at all. Yet, overall health, as Charles Woodmason found on the frontier, was surprisingly good. Compared with the incoming English troops, even poorer Americans were tall (probably about 5 feet 7 inches would be the average for an American soldier, about 2 inches taller than a British soldier), sturdy, and well built, because, poor as they were, Americans ate and were housed better than Englishmen.

The colonists paid a price for what they acquired from England. By the time of the Revolution, southern planters were in debt to London merchants for 4.5 million pounds, or roughly as much as the yearly service on the British public debt. Thomas Jefferson commented, "These debts had become hereditary from father to son for many generations, so that the planters were a species of property, annexed to certain mercantile houses in London." Oliver Wolcott, a militia general, signer of the Declaration of Independence, and later governor of Connecticut, believed it was this debt that converted the generally conservative gentry of Virginia into rebels.

Certainly the English merchants gouged their colonial customers. Not only did they supply, as George Washington mildly complained, inferior or shopworn goods, but, acting as agents, they probably manipulated the prices of the tobacco they sold on behalf of growers. More significant was the cost of transporting tobacco. Since a ship could make only one round trip a year, transport averaged nearly 20 percent of the sale price. Producers were lucky to end up with one-third of the gross. Market forces then as today were against "primary products."

Prices of primary products fell disastrously during the seventeenth century. When the colonists at Jamestown first began to grow tobacco, it sold for 5 shillings a pound. A decade later, the price had fallen 98 percent. Horrified, the fledging colonial government of Virginia tried to limit production, but its efforts failed. Marylanders jumped in to flood the market with their tobacco, setting off a century-long conflict, the first of many among the colonies, as Maryland and Virginia fought for market share. North Carolina then joined in, and Virginia retaliated by closing its ports to North Carolina's shippers for half a century. Tobacco prices never recovered, despite a vigorous and questionable advertising campaign. There is "a singular virtue against the Plague in fresh, strong, quick-sented Tobacco," wrote William Byrd, a large producer in Virginia, in a pamphlet published in 1721 in Virginia's most important market, London. "The sprightly effluvia sent forth from this vegetable, after it is rightly cur'd, are by nature peculiarly adapted to encounter and dissipate the pestilential taint, beyond all the antidotes that have been yet discover'd."

Americans were slow to learn that distributors rather than producers controlled the market and gained the most from trade. So most American farmers continued to borrow in order to buy. Borrowing from London was a taste most indulged but a luxury few could afford, and many went bankrupt. Admiration of "every thing that is *British*" sometimes produced humorous results. One activity stood out: the aristocratic sport of fox hunting. Unfortunately, Americans did not have the key ingredient, a suitable fox. So sometime around 1680, they began to import foxes from England, and the hunt was on.

Meanwhile in cities, streets were nearly impassable for pedestrians. When dry, they were dust heaps that became dust storms in a breeze; in times of rain "the wheels of heavy carriages plough'd them into a quagmire, so that it is difficult to cross them." Always, they were running sewers. The stench in summertime was almost overpowering. Underground sewers were eventually built in parts of Boston and New York in the eighteenth century, but they were uncommon throughout the colonial period. Because people dumped their garbage outside their doors, drains (if there were any) were often clogged. Little progress was made with paving until the middle of the eighteenth century. Such cleaning as was done was usually the work of vultures or feral pigs. In Philadelphia, in one of his many services to Americans, Benjamin Franklin began a project of street cleaning, paid for by subscription of the residents.

To avoid the muck, wealthy people arranged to be carried in horse-drawn light "chairs." Sophisticated Massachusetts at mid-century boasted twenty coaches and 1,200 one-horse "chaises." Those who could afford it let other men's or animals' feet trample the morass.

The morass was a breeding ground for all sorts of pests and pathogens. Mosquitoes were considered just a nuisance, since their role in spreading malaria was not understood. Some relief from malaria was gained by another borrowing from native Americans, "Peruvian bark," a tropical plant of the madder family that yielded quinine. Then, shortly after 1722, experiments began to be made in the colonies on another borrowing, this time from the Ottoman Turks: an inoculation against smallpox. Although this inoculation came from the Ottoman Empire, as I pointed out in chapter 6, its main sponsor, the Reverend Cotton Mather of Boston, had actually heard about it from a slave he owned. Onesimus, as this slave was called, came from the Gold Coast and explained to Mather that it was common there. Before being enslaved, "he had undergone an Operation, which had given him something of the *Small-Pox,* & would forever praeserve him from it: adding, That it was often used among the Guramantese"—[that is, the Coromantee, the Akan or Twi of the Gold Coast]. Smallpox was then killing about one in four Englishmen and even more Indians.

Lady Mary Montagu, the wife of the British ambassador to Constantinople, had been severely scarred by smallpox, and she noted how the Turks prevented the worst of its ravages. In England, she arranged to have inoculation tested on six condemned prisoners and on children in an orphanage. Then, in absolute terror, the princess of Wales tried it, with success. With her endorsement, it was widely adopted. Mather had received word of the experiment from members of the Royal Society and wrote to tell them that he already knew about it from his slave. There was much resistance to accepting a foreign technique, especially from a Turk or an African slave, but the situation was desperate both in London and in Boston. During the outbreak of 1752, one in three Bostonians contracted smallpox and 514 died. Inoculation was then grudgingly accepted, but those who could took to their heels when epidemics broke out. No one wanted the refugees, though: when apprehended, neighboring colonies treated them as "wicked and indiscreet persons," fined them heavily, and sent them home.

Not just the sick, but rich and poor, native and foreign, men and women, were sharply divided. Despite what we would like to believe, those who had so recently suffered from intolerance in Europe demonstrated in America little tolerance for anyone else. Although sex was not so suppressive as race or religion, it still restricted women's lives. Women had few property rights; divorce required an act of the legislature; and there were practically no independent job opportunities for women except prostitution. Married women had much to do: cooking; cleaning; repairing or making clothes; planting, weeding, and harvesting vegetables; and above all, producing and taking care of swarms of children. The Reverend Charles Woodmason remarked, "There's not a Cabbin but has 10 or 12 Young Children in it." Even well-to-do women produced nearly a child a year.

Except for the recent immigrants and black slaves, Americans tended to be better-educated than Europeans of comparable means. The standard of education was not high, but even unskilled coastal laborers often could read and write understandably; almost none of their English counterparts could. On the frontier, however, Charles Woodmason reported, "I have not yet met with one literate, or travel'd Person." Schooling was an urban activity. More schooling was offered in Boston than elsewhere; its schools were generously supported by the annual town meetings and enrolled perhaps as many as 700 students. Colleges, which were more or less the equivalent to modern high schools, were opened on the East Coast; they included Harvard, the College of New Jersey (later called Princeton), King's College (later called Columbia), and William and Mary. For advanced education, particularly in law and medicine, the more affluent people sent their sons back to England or Scotland. Philadelphia had twenty-six lawyers who had read law there, and even the more rural southern states were well represented. Fifty-eight students went to England or Scotland from South Carolina, forty-three from Virginia, thirty-three from Maryland, eleven from Georgia and eight from North Carolina. Between 1760 and the outbreak of the Revolution, a total of 115 young Americans went to London to read law at the Inns of Court. Some, including the great Pennsylvanian writer, lawyer, and statesman John Dickinson, later played an important part in the events leading to the Revolution.

In the more settled areas, even modest households boasted (usually at the time of death in wills) "libraries" that might contain as many as fifteen or twenty books. By mid-century, bookselling was a prosperous trade. Philadelphia and Boston were the leaders, with about fifty bookstores each. Books on theology were common, but "almanacks" were also widely read. Families were often isolated and had to rely upon themselves, so "how to do it" books were as popular then as today. Since doctors were few and the nearest doctor might be days of horseback riding away, what to do about aches and pains could become literally a matter of life or death for the patient. As literacy increased and the relationship with England was debated, histories and political tracts were also widely circulated.

Books were expensive. So, in 1731, Benjamin Franklin led the way to form lending libraries. Already popular in England, their worth slowly came to be recognized in America. Other private initiatives were taken to spread information and opinion to inform. In 1772, the Boston bookseller and printer John Boyle offered an edition of Montesquieu's *Spirit of the Laws* at a 40 percent discount because he thought it "ought to be in

Everyman's Hands." Even at that discount, the book still cost a laborer's wage for several days' work.

The first newspaper in the colonies—the *Boston News-Letter,* published by the postmaster John Campbell—did not appear until 1704. New York City got its first newspaper in 1725. Thereafter, newspapers multiplied, and broadsheets became common, giving a new impetus to reading, particularly as the exciting events that led to the Revolution captured public interest. Pamphlets and books published in London were circulated and sometimes reprinted in Philadelphia, Boston, and New York. The range of quotations from European literature is striking in American political writings.

Reading not only required intellectual skill but also posed a physical challenge. With long working hours, the logical time to read was at night, but lighting was provided only by tallow candles and the blaze of fireplace logs. Since the frugal small rooms of even the more opulent houses could accommodate only a few people, and poor people's rooms were often crowded with adults, children, and animals, taverns were oases of culture and conviviality. Little wonder that men in search of communication often escaped to them. As in contemporary London, taverns became academies of politics.

Culture and conviviality were not widely shared. Dr. Alexander Hamilton, one of the most-traveled Americans of the time, does speak of a "marvelous mixture of Scots, English, Dutch, Germans, and Irish [comprising] Roman Catholicks, Churchmen, Presbyterians, Quakers, New-lighters, Methodists, Seventh daymen, Moravians, Anabaptists, and one Jew" in a meeting hall in Philadelphia in 1744; but he goes on to say that they divided themselves "into committees in conversation." People tended to cluster in small groups with those they considered like-minded. Race, religion, language, wealth, class, and geography were formidable barriers, insisted upon by even the most "liberal." Within any single group, those who were more established and free tended to fear those who were recent, and frequently indentured, arrivals. In 1751, noting a rise in violent crime, the *Virginia Gazette* attributed it (as did Benjamin Franklin) to England's practice of shipping felons to the colonies. About 30,000 criminals were sent to America in the eighteenth century. "In what," the *Gazette* asked, "can Britain show a more sovereign contempt for us than by emptying their jails into our settlements; unless they would likewise empty their jakes on our tables!"

This "sovereign contempt" was one of the things that infuriated the already established Americans. Increasingly, they were asserting social as well as cultural equality with Englishmen. After his satirical offer, in 1751, of shipments of American rattlesnakes in return for English felons, Benjamin Franklin again entered the fray in a letter to the London *Chronicle* on May 9, 1759. Calling himself "A New Englandman," he angrily wrote:

> We call *Britain* the *mother* country; but what good mother besides, would introduce thieves and criminals into the company of children, to corrupt and disgrace them?—And how cruel is it, to force, by the high hand of power, a particular country of your subjects, who have not deserv'd such usage, to receive your outcasts, repealing all the laws they make to prevent their admission, and then reproach them with the mixture you have made."

Whether or not it was merely a dumping ground, America had already become a mixture by the time of the Revolution. No longer almost entirely English, it had begun the process of becoming a multiracial, multicultural society. In addition to the English, there were the Scots, both Highlanders and "Scotch-Irish"; Germans; Dutch; Huguenot French; and Bohemians. These linguistic and ethnic divisions were compounded by religious beliefs sufficiently intense to have caused their adherents to emigrate from Europe. In America they acquired new interests that separated them still further. Widely scattered along the seafront of the vast new continent, they developed virulent local antagonisms to one another—Marylanders fought Virginians, New Yorkers fought Pennsylvanians, and so on. Quarrelsome and opinionated, they flouted English regulations, and, when pressed, they evaded government control by moving inland. Over the century scores of thousands did just that.

On the eve of the Revolution, the overall population was about 2.5 million, of whom roughly one in four was German or Scots, one in six was black, and one in ten lived on the frontier. The population was not only doubling each generation but becoming more diverse. An Anglican, Reverend William Smith, who was an early leader in the attempt to persuade non-English immigrants to speak English, wrote, "We are a people, thrown together from various quarters of the world, differing in all things—language, manners and sentiments." In 1767, a new ingredient was added to the mixture when some 1,500 Greeks, Italians, and Minorcans arrived to raise indigo and sugarcane at New Smyrna, on the Carolina coast. It would be a long time before the concept of a "melting pot" was suggested; few of the ingredients wanted to mix, and marriage records and genealogies show that religious "like" married like. But at least some people wanted to create a sort of unity. The first "cook" for America's melting pot was certainly Benjamin Franklin.

As Franklin well knew, bringing about some degree of unity would be a Herculean task. Until the 1760s, people in the different colonies hardly knew one another. A Harvard College alumnus demonstrated how little information was exchanged, even between cities. Tom Bell, "the most traveled, most notorious, and probably best-known of all the colonists either on the continent or in the West Indies," was a confidence man, robber, horse thief, and counterfeiter—a jack-of-all-illegal-trades who parlayed his gentlemanly airs, his Harvard education, and his wit into a colorful career. Exposed in one colony, he moved effortlessly and rapidly to the next for a decade between 1743 and 1752. Despite jail terms, escapes, and a death sentence, he operated not only easily but in full public view, apparently without any fear that his reputation would follow him. After he had exhausted the possibilities in the colonial urban scene, his stories merged into a shared legend; so, inadvertently, he did as much to foster a sense of (outraged) unity among the colonies as any other man with the possible exception of King George III.

The colonies were divided by obstacles then almost unbridgeable. Movement between towns was slow and dangerous. Even a generation after the Revolution, in 1801, Thomas Jefferson remarked that to get from Monticello to the new capital at Washington, D.C., he had to ford five rivers that had neither bridges nor ferries. Throughout the colonial period and for years thereafter, any stream could be a formidable barrier in winter or the cause of a good dunking in summer. Flat-bottom ferries frequently were capsized by frightened animals or storms, but they were, relatively speaking, comfortable and in many places essential. Often a complex set of means of travel was required. To trek from New York to Philadelphia in the middle of the eighteenth century, the traveler began on a Hudson River ferry, then changed to a stagecoach, and finally floated on a bateau down the Delaware from just below the site of modern Trenton. Such travel took organization, was subject to frequent delays, and was expensive. What was true for people was also true of freight. Even more than half a century later, in the 1830s, after much development of transportation infrastructure, carrying a ton of goods 30 miles inland from any of the coastal cities cost about the same as shipping it from London or Liverpool to Boston.

Traveling was always an exhausting experience and sometimes a lethal one. First of all, there was no such thing as waterproof covering. Every drizzle was a shower bath, often to be followed by a cold or influenza, for which there was no treatment except perhaps a bloodletting that might be more dangerous than the disease. Then there was an arduous, jolting, lurching carriage or the tedium of riding on a hard saddle. Few parts of the countryside had "ordinaries" (inns), and none had any equivalent of hotels; the traveler was lucky to find a complaisant farmer with a spare pallet or a dry floor. Travelers were more likely to have to sleep under a tree or on a haystack—and even haystacks were still rare.

Even for hardened post riders, travel was difficult. In the relatively well-populated East during the mid-eighteenth century, New York had only 57 miles of "post roads": that is more or less leveled dirt trails wide enough for a coach or wagon and generally free of rocks or fallen trees. Well into the 1780s, European visitors were shocked at how primitive these roads were. Pennsylvania had seventy-eight more or less cleared tracks, and Massachusetts had 143. Between Boston and New York, the mail moved only once a week in the summer or once every two weeks in winter and took four or five days. These were "best cases." Elsewhere, conditions were much worse. Governor Dobbs of North Carolina complained in 1762 that letters coming to him from London were usually three to four months en route and sometimes as long as twelve months; Governor Robert Dinwiddie of Virginia said that the post from New York often took five to six weeks. By the time letters went to and fro, the sick had died or recovered, attacks had succeeded or been repulsed, couples had been married or separated; in short, whatever was described in a letter had probably long since changed.

And mail was not cheap. Even after Benjamin Franklin, as "co-deputy postmaster general" for the colonies from 1753 on, had much improved the post roads, made service more frequent, and persuaded Parliament to reduce the charges, letters mailed in 1765 cost fourpence sterling for the first 60 miles, sixpence for 100 miles, eightpence for 200 miles, and eightpence for each additional 100 miles. To note the cost of mail, however, tells us little unless we relate the figures to prices and income: a soldier in the British army at that time was expected to support himself and his family on less than sixpence a day; a civilian worker earned perhaps twice that amount. Urban Americans, on average, earned more but paid more for food. A loaf of bread in New York City cost about twopence a pound.

It was expensive to keep in touch. Even if people in, say, Boston and Charles Town (later Charleston), South Carolina, knew one another well enough to write, which was not likely, the cost of sending a letter in 1756, when the first overland post was begun, amounted to the wage of a laborer for a week. Not surprisingly, one of the most crucial tasks the American radicals, led by Samuel Adams of Boston, set themselves a decade later was to organize "committees of correspondence" to introduce the colonists to one another. At the first meeting of the Continental Congress in 1774, even the sophisticated and active leaders had not met each other. When they did meet, they were so worried about their "diversity of religions, educations, manners, interests, such as it would seem almost impossible to unite in one plan of conduct," in the words of John Adams, that they pledged secrecy in their discussions. Americans were far from being at ease with one another.

At ease or not, the colonists were learning to manage their own lives. From the model of Parliament, they adapted their legislatures; from English courts, they freely borrowed their legal system, with its tradition of decisions and concepts of evidence and justice.

The colonists adapted but rarely invented. So they did not establish police forces. Like the English, they considered what we think of as police work to be the duty of every citizen. Hearing a "hue and cry"—the antiquity of the custom is shown by the origin of the expression, the Anglo-Norman *hu-e-cri*—every nearby person was obliged to respond. When malefactors were caught in the act, a mob or posse—from the even older Latin phrase *posse comitatus*—was immediately called into being. Those who refused to help run down the suspect became liable. The person the posse caught was often summarily punished; otherwise, he was turned over to a legal system manned by appointed or elected colonials. While these officials theoretically administered English law, they did so in ways that reflected local social and religious custom and popular opinion. They were, in short, another but less material manifestation of the ceiling in the colonial house I described earlier, *almost* but *not quite* English.

If the officials were not vigorous enough, mobs often took the law into their hands. For example, when the spread of prostitution infuriated the straight-laced inhabitants of Boston, mobs rose to destroy the whorehouses and to enforce bans on unseemly conduct by prostitutes. They did this time after time, in 1734, 1737, and 1771. Again, concerned by threats of famine, Boston mobs rose in 1710, 1713, and 1729 to prevent food from being exported from the colony. And, frightened by an epidemic of smallpox, inhabitants of a number of towns and villages similarly took control of measures of public health in the 1770s. One riot in 1747 lasted three days; it and another in 1764 aimed to stop the Royal Navy from impressing local sailors. Mobs in Philadelphia and other cities similarly rioted over local infringements of what they considered right and proper.

Because the British did not prevent such popular action and allowed or encouraged the colonists to develop and man their own legal system, elected colonial officials became conscious of their power—and, equally important, of their right to act in matters of public interest. By the decade before the Revolution, they had acquired a status, even admitted by the British, that enabled them to exercise a veto over British decrees and to control royal and even military officials. One reason they could do this was that there were so many of them. Each generation brought over a new complement of indentured servants who swelled the ranks of the soon-to-be American public.

Early on, the colonists realized, as John Pory wrote from Virginia in 1619, that "our principall wealth . . . consisteth in servants." That remained true throughout the colonial period. In 1755 Governor Sharpe of Maryland wrote, "The Planters Fortunes here consist in the number of their Servants (who are purchased at high Rates) much as the Estates of an English Farmer do in the Multitude of Cattle."

With few Indian or black slaves and with free labor at such a premium, animals should have been avidly sought; yet in 1616 only six horses were reported in the colony at Jamestown. A few years later, the colonists asked that twenty mares be sent over from England. Horses remained so rare and so valuable that the government in 1626 mandated the death penalty for stealing one. Other than horses, colonists in Virginia initially had few domesticated animals. Ships' crews and passengers customarily carried some animals on voyages, to eat along the way. They favored small animals—pigs, goats, chickens, or geese—since cattle and horses were difficult to keep on the tiny sailing ships and would often suffer broken legs in storms. Gradually, however, the survivors multiplied. Usually they were allowed to roam free in the forest surrounding settlements until the colonists found the time and energy to build "snake" fences—interlaced logs—around their plots both to protect the crops and control the animals.

What made indentured servants so attractive was that they could be obtained for about 10 pounds—it cost only about 6 or 7 pounds to bring them over—and an individual laborer could produce 49 or 50 pounds' worth of tobacco yearly. The economics were compelling: colonists knew that unskilled workers in England earned what seemed to them a fair wage of about a shilling a day. But free laborers in America expected two or three times that rate, and skilled craftsmen might demand eight or nine times as much. To feed an indentured servant cost only about a penny a day. Also attractive to employers was the fact that whereas free laborers were likely to strike out on their own as soon as they had a small amount of capital, indentured servants were bound by their contracts. If these servants tried to flee, an employer could call on the colonial government to have them hunted down.

Employers in the southern colonies often complained about the quality of the servants they got from England, particularly exiled criminals; but they needed labor so urgently that they took whatever was sent. In New England, where plantation agriculture was less common than in the South, white Englishmen and Irishmen were less welcome than obviously distinct, and so segregatable, black slaves. The first blacks in Jamestown came off a Dutch ship in 1619, but it would be decades before many others arrived. In 1624–1625, only twelve black men and eleven black women are mentioned as living in English-controlled Virginia.

Many white servants had been convicts. Although Virginia had passed a law in 1670, and Maryland in 1676, forbidding convicts from entering, an Act of Parliament in 1717 opened both colonies to convicts; and, as Abbot Emerson Smith wrote, "no one on either side of the Atlantic seems to have doubted that a colonial law flatly prohibiting their importation could not stand against the new parliamentary statute." The colonial legislatures tried a subtle way to avoid a confrontation with Parliament. Realizing that they could not prevail in a head-to-head dispute, they enacted laws that made the importation of convicts unprofitable. Fines were to be levied for various acts of negligence, and impossibly high bonds were to be provided against escape or subsequent misbehavior. In 1723 the contractors appealed directly to the Board of Trade, a committee of the Privy Council that had taken over from the Lords of Trade in 1696 to administer the colonies. The contractors claimed, rightly, that the intent of the acts was to overrule Parliament and render the transportation of felons impracticable. The board agreed, and with unusual speed the Privy Council disallowed the acts. There were subsequent attempts, but these were finally abandoned just a generation before the Revolution broke out. In one instance, the king personally instructed Governor Culpeper of Virginia "to allow the entry of [the merchant Ralph] Williamson's prisoners, any law, order, or custom of Virginia to the contrary not withstanding. Thus protected against any colonial disposition to exclude the Scots by virtue of the act against convicts, Williamson loaded sixty-eight captives." Private contractors are recorded as having taken 17,740 felons from prisons in the area around London and probably landed at least 20,000 in Virginia and Maryland in the thirty years before the Revolution.

Other than prisoners and those who wished to emigrate, individuals were kidnapped, drugged, plied with drink, or overwhelmed as they ate or drank in taverns or walked London's dark streets and alleys. Once in the hands of "spirits" or "crimps," a person had little chance to escape, as he was usually put in a private jail until he could be bundled abroad a ship. The practice was recognized and deplored, but Parliament repeatedly refused to legislate against it—even though one kidnapper turned king's evidence and described the system in graphic detail. These whites were only marginally better off than the blacks then being imported from Africa. It is to the plight of the blacks I now turn.

CHAPTER 10

Blacks in America

Like a giant whirlpool, the transatlantic traffic sucked out of Africa perhaps as many as 12 million people. Of them, about half a million were sent to North America as slaves. The objective of this chapter is to show how they were transformed from Africans into Afro-Americans and how their presence altered white society.

This process was begun in Africa by African slave dealers; the captives were then taken over by ships' crews who carried forward what had begun in Africa and delivered their "black cargoes" to slave markets in the New World, where the Africans were parceled out to white Americans. At each stage, we today are tempted to stop to express our astonishment—and horror—at how otherwise decent human beings, even clergymen like Revenend Cotton Mather of Boston, could participate in such a barbaric, inhuman, vicious enterprise. Here I will try to understand. I begin with the context in which black slavery fit.

As we have seen, large numbers of whites who became Americans arrived from Europe in bondage, either having been expelled or having traded such "liberty" as they had for an escape from poverty, oppression, or insecurity. Difficult as their position sometimes was, however, it was alleviated by three considerations: the condition was temporary, they had some legal safeguards, and no one suggested that they were not fellow human beings. None of these considerations pertained either to the native Indians or to the arriving Africans. Both groups were regarded (and treated) as nonhuman, but the way they were treated differed. After initial attempts to use Indians as slaves, the whites drove them away or killed them. By contrast, they kept the Africans under tight control and put them to work. In effect, they treated the Africans as a species of farm animal.

Treating enslaved people as animals has a long history. The ancient Assyrians spoke of slavery as *eli erbi ritti pasalu,* "walking on all fours," that is, behaving like a domesticated animal. Most primitive and ancient peoples referred to themselves as "*The* People." That is the meaning of many of what we think of as American Indian tribal names: Lenni Lenape (Delawares), Ongwe Honwe (Iroquois), Illinois, Navaho or Nemene (Apache). In Africa, there are the Nuer or Dinka and scores more. A stranger was not of *the people,* not one of us; by definition, he was not a fellow human being. The word "slave" in all Indo-European languages is derived from the concept "captive alien"; that was also the meaning of *lu* in Tang Chinese. Terms for aliens in many languages are derived from or formed on analogies to wild beasts: the Aryan invaders of India called the Dravidian aboriginals *dasa,* which means something like "beasts of the forest"; the Bushmen of the Kalahari called aliens "animals without hooves." Like animals, aliens could not speak (our) language: they were to the Greeks, *barbaros;* to the Arabs, *barbarri;* to the Sanskrit-speaking Aryans, *me-luh-ha.* To look or smell different was, in itself, a justification for being regarded as not human. The wall friezes of the Assyrians and Egyptians minutely detail the outlandish dress, unfamiliar facial features, and bizarre hairstyles of foreigners who were destined for slavery.

An echo of this attitude can be heard in Thomas Jefferson's strictures on the "very strong and disagreeable odour" of Negroes who are, in sexual preference, close to apes; and in George Washington's description of the hair of his black manservant as "wool." Terms applied to blacks and their treatment always carry at least a hint that the white thinks of the black as a nonhuman animal. An American court made this explicit, ruling that the word "person" did not apply to a slave. Black people's color, their facial features, their hair, and even—as Jefferson remarked—their smell set them aside. The accuracy, morality, and legality of these attitudes were unquestioned in the sixteenth and seventeenth centuries and were rarely challenged in the eighteenth century.

Later, pseudoscientific explanations were offered. Dr. Samuel W. Cartwright of Louisiana claimed that surface attributes—color, hair, and smell—were merely the visible manifestations of much deeper differences in "the membranes, the muscles, the tendons, and . . . all the fluids and secretions. Even the negro's brain and nerves, the chyle and all the humors, are tinctured with a shade of the pervading darkness." If the master believed that Africans were not of the same species, were indeed animals, how should he handle them? By analogy, the answer was clear: he had to domesticate them. The key behavioral characteristic of the domesticated animal as opposed to the wild animal is that it is "broken," "tamed," or "trained" so that it becomes subservient to the will of the master. That is the essence of slavery.

Being "broken" began with capture at interior African villages and a march to a port. The captives were at least traumatized and often also wounded; all their possessions, usually including their clothing, had been taken away from them; they had seen family members killed or abandoned; and they had no idea what would happen to them. Then, awaiting transfer to a ship, they were chained and penned in what amounted to dungeons and were almost starved. When they were examined by ships' mates, usually the first white men they had ever seen, they suspected that they were being prepared for slaughter to be eaten. Then they were "burned," as branding was called. The whole experience to that point must have resembled what happens as people are taken to concentration camps or as "enemy combatants" are hooded to disorient them. Such vestiges of humanity as the captives had left were stripped from them. Many died, as the surgeon of one ship put it, simply of "melancholy." What he called melancholy was known in the Nazi concentration camps as becoming a "zombie." If not already in the barracoon, then certainly during the voyage, the captives were already well on the way to being broken in spirit.

It was not only the spirit that was broken. The prisoners were loaded onto waiting ships, where they were packed, as one account reports, like "books on a shelf," fed only enough to keep them barely alive, and clamped two by two in leg irons. Thus their bodies were also wasted. "Walking skeletons covered over with a piece of tanned leather" was one description. Jammed with the sick and dying, with no facilities for hygiene, a slave ship was a floating cesspool. Smallpox, diarrhea, "the bloody flux," and other diseases swept through the ships like wildfire; so bad were the conditions aboard that about 15 of 100 captives died on the voyage. Often the death rate was much higher. On one Dutch ship from Elmina on the Ghana coast in 1731, at least 150 of 753 slaves died of scurvy. Slaves who fell ill and did not recover from their ailments before the end of the voyage were worthless as commercial goods and were often thrown to the sharks. When water ran short, as it sometimes did when a ship was becalmed, the captain would routinely "dispose" of the cargo. We know this primarily because some captains tried to collect insurance. In 1781, for example, the slaving ship *Zong* was running low on water, so the captain ordered the crew to handcuff 132 slaves and throw them into the sea. When he arrived in England, he claimed 30 pounds each for them under his maritime insurance. The underwriters took the case to court and lost: the jury held that "the case of the slaves was the same as if horses had been thrown overboard." If nothing else could be done with the captives, they might be simply dropped ashore, as Captain John Lovell did with "ninety-two pieces of Blacks, all old, and very sick and thin." Slaving was a brutalizing business.

It was also dangerous not only because of fevers but also because of the captives. Captured Africans were a hazardous cargo and were treated as such by the sailors. Many came from warlike societies, and those who were not completely cowed or totally immobilized were sometimes driven to desperate acts. Despite having all odds against them, hundreds of captives tried to overwhelm the crews. To actual or anticipated rebellions, the crews responded with fury, whipping the ringleaders, hanging them, chopping them to pieces with axes, or throwing them overboard. Yet, as David Eltis has pointed out, "in the eighteenth century alone, resistance ensured that half a million Africans avoided the plantations of the Americas. . . . Africans who died resisting the slave traders, as well as those who resisted unsuccessfully but survived to work on the plantations of the Americas, saved others from the middle passage." Although the records are incomplete, 392 revolts by slaves on shipboard are now known.

Olaudah Equiano later wrote a record of his capture as a child and of being taken aboard a slave ship. When he saw "a multitude of black people of every description chained together, every one of their countenances expressing dejection and sorrow," he said, "I no longer doubted my fate; and quite overpowered with horror and anguish, I fell motionless on the deck and fainted." Seeing a huge copper pot boiling, he thought he was to be eaten. As a contemporary witness who took the trouble to interview some slaves remarked, they believed that "upon their arrival they would be made into oil and eaten."

Under sail, Equiano wrote, conditions were

> absolutely pestilential. The closeness of the place, and the heat of the climate, added to the number in the ship, being so crowded that each had scarcely room to turn himself, almost suffocated us. This produced copious perspirations, so that the air soon became unfit for respiration, from a variety of loathsome smells, and brought on a sickness among the slaves, of which many died. . . . This deplorable situation was again aggravated by the galling of the chains, now become insupportable; and the filth of necessary tubs [toilets] into which the children often fell, and were almost suffocated. The shrieks of the women, and the groans of the dying, rendered it a scene of horror almost inconceivable.

One day, he continues, when the sea was calm, the chained slaves were brought onto the deck, and two managed to break through the netting on the sides and jumped into the sea, "preferring death to such a life of misery." In slavery, life itself was just another form of suicide.

The first blacks bought by English colonists arrived in Virginia one year before the *Mayflower*. As John Rolfe reported from Jamestown in 1619, "a dutch man of warre . . . sold us twenty Negars." How the colonists in Virginia understood the transaction is not known. Virginia then had no law dealing with slavery—not until forty years later would slavery be legalized—and the only local practice under which the blacks could have been categorized was indenture. Without question, however, the colonists put the new arrivals to work as slaves.

Slaves are first mentioned in New Netherland in 1628, a decade later than in Virginia. The following year the Dutch West India Company promised to furnish each landowner (*patroon*) "with twelve Black men and women out of the prizes in which Negroes shall be found (raids on Spanish slave ships) for the advancement of the colonies in New Netherland." In 1638, the first black slaves were imported into Massachusetts, transshipped from Bermuda. Rhode Islanders joined the parade in 1647 and soon were among America's foremost slavers. Soon slavery was common in "British America."

Curiously, given the horror of enslavement and transport to the New World, the colonists almost immediately thought of the blacks as their defenders. Slave men were frequently enrolled alongside whites in the early militias. Unconsciously, the colonists and their British governors were adopting a very ancient practice: the Egyptians used Nubian slaves as soldiers; the Athenians used Scythian slaves as policemen; wealthy Romans often armed large bands of their Balkan slaves as bodyguards; and the later Islamic states assembled whole armies of Turkish and Circassian slaves. In America, necessity was the mother of re-invention. Out of fear of the still-free Indian tribes on their frontiers, whites armed blacks to fight against them. In 1641, the Dutch governor of New Netherland armed "the strongest and fleetest Negroes." In 1652, the General Court in Massachusetts ordered all inhabitants including "Scots and Negroes" to be armed and trained to fight. South Carolina and Rhode Island adopted the practice and continued it into the first decade of the eighteenth century.

Other black slaves were employed to build fortifications because they were cheaper than free whites, who had to be paid. In 1699, the earl of Belmont wrote to the Lords of Trade in London, urging them to send more "negroes from Guinea, which I understand are brought hither, all charges whatever being borne, for ten pounds apiece, New York money, and I can clothe and feed them very comfortably for nine pence apiece per day sterling money, which is threepence per day less than I require for [white] soldiers." Despite the saving in cost, the experiment was soon dropped. Massachusetts led the way in 1656 and Virginia followed in 1680. Eventually, all colonial governments acted to curb what they feared might become "black power."

Probably this sequence of enrolling blacks and then curbing them was a function of the growing size of the black population. As long as there were few blacks in the American colonies, they posed no threat of revolt even if armed; as their numbers increased, the colonists grew apprehensive and ceased to arm them. Between June 1699 and October 1708, thirty-nine ships brought to Virginia some 6,607 blacks, of whom 679 were imported by the recently established Royal African Company of London and 5,692 by private traders. The number again multiplied: between 1727 and 1747 it reached 45,440; and from 1741 to 1760 it reached over 100,000. Ironically, in the 1770s, on the eve of the Revolution, the importation of slaves dropped to only 3,338 because of the colonists' boycott of British goods.

Modern Americans associate slavery with cotton; but a century before cotton became a major crop, sugar employed the most African slaves. It was mainly the growth of sugar plantations that brought about the growth of slavery. Then, in America in the eighteenth century, the spread of rice and tobacco raised the value of slaves on the African coast fivefold from 1730 to 1780. Still only about six of 100 Africans came to the British American colonies. Most of these were sold to whites in the colonial South, but some were sold in the North. In the North, no "plantation" crops were grown; but in a site excavated in the 1990s on Shelter Island, there is evidence that slaves in New York produced provisions and timber for a sugar-producing plantation on Barbados which was owned by the same family that owned the Shelter Island plantation.

Getting Africans into the hands of white buyers was usually accomplished by advertisements like the following in the South Carolina *Gazette* of May 25, 1762:

> Just imported from the River Gambia in the Schooner Sally, Barnard Badger, Master, and to be sold at the Upper Ferry (called Benjamin Cooper's Ferry), opposite to this City, a Parcel of likely Men and Women
>
> SLAVES
>
> With some Boys and Girls of different Ages. Attendance will be given from the Hours of nine to twelve. . . . It is generally allowed that the Gambia Slaves are much more robust and tractable than any other Slaves from the Coast of Guinea and more capable of undergoing the Severity of the Winter Seasons in the North-American Colonies, which occasions their being vastly more esteemed and coveted in this Province and those to the Northward than any other Slaves whatsoever.

Many descriptions of what happened next are available. Mary Prince, who wrote one of the few narratives of this period, remarks that when she

TO BE SOLD, on board the Ship *Bance-Island*, on tuesday the 6th of *May* next, at *Ashley-Ferry*; a choice cargo of about 250 fine healthy

NEGROES,

just arrived from the Windward & Rice Coast. —The utmost care has already been taken, and shall be continued, to keep them free from the least danger of being infected with the SMALL-POX, no boat having been on board, and all other communication with people from *Charles-Town* prevented.

Austin, Laurens, & Appleby.

N. B. Full one Half of the above Negroes have had the SMALL-POX in their own Country.

was sold, blacks were treated "like cattle or sheep." Generally, as in any livestock show, the sellers tried to loosen the purses of the buyers by providing food and drink while they paraded their offerings. The buyers reacted as they might at a sale of horses, prying open the mouths of the slaves to examine their teeth, and rubbing their skin to check for sores that might be hidden by paint, powder, or lead dust. Then the Africans were auctioned individually, so that relatives were frequently separated. This happened not only because a buyer might need only one or a few slaves but also because it was thought that if slaves were cut off from relatives, friends, or speakers of the same language, they would have less opportunity for revolt or flight. This moved forward the homogenization process that was to be the hallmark of the creation of Afro-Americans.

When they finally reached their destination, slaves were "broken to the bit" and so transformed from "outlandish" Africans to "new Negroes." Not yet knowing English, they were unable to understand orders from their new masters. Against what the whites regarded as "insolence," the whip was the main educational tool. Some rudimentary skills were also taught, but some of the newly enslaved people came from agricultural societies in which they had been raising similar crops; so, particularly in the South Carolina rice belt, farming techniques and tools can often be traced to their African origin.

Above all in the coastal rice-growing areas, but by no means uniquely there, little human warmth could be found in the relationship of whites and blacks. As in the breaking of a mustang, the slave drivers believed that they had to prove beyond any challenge their absolute power. So most of the newly arrived blacks were immediately put to work as field hands under gang bosses. Like domesticated animals, they were fed enough to enable them to perform the work for which they had been bought, but they were certainly not given more. Their staple food was corn bread, known as "pone" (Algonquian, *appone*). Insofar as records exist, they suggest that field hands were rarely given meat to supplement their carbohydrate diet; consequently, various dietary-deficiency sicknesses (pellagra, beriberi, and scurvy) were as common among them as among the inhabitants of the Spanish mission towns I described in chapter 3. A "kind" master was defined as one who allocated small patches of land to his slaves on which they could work during their short periods of time off, usually part of a Sunday, to raise vegetables and, perhaps, chickens or even pigs. A few slaves, particularly in the tobacco area of the Chesapeake, managed to acquire more, but, at least in the southern colonies, they were a tiny minority.

Regarding the slaves, particularly the field hands, as virtually animals, owners usually provided them with neither plates nor eating utensils; rather, the slaves ate from a shared wooden tub. Frederick Douglass remembered his childhood on a farm in Maryland. Although he wrote well after the Revolution, conditions then were essentially the same as in the pre-Revolutionary period; I cite him because he was one of the few to give a graphic description of what must have been the experience of thousands of blacks in America generation after generation. When dinner was called, he wrote, a cornmeal mush was poured onto large wooden trays; then the little children "like so many pigs would come, and literally devour the mush—some with oyster shells, some with pieces of shingles, and none with spoons."

Clothing was as scanty as food. Adult slaves often went virtually naked, and children almost always. Notices of rewards for runaways might describe the slave as wearing "only an Arse-Cloth," or having "nothing on but an old rag about his middle," or having "nothing on but a piece of check linen about her middle," or "a Clout round his Loins." Such clothing as was available was almost entirely locally made, often by the slaves themselves, from "Negro cloth," described as cotton "calicoes, nankeens, osnaburgs, tows, linsey-woolseys, cassimeres, ducks, kerseys, and Kentucky jeans." Most of these terms will mean nothing to modern, affluent, free readers, but all the fabrics were thin and offered little protection against cold. Moreover, they were usually in short supply. Frederick Douglass notes,

> I suffered much from hunger, but much more from cold. In hottest summer and coldest winter, I was kept almost naked—no shoes, no stockings, no jacket, no trousers, nothing on but a coarse tow linen shirt, reaching only to my knees. I had no bed. I must have perished with cold, but that, the coldest nights, I used to steal a bag which was used for carrying corn to the mill. I would crawl into this bag, and there sleep on the cold,

> damp, clay floor, with my head in and feet out. My feet have been so cracked with the frost, that the pen with which I am writing might be laid in the gashes.

Housing and furniture were, as Douglass experienced, rudimentary in the extreme. Mary Prince, the first black woman to tell her story, commented that she and other slaves "slept in a long shed, divided into narrow slips, like the stalls used for cattle." Most slave quarters were small shacks, generally a single room. Some were built by the slaves, who followed patterns they remembered from Africa—that is, a wicker frame daubed with clay or mud. At best, these quarters were protection against rain and wind. While they were no worse than the lean-tos lived in initially by white "Indian traders" on the frontier or by Indians, they were often densely crowded. The slave quarters had no furniture and no bathing or toilet facilities. Not surprisingly, black slaves had the "strong and disagreeable odour" that offended Thomas Jefferson.

What I have just described may be taken as the condition of the majority of blacks who were field slaves. But three other categories of blacks came into existence during the late seventeenth century, and for them life was more ample and less arduous. They were household servants; skilled slaves who were allowed to work for their own profit; and free individuals. For purely practical reasons and in singular ways, they were treated differently from the field hands.

First, consider the household slaves. Household slavery, wrote Michael Mullin, was "the dimension of American Negro slavery that changed least over time." Most accounts date from later, but two reports come from the eighteenth century, when much of the South was already a plantation society. In the great houses of that society, contact with household slaves was often intimate. Since some masters saw household slaves, like horses and carriages, as a public manifestation of their social status, they tended to clothe these slaves better and to give them the time and facilities necessary to keep relatively clean.

Male slaves, who often served as butlers and coachmen, were frequently dressed in an American version of the livery in which Americans thought English aristocrats dressed their household servants. As one rice planter in South Carolina wrote of his household staff, they were "supplied *without limit* to insure a genteel and comfortable appearance."

Foreign visitors recorded instances of whites and blacks eating together, maintaining a sort of "joking relationship," and even, on occasion, exhibiting sincere affection. But there were also, certainly, numerous if seldom recorded instances of casual cruelty by whites. William Byrd, in diaries that he intended as personal and secret, often mentions having slaves whipped. Although Landon Carter does not seem to have done that, he did punish his slaves in other painful and humiliating ways; and in general whipping, was not only common but horrible. In an account published after the Revolution, but referring to an earlier time, the black Baptist minister David George, who says he was from Sierra Leone but was actually born in Virginia about 1740, portrays one of the ugliest realities of slavery:

> My oldest sister was called Patty: I have seen her several times so whipped that her back has been all corruption, as though it would rot. My brother Dick ran away, but they caught him. . . . they hung him to a cherry-tree in the yard, by his two hands, quite naked, except his breeches, with his feet about half a yard from the ground. They tied his legs close together, and put a pole between them, at one of which one of the owner's sons sat, to keep him down, and another son at the other. After he had received 500 lashes, or more, they washed his back with salt water, and whipped it in, as well as rubbed it in with a rag; and then directly sent him to work pulling off the suckers of tobacco. I also have been whipped many a time on my naked skin, and sometimes till the blood has run down over my waist band; but the greatest grief I then had was to see them whip my mother, and to hear her, on her knees, begging for mercy.

Still less recorded but also common was sexual exploitation of slaves by masters and their sons. Thomas Jefferson was concerned about what he saw as the cultural mold slavery created:

> The whole commerce between master and slave is a perpetual exercise of the most boisterous passions, the most unremitting despotism on the one part, and degrading submissions on the other. Our children see this, and

> learn to imitate it. . . . The man must be a prodigy who can retain his manners and morals undepraved in such circumstances.

The jury is still out over Jefferson's own relationship with his slave Sally Hemings.

Various attempts were made to prevent "abominable Mixture" and its "spurious issue," as a law of 1741 in North Carolina described a sexual union between a black and a white. As early as 1630 a court in Virginia ordered a white man whipped for "defiling his body in lying with a Negro." In 1664 Maryland prohibited mixed marriages; Virginia followed suit in 1691. The legislature of Virginia, in Act 14, provided that "for the prevention of that abominable mixture and spurious issue which hereafter may encrease in this dominion . . . negroes, mulattoes, and Indians intermarrying with English, or other white women . . . shall within three months after such marriage be banished and removed from this dominion forever." In 1715 a statute in North Carolina specified that "no White man or woman shall intermarry with any Negro, Mulatto or Indyan Man or Woman under the penalty of Fifty Pounds for each White man or woman."

Law after law was passed; but in a situation of "most unremitting despotism" it remained too easy for white men to satisfy their lust with black women. Not only legislation but also court cases and folklore reveal that many white masters found black women sexually attractive. By the middle of the eighteenth century, many slaves were the offspring of such unions, and mulattoes were quite common. The General Assembly of North Carolina complained in 1723 that "great Numbers of Free Negroes, Mulattoes, and other persons of mixt Blood . . . have intermarried with white Inhabitants of this Province." "Intermarrying" remained common. When, in 1773, the lawyer Josiah Quincy of Boston went to the South to encourage support for Massachusetts's stand against Britain, he found that "enjoyment of a negro or mulatto woman is spoken of as quite a common thing: no reluctance, delicacy or shame is made about the matter." As is evident in the proliferation of the mulatto population, attempts to restrain white masters were to no avail.

The American South unknowingly followed a pattern common in colonial territories. In Bengal, then Britain's most important colony, Englishmen quite openly kept zenanas (harems) of dark-skinned women. But when the wives and families of the Englishmen began to arrive in India toward the end of the eighteenth century, the attitude toward interracial sex underwent a reversal: whereas earlier white men had enjoyed relations with dark-skinned women, suddenly there was a perceived threat that white women might be violated by dark-skinned men. In India, as in America at about the same time, whites who engaged in sexual relations with "people of color" began to risk public censure: forms of apartheid became the norm. Most of what is known about sexual unions in America, therefore, is inferred from the large number of mulattoes mentioned in documents, and from occasional court records.

"Bastardy lists" suggest that not only the lust of masters and their sons produced mulatto children. Another cause was the custom, already common in the seventeenth century, of mixing white indentured servants with black slaves. Sharing living quarters, sharing workplaces, and eating together, they formed attachments. Nor were the offspring always those of a white father and a black mother. Little evidence exists, but it appears likely that about one illegitimate child in three or four had a black father and a white mother. This is apparent from divorce petitions filed by white husbands whose wives had produced mulatto babies.

Although sexual relations among the races in the Old South probably cannot ever be adequately studied, some can be inferred. In his diary, William Byrd several times mentions his annoyance that his wife was particularly cruel and vindictive to an attractive household slave girl. Such tension must often have nearly wrecked both white and black families.

A second category of slaves who fared better than field hands included those who managed to distance themselves from a master or to place themselves among several "virtual" masters so as to achieve a degree of autonomy. This was true not only in British America but also in New Netherland. For example, a slave might be permitted to farm a plot of land so as to buy his freedom, or a blacksmith might be shared among several plantations or might be allowed to set up shop on his own in a town. Those who acquired the most freedom were probably sailors. Drawing on skills they had learned in Africa—along the Atlantic and on the African rivers, they got even farther away from the plantations. There were a surprisingly large number of such sailors. They included perhaps one in four slaves in coastal Massachusetts in the middle of the eighteenth century.

Our third additional category was made up of blacks who in one way or another had become free. Achieving freedom was less common in the seventeenth or eighteenth century than in the nineteenth, but it occasionally happened. Manumission was first recorded in New Amsterdam in 1643, when the Dutch West India Company freed some slaves in return for a sort of quitrent to be paid to the company for the rest of the person's life. The newly freed men and women could not, of course, pay quitrent unless they had some way to earn money, so the company gave them land to farm. When the English took over the colony in 1664, they did not interfere with the already established black farmers; but the earlier custom does not seem to have been continued by the English.

Some blacks, particularly at the end of the eighteenth century and during the nineteenth century, were able to buy their freedom by saving what they earned from handicrafts or from selling produce from their kitchen gardens.

Manumission by owners, whether by purchase or by gift, was illegal in many areas and was not common anywhere until after the Revolution. Nowhere was it automatic. Often, the owner was required to post a bond to ensure that the manumitted person would not become a burden on the community, since some owners used this manumission to get rid of slaves who were sick or too old to work.

Fugitive slaves, like the man Sandy for whom Thomas Jefferson advertised as a "run-away," might manage to become and even to stay free, but escape was always difficult. In Virginia, the legislature provided in an act of 1691 "for suppressing outlying Slaves" that:

> in case any negroes, mulattoes or other slave or slaves lying out as foresaid shall resist, runaway, or refuse to deliver and surrender him or themselves. . . . It shall and may be lawfull for [deputies] to kill and distroy such negroes, mulattoes, and other slave or slaves by gunn or any otherwaise whatsoever.

The slaves who were most likely to succeed in getting away were artisans who had a skill by which they could support themselves, were more likely to know enough English to pass as free in a town or city, and were accustomed to working more or less on their own. Like the household slaves, who also frequently tried to run away, they would have had access to clothing that would not immediately give them away. "Slave clothing," by contrast, was almost as obvious as a modern prison uniform. A reward notice in 1771, for example, notes that a "new" Negro named Step had absconded in "a white Plains Waistcoat and Breeches, Osnaburg shirt and a tolerable good bound Hat." Field hands, unlike the artisans and house servants, typically had no passable clothes, no marketable skills, no possessions, and no good command of English, so the odds were against them.

The field hands, at least during the seventeenth and eighteenth centuries, were mostly "outlandish" blacks, many of whom bore the facial scars (called "country marks") that in some African tribal societies were a mark of the coming-of-age ceremony. They were therefore easy to distinguish. Moreover, since few of them had been born in America or had lived long among English-speakers, they were—as noted above—unlikely to have a command of English sufficient to let them pose as free blacks in towns or cities. And because they had little or no access to geographical information, few could have had any idea exactly where they might go.

Still, what an individual probably could rarely do, a group might accomplish; so when field hands bolted for freedom, they tended to do so in groups in which one might know English and another might have some idea where they could hide. Hiding meant avoiding towns and heading for the nearest equivalent of the African bush. The Spaniards in the New World called an escaped person—or an escaped domestic animal—a *cimarrón,* from which comes the English word "maroon."

By the end of the seventeenth century scores of runaway slave communities had been established. Close to Africa, on the tiny island of Principe, escaped slaves managed to set up a free colony. In Surinam such communities became much more numerous and quite powerful. Those in the inaccessible areas of Jamaica were known as *macambos.* Others were established in La Florida, in hidden valleys among the Appalachians, and in the then virtually uninhabited Great Dismal Swamp that extended to the south from Norfolk, Virginia.

White Americans saw the runaway blacks as a constant threat. Since combating fugitives was dangerous and expensive, whites hit upon the plan of using Indians to do the job for them. But, even tracked by forest-wise Indians, some *cimarrón,* or maroon, communities managed to survive for many years. One colony on the border between New Jersey and New York, composed of a mixture of blacks, Indians, and whites, now known as the Ramapo Mountain People, still exists. But maroon communities were, of course, the exception, not the rule; and only a tiny portion of slaves managed to escape to them.

Since a slave was valuable property, owners did what they could to prevent slaves from escaping and created agencies to get fugitives back. The New England Puritans set out as a principal objective in their Confederation of New England apprehending both white and black runaways, and this task was assigned throughout the colonies to the militias. Lieutenant Governor Sir William Gooch of Virginia made this clear in a dispatch to the Board of Trade in London in 1729. After describing a runaway community that had been overwhelmed by an armed group of settlers, he wrote:

> Tho' this attempt has happily been defeated, it ought nevertheless to awaken us into some effectual measures for preventing the like hereafter, it being certain that a very small number of Negroes once settled in those Parts [the mountains on the frontier], would very soon be encreas'd by the Accession of other Runaways and prove dangerous Neighbours to our frontier Inhabitants. To prevent this and many other Mischiefs I am training and exercising the Militia in the several counties as the best means to deter our Slaves from endeavouring to make their Escape, and to suppress them if they should.

The role of the Indians in escapes by blacks varied from group to group. Indians—such as the Yamassee and Lower Creeks in particular—who were themselves still free and more or less out of reach often took in and sheltered runaway blacks. But nearby Indian groups, particularly those who had become dependent upon white society, frequently caught and returned or killed black runaways. Recognizing that such "inner" Indians could form a cordon along the frontier, the legislature in South Carolina was anxious to encourage them. When they aided in putting down a black uprising, it proposed "that the sd Indians be severally rewarded with a Coat, a Flap, a Hat, a pair of Indian Stockings, a Gun, 2 Pounds of Powder & 8 Pounds of Bullets." When runaways were caught and killed, cooperating Indians were to receive 20 pounds "for every scalp of a grown negro slave, with the two ears." This early alliance with Indian groups, notably the Catawba and the Chickasaws in the Carolinas, was later extended and used all along the frontier to pursue runaway blacks.

Blacks did not always run away. In a few instances—surprisingly few, and sometimes tragic—they endeavored to strike back at their oppressors. The most famous example in colonial America was an uprising in Stono, South Carolina, on September 9, 1739. In comparison with the slave revolts in the Roman Empire, which involved thousands of people, Stono was on a very small scale, initially just twenty men; but it terrified the owners of the rice plantations.

At Stono, on a Sunday, when plantation owners gave the blacks time to tend their kitchen gardens, a group identified as being from Angola broke into a warehouse containing muskets, powder, and bullets. There they killed two white men who were presumably guarding the building. Then, armed, they broke into the houses of families they identified as being particularly cruel to slaves and killed the inhabitants. The slaves obviously were evaluating the actions of the whites, because they spared at least one white tavern owner whom they regarded as a kind man. Making no attempt to run away or even to conceal themselves, and "calling out Liberty," they "marched on with Colours displayed, and two Drums beating." As they proceeded, a few more blacks, perhaps as many as forty or fifty, joined them. But after they had gone about 10 miles, apparently without any clear objective, they were overtaken by a number of planters and by the local militia. Practically all of the rebels were immediately killed. When the short, bloody foray ended, the casualties were said to be about forty blacks and twenty whites.

For most blacks before the Revolution, life was governed ultimately by statutes passed by the colonial legislatures and administered by the courts and local sheriffs. How to regulate the growing number of existing slaves and new arrivals was not a major issue until nearly the last quarter of the seventeenth century. Piecemeal regulations were passed that seemed useful or necessary to the whites. By about 1680, however, the increase in the black population seemed to Virginia's white colonists to demand a comprehensive and harsh set of laws. These became the model throughout the South for the next two centuries. Their main thrust was to deny blacks the limited rights of white indentured servants, which included the right to take a master to court for brutal treatment and the right to own property. The clear intent of the law was to separate all whites from blacks, mulattoes, and Indians, who were to have no rights at all. As much force as was required to keep them in submission was declared legal. As the black population increased, the regulations became more codified, more proactive, and more harsh.

But even these harsh codes were deemed insufficient. Several colonies tried to limit the importation of slaves. Largely for the sake of security, in 1738 the British authorities in the new colony of Georgia, which was considered dangerously close to Spanish Florida, banned slavery. After the Stono Rebellion, South Carolina also considered banning slavery. But its prosperity was based on rice and indigo, and, perhaps even more important, the planters' lifestyle depended on a servile class, so the South Carolinians continued to import great numbers of slaves. In the most populated areas, around Charleston, only one person in five was white. To be sure of retaining the preponderance of power, the whites passed a code in 1740, which, among other restrictions, prevented blacks from being taught to read. As an American judge has ironically commented, "The reward for killing a runaway slave was far less than the fine for teaching him to write."

Fearing rebellion, whites in both the North and the South instituted draconian punishments. "Raising a hand" against a white, even to ward off a blow, was a serious criminal offense. Without reproach, the southern apologist historian Ulrich B. Phillips noted that when, in 1755, two slaves were convicted of poisoning their master, the woman was burned at a stake and the man was hanged and his body left suspended in chains on the Charleston common. This was macabre but not unique. L. Michael Kay and Lorin Lee Cary found records of how fifty-six slaves were executed by the authorities in North Carolina between 1748 and 1772. Of those whose fate is known, two died after being castrated; six were burned at the stake; twenty-four were hanged. Kay and Cary concluded that "brutal slave punishments were both common and normative forms of punishment, encouraged by statutory law in order to cow the slaves into submission."

To ensure punishment without delay and without harm to the owner, slaves were subjected to special procedures. In North Carolina, from at least 1715 on, they were tried without a jury, in special courts. If a slave was condemned and executed, the owner was compensated from public funds. The owner, however, had to establish that he had acted reasonably in the affair. Under a law passed in 1753 in North Carolina, he would not be compensated for an executed slave unless he stated, as of course he always did, that for the previous year, the slave had been properly clothed and provided with a quart of corn each day as food. Some owners found in the law a means of getting rid of a surplus or refractory slave at a higher price than they could get on the open market. It was reported that "many persons by cruel treatment of their slaves cause them to commit crimes for which many of said slaves are executed"—so many, in fact, that compensation was abolished after the Revolution, in 1786.

Except for the money involved, masters had little need to take a slave to court. The "casuall killings of slaves" was legalized by the Virginia legislature in 1669 and reaffirmed in a further statute of 1705, which specified that the master "shall be free and acquit of all punishment and accusation for the same, as if such accident had never happened." It was not until 1774 that the "malicious and wilful killing of a slave" would result in one year's imprisonment for a first offense.

Falling afoul the law or a master was virtually a certainty for newly arrived, or "outlandish," Africans. Traumatized by their capture and the ordeal of the slave ships, confused by the life into which they had been plunged, and bewildered by the commands they received in a language they did not understand, they frequently failed to obey. Moreover, being often unable to communicate with one another, both because of the variety of the languages they spoke and because whites tried to prevent them from "conspiring," they could not give one another guidance. To hoe and cut cane or pick cotton, only a primitive level of skill was sufficient.

Sufficient was not efficient. Many whites recognized this. However, during the eighteenth century few whites tried to "upgrade" their slaves even though doing so would have increased the slaves' value. In South Carolina in 1740, teaching a slave to write was punishable by a fine of 100 pounds. Attempts to Christianize slaves were also frowned upon. In this, the English Protestants were far less liberal than the Catholic Spaniards. In 1667 the Virginia legislature attempted to forestall any thought that conversion of slaves would make them candidates for freedom. By 1706 Virginia's lead had been followed by five other colonies. The South Carolina Society for the Propagation of the Gospel complained in 1713 that masters

> are generally of [the] Opinion that a Slave grows worse by being a Christian; and therefore instead of instructing them in the principles of Christianity which is undoubtedly their Duty, they malign and traduce those that Attempt it [and in the Legislature also it is] thought inconsistent with the planters' secular interest and advantage; and it is they that make up the bulk of our assembly.

This remained true during all of the seventeenth century and most of the eighteenth.

Why such opposition? If Karl Marx (much later, of course) was right in saying that "religion is the opiate of the people," it would seem to have been more to the interest of the masters to promote Christianity among slaves. But the masters did not see it that way, for both economic and security reasons. In the first place, giving Sunday to the slaves as a day to grow some of their own food cut down on expenses. They could not do that if they went off to church. Also—particularly during the planting and harvesting seasons, when men, women, and children often worked as much as eighteen hours a day, every day in the week—going to church would be a distraction; moreover, it could, conceivably, grow into a demanded "right." Second, when slaves attended gatherings of any sort, even (or perhaps especially) religious meetings, they could "conspire," be encouraged to rebel, or plan to escape. In 1715, in "An Act Concerning Servants and Slaves," the legislature of North Carolina forbade gatherings of any sort, including attendance at church. It was safer to keep slaves isolated in the small groups that worked on each separate plantation.

In addition to these evident reasons, slave owners were growing sensitive to critiques of the "peculiar institution." When religious leaders such as John Wesley (who denounced slavery as evil), Francis Asbury, and George Whitefield visited slaveholding areas, attempts were made to boycott them. One plantation owner, Hugh Bryan in Georgia, was arrested in 1742 for trying to carry an evangelical message to the slaves and for criticizing what the whites were doing.

How did religion spread in the face of this formidable opposition, particularly when so few slaves could read? The fact is that no one knows. Even more curious is the desire of the slaves to adopt what they must have thought of as the religion of the masters, particularly a religion that seemed to justify slavery. We are on safer ground in considering what the blacks did with Christianity.

Two points stand out. First, blacks chose from Christianity those elements that spoke to their condition. They found, particularly in the Old Testament, stories that were parables for their own tribulations. They generally identified themselves with the "children of Israel" and their masters with "old Pharaoh." Second, and more generally, by portraying themselves as "children of God" slaves overcame the central problem of their existence, being nonpeople. As slaves became fellow Christians, avoidance of the question of their humanity became more difficult for any but the most obdurate whites. Yet it was not until nearly the eve of the Revolution that whites came to tolerate missionaries' activities or even limited education for blacks. Perhaps when they did, it was because neither missions nor schooling seemed to be leading toward emancipation.

Virginia, the state with the largest black population, was the scene of the first emancipation proclamation. How it came about is one of the ironies of the American struggle for freedom. On November 7, 1775, in his final act in office, the last royal governor, Lord Dunmore, set out to enroll blacks and indentured whites to fight the rebellious colonists. Some 800 blacks were enlisted, but they were little used by the British except to support Scottish and English raiders on the Virginia coast. The main effect of Dunmore's proclamation was to consolidate and energize the southern rebels and to push an unknown number of wavering whites, who feared that British policy would provoke a general slave rebellion, into the rebel camp.

In the ensuing Revolution, blacks fought on both the British and the rebel side; but much more common in the confusion of the times was flight. Herbert Aptheker has estimated that 80,000 to 100,000 blacks took the opportunity to flee. John Hope Franklin quotes Thomas Jefferson as saying that "in 1778 alone more than 30,000 Virginia slaves ran away." Franklin adds that David Ramsay, a contemporary South Carolinian historian, "asserted that between 1775 and 1783 his state lost at least 25,000 Negroes." Neighboring Georgia, upon joining the Continental Association in January 1775, declared slavery to be an

> unnatural practice . . . founded in injustice and cruelty, and highly dangerous to our liberty (as well as our lives) debasing part of our fellow creatures below men, and corrupting the virtue and morals of the rest, and is laying the basis of that liberty we contend for . . . upon a very wrong foundation.

The blacks certainly agreed with this sentiment. As many as 11,000 of the 15,000 held as slaves in Georgia used the outbreak of war as a chance to run for freedom.

It is evident that—like the Indians, to whom I turn next—blacks were repelled by the American colonists, disregarded what they knew of the colonists' ringing pronouncements on liberty, and thought their own best hope lay with the British. Their actions on the eve of the Revolution certainly make clear that, despite the tragedy of their early residence in America, "This celestial spark [of liberty was] not extinguished in the bosom of the slave."

CHAPTER 11

Whites, Indians, and Land

"Your coming is . . . to invade my people, and possess my Country." Those words from an Indian "emperor" whom the colonists knew as the Powhatan in 1609, just two years after the founding of Jamestown, might have been said by any Indian from then until the end of the nineteenth century. As the frontier moved westward, each newly exposed group learned anew the meaning of the arrival of whites. Only rarely were Indians given respite from settlers' relentless push into their lands, and only rarely could the Indian peoples coordinate their activities to defend themselves.

Those Indians who met the Spaniards in La Florida and in the American southwest, *Nueva Andalucia,* had a different tale to tell: the relatively few Spaniards who came among them sought not to expel them but to incorporate them into a limited version of Spanish society. Unlike the English settlers, Spaniards did not intend to farm the land themselves. Rather, they wanted to "reduce" the Indians to urban life, destroy their beliefs, and demolish their culture. It was thus not land but subversion that caused the Apalachee and Timucua of La Florida to begin what would become a sequence of revolts that gained in intensity later in the southwest among the Pueblo people.

The French were no more colonists than the Spaniards. They never amounted to more than a fraction of the numbers of the English settlers. Land was not an important objective of their rule. Rather, they were traders. To trade successfully and even to defend themselves from the Iroquois, whom they made into enemies almost from their entry into the Saint Lawrence, the French needed Indian allies. They therefore sent French

By the KING,

A PROCLAMATION,

For suppressing Rebellion and Sedition.

GEORGE R.

WHEREAS many of Our Subjects in divers Parts of Our Colonies and Plantations in *North America*, misled by dangerous and ill-designing Men, and forgetting the Allegiance which they owe to the Power that has protected and sustained them, after various disorderly Acts committed in Disturbance of the Publick Peace, to the Obstruction of lawful Commerce, and to the Oppression of Our loyal Subjects carrying on the same, have at length proceeded to an open and avowed Rebellion, by arraying themselves in hostile Manner to withstand the Execution of the Law, and traitorously preparing, ordering, and levying War against Us: And whereas there is Reason to apprehend that such Rebellion hath been much promoted and encouraged by the traitorous Correspondence, Counsels, and Comfort of divers wicked and desperate Persons within this Realm: To the End therefore that none of Our Subjects may neglect or violate their Duty through Ignorance thereof, or through any Doubt of the Protection which the Law will afford to their Loyalty and Zeal; We have thought fit, by and with the Advice of Our Privy Council, to issue this Our Royal Proclamation, hereby declaring that not only all Our Officers Civil and Military are obliged to exert their utmost Endeavours to suppress such Rebellion, and to bring the Traitors to Justice; but that all Our Subjects of this Realm and the Dominions thereunto belonging are bound by Law to be aiding and assisting in the Suppression of such Rebellion, and to disclose and make known all traitorous Conspiracies and Attempts against Us, Our Crown and Dignity; And We do accordingly strictly charge and command all Our Officers as well Civil as Military, and all other Our obedient and loyal Subjects, to use their utmost Endeavours to withstand and suppress such Rebellion, and to disclose and make known all Treasons and traitorous Conspiracies which they shall know to be against Us, Our Crown and Dignity; and for that Purpose, that they transmit to One of Our Principal Secretaries of State, or other proper Officer, due and full Information of all Persons who shall be found carrying on Correspondence with, or in any Manner or Degree aiding or abetting the Persons now in open Arms and Rebellion against Our Government within any of Our Colonies and Plantations in *North America*, in order to bring to condign Punishment the Authors, Perpetrators, and Abettors of such traitorous Designs.

Given at Our Court at St. *James's*, the Twenty-third Day of *August*, One thousand seven hundred and seventy-five, in the Fifteenth Year of Our Reign.

God save the King.

LONDON:
Printed by *Charles Eyre* and *William Strahan*, Printers to the King's most Excellent Majesty. 1775.

Catholic missionaries to convert the Indians and, like the Spaniards, created settled mission communities. The missionaries were crucial to French imperialism, but above all it was the Indian trader, the *coureur de bois,* who was the Frenchman best known to the Indians.

Because the English most affected the Indians and because land became the cause of the destruction of so much of Indian society, I will focus here on what the English did to the Indians and on how the Indians attempted to meet the challenge.

The relationship between the incoming colonists and the Indians usually began in peace and friendship and was often characterized by the settlers' hunger and the Indians' generosity, but it invariably turned sour. Hostility arose faster when the incoming colonists stole Indians' food caches (as in Virginia and Massachusetts) or kidnapped Indian children (as in Virginia, Massachusetts, and Carolina), and more slowly when the colonists were at least polite (as in Maryland) or reasonably honest (as in Pennsylvania). In each case, though, an eventual clash seemed unavoidable.

This was because colonialism is about land: those who live on it want to keep it while those who come among them want to take it. The inhabitants assert their rights and in desperation sometimes fight. The newcomers feel a need to cloak their aggression in more exalted terms than simple greed, and readily find a justification: the natives do not efficiently use the land; they are gypsies, nomads, or drifters, whose land means nothing to them; they are treacherous and murderous; they do not respect the true religion (that is, the colonists' religion), so God has ordained the colonists' success; and, finally, the natives are little better than wild animals.

Such arguments have been advanced by various colonizing groups not only in America but also in Ireland, the Canaries, Central Asia, South Africa, the Congo, and Algeria from the sixteenth century to our own time. In each of these places, the technologically advanced people saw the natives as merely a hindrance to progress. Soon the newcomers come to look upon the natives as the intruders although as John Lawson remarked of the American colonists in 1701, it is "we [who] have abandoned our own Native Soil, to drive them out, and possess theirs."

When the first English settlers began to arrive in the Chesapeake in

1607, their numbers were small and their skills few. Recognizing their vulnerability, the Virginia Company of London instructed them to "keep aloof and not to show that they were merely human." Otherwise, the company warned, the Indians "will make many adventures upon you." In contrast to the whites, the native population was numerous. The first society with whom the colonists came into contact numbered about 13,000, whereas at one point the population of Jamestown dipped to less than 100. After the Pilgrims in Plymouth had suffered through their first winter, "scarce fifty remained" of the original 102; their Indian neighbors numbered at least 10,000. In Carolina, colonists were even fewer in comparison with the Indian inhabitants. Yet the shrewder of the Indian leaders realized that the few Europeans they saw represented not only a more technologically advanced but probably a much more numerous civilization. Captain John Smith relates how Powhatan attempted to find out what the odds were.

When Powhatan's daughter Pocahontas, renamed Lady Rebecca, went to London with her new English husband, John Rolfe, in 1616, Powhatan used the occasion to send along with her a man by the name of Uttamatomakkin. As Smith wrote,

> This favage . . . was one of *Powhatan's* Council, and was, amongft them, held to be an underftanding fellow. The King fent him, as they fay, purpofely to number the people here, and to inform him well what we were and what was our ftate. Arriving at *Plymouth,* according to his directions, he got a long ftick, whereon by notches he did think to have kept the number of all the men he could fee, but he quickly wearied of that tafk.

In England, people were too numerous to be counted; soon they would be innumerable in the New World as well.

By the middle of the seventeenth century, 17,000 colonists had arrived in Massachusetts; settlers in what became Connecticut reached 10,000; Plymouth, where the Pilgrims were established, reached a population of 5,000; Rhode Island, 5,000; Virginia in 1670, 40,000; and Maryland, 20,000. The number of white colonists was almost doubling every twenty-five years. Meanwhile, Indian societies declined drastically. The Spaniards first brought diseases, and Sir Francis Drake's crews brought what appears to have been typhus in 1585. Smallpox, mumps, typhus, diphtheria, and whooping cough were brought by Europeans; and hookworm, dengue, blackwater fever, yellow fever, elephantiasis, leprosy, and yaws by Africans. Travelers in the Carolinas found whole towns abandoned and overgrown with weeds. "Two and a half centuries after contact with the Spaniards, all of Florida's original Indian people were gone." Farther north, shortly after the English established themselves on Roanoke, large numbers of their Indian neighbors sickened and died. A Catholic priest in Louisiana speculated that because the pores of the skin were more open in Indians than in whites, diseases could more easily enter Indians' bodies; but knowing nothing of immunities, both whites and Indians ascribed the devastation to some divine force. As Thomas Hariot reported in 1587, within a few days of the departure of the English from any Indian village,

> the people began to die very fast, and many in short space, in some Townes about twentie, in some fourtie, and in one sixe score. . . . The disease also was so strange, that they neither knewe what it was, nor how to cure it, the like by report of the oldest men in the Countrey never happened before, time out of minde. . . . They were perswaded that it was the worke of our God through our meanes, and that we by him might kill and slay whom we would without weapons, and not come neere them.

In Massachusetts, the Pilgrims agreed that devastation of the Indians was the work of God. As Governor William Bradford recorded in 1634, "it pleased God to visit the Indians with a great sickness, and such a mortality that of a thousand, above nine and a half hundred of them died." For many, death was horrible. Bradford wrote:

> Indians that lived about their trading house there fell sick of the small pox and died most miserably; for a sorer disease cannot befall them, they fear it more than the plague. For usually they that have this disease have them in abundance, and for want of bedding and linen and other helps they fall into a lamentable condition as they lie on their hard mats, the pox breaking and mattering and running one into another, their skin cleaving by reason thereof to the mats they lie on. When they turn them, a

> whole side will flay off at once, as it were, and they will be all of a gore blood, most fearful to behold. And then being very sore, what with cold and other distempers, they die like rotten sheep.

Once thriving and powerful Indian communities collapsed. Casualties ranged upward from five out of ten inhabitants. Some populations may have declined by a factor of twenty to one. In Virginia, the Indian population fell from more than 30,000 when the colonists arrived to about 3,000 divided among nineteen tribes in 1670. Everywhere the whites ventured, such disasters were repeated: The population of the Pequots of southern Connecticut dropped from 13,000 to 3,000; and that of the Kanienkahaka (known to their enemies as Mohawks) of eastern New York from 8,000 to less than 3,000. Half the Cherokees, Creeks, Choctaws, and Chickasaws were gone by 1738, and about the same proportion of a group of tribes known as the Catawbas in the Carolinas in 1759. Florida, in Amy Turner Bushnell's arresting phrase, had become "a hollow peninsula." As Colin Calloway noted, Indian America was turned "into a graveyard."

Large as they are, the numbers do not tell the whole story. Because older people were less likely to survive, and because in communities without written records elders were the guardians of heritage, society after society in the Indian world foundered culturally as well as demographically. Thus, providentially, much of the land was cleared for the incoming whites.

Before the arrival of the whites, warfare was common among Indian societies: the powerful often attacked weaker neighbors, kidnapping them, taking their possessions, and sometimes driving them off their lands. Warfare was not a sport, but as many early European visitors described it, it was more ceremonial than lethal, more a matter of "violent theater" than true combat. With the coming of the whites, the scale and intensity of conflict changed. As the Indians began to learn to use muskets, they found that the soft lead slug of a 50- or 60-caliber musket inflicted often unhealable wounds. For the first time, Indian societies could destroy one another. Those who had guns won. Getting guns became absolutely vital. By 1643, the Mohawks had nearly 300 arquebuses. Their enemies had to have this weapon also.

As we have seen in chapter 6, roughly the same process was at work in

Africa, impelled by the same forces—societies fighting one another and outsiders eager to give them the means. In America, colonists were delighted to see Indians kill one another, but they feared that the guns would also be turned against them. So as early as 1619 they sought to prevent merchants from selling muskets, shot, or powder to Indians "upon pain of being held a traitor to the colony and of being hanged as soon as the fact is proved, without all redemption." But even government officials violated the restrictions. The incentive to acquire the fur pelts that the Indians offered was too great to resist, so guns were sold despite the ban.

Guns also made possible, and purchasing them made necessary, massive hunting of animals, particularly deer and beaver, for their pelts. Whereas previously Indians had lived in balance with their environment, they soon hunted some animals to extinction. Widespread famine followed in due course.

Guns were not the only cause of the collapse of Indian societies; drunkenness was at least as lethal. Rum came with the whites' incursion. In early Maryland, before extensive contacts with the whites, Indians were "very temperate from wines and hote waters, and will hardly taste them, save those whome our English have corrupted." So wrote Father Andrew White, S.J., in 1634. Later, liquor not only became the scourge of the Indians but also the principal means by which Indian traders acquired furs. Why the Indians were so incapable of abstaining from or withstanding "demon rum" has long been debated. No certain answer is forthcoming, but two possibilities merit consideration.

First, as we may observe today in Peru, people who have traditionally used narcotics have developed a tolerance that outsiders lack. Perhaps something similar affected the East Coast Indians, who were products of an evolution different from that of Europeans. One indication of this difference is that no Indians had type A or type B blood from the ABO blood group; they were all type O. This may be unrelated to immunity, but perhaps some other, as yet unidentified, biological characteristic or genetic trigger made them potential alcoholics. It is certain that alcohol creates an imbalance of sugar, for which Indians had no genetic tolerance. Among other peoples, this has led obesity, heart disease, and dependence. Never having had liquor before, Indians were unable to control it.

There is a second, and to me more persuasive, reason for Indians' drunkenness. It is their attempt to "escape" from facing the devastation of their societies. Like the shocked survivors of the black death in fourteenth-century Europe, or the distraught people Hogarth drew in seventeenth-century London's "Gin Alley," or like some heroin addicts today, Indians found that their lives had become nearly unbearable. We have no means to judge the psychological impact on the Indians, at least from their own words, but we would probably not be wrong in assuming that the survivors were so traumatized, disheartened, and disorganized that they found it difficult to continue. In modern terms, their experience would be like surviving a nuclear war in which family, friends, and whole communities perished, and all that had made life worthwhile was gone.

Whatever the reason for their susceptibility, rum was almost as lethal to the Indians as smallpox. "Rum-debauched, Trader-corrupted," was Benjamin Franklin's summation of their predicament. Occasionally, colonial governments tried to curtail sale of liquor. Bans were enacted, violated, and then lifted. The brutal reality was that sober Indians were not so willing as drunken Indians to give up land, and that addicted Indians were willing to give up everything. Liquor was a tool of colonialism. One official who openly acknowledged this was the British Indian agent Sir William Johnson; he advocated lifting one recently imposed ban on the sale of liquor, figuring that, drunk, Indians would destroy themselves. The Board of Trade agreed and provided tens of thousands of gallons of rum, which traders almost literally poured down the Indians' throats.

"Indian traders," as the English called them, or *coureurs de bois,* led the penetration into Indian lands. As "advance men," they peddled cloth, guns, and rum for pelts; scouted the land; made contacts; learned local customs and languages; and unintentionally prepared the way for settlers. The first known Indian trader, Ensign Thomas Savage, came to the fore shortly after Jamestown was established. Having learned a dialect of Algonquian, Savage became an important intermediary. Knowledge of Indian languages was to become the great virtue of his successors; their great vice was spreading the curse of liquor.

Not only Indian traders were involved in swapping rum for furs; although less openly, many coastal plantation owners built their fortunes selling rum to the Indians. The trade, once denounced by a colonial legislature as "abominable filthyness," was actually abominably profitable. Rum was a by-product of the molasses made on the West Indies sugar plantations and became one of the earliest and most successful colonial industries. At least 140 distilleries were converting molasses into more than 25,000 tons of rum a year by 1770.

Subversion of the Indians was a subtle process involving not only guns and rum but a variety of other trade goods on which the Indians became dependent. Fur capes were replaced by imported cloth, stone tools by steel, and clay pots by iron. As each new item became standard, the Indians forgot how to make the things that had formed their traditional material culture. Finally, as Skiagunsta, the war chief of the Lower Cherokee, said, "The clothes we wear we cannot make ourselves. They are made for us. We use their ammunition with which to kill deer. We cannot make our guns. Every necessary of life we must have from the white people." The Mohawks coined a name for Europeans, *asseroni,* or "ax-maker"—and sadly concluded that they could no longer dress themselves in skins but required English "strouds" or Belgian "duffles."

This process, of course, was slow, but the colonists were in a hurry. Since more distant Indian societies had not yet collapsed, they rediscovered a strategy that had been developed 2,000 years earlier in China. Much as the American colonists considered Indians barbarians, the Chinese considered all non-Chinese barbarians, but they recognized that barbarians could be useful. Those who lived closest, the "inner barbarians," had as much reason as the settled people to fear the wilder "outer barbarians." So the Chinese used them as a human wall. Much the same strategy was adopted by the Romans, who enrolled the relatively pacified Gauls to keep the wilder Germans at bay. Faced with a similar challenge in the seventeenth century, American colonists reinvented the ancient strategy. An act by the legislature of Virginia in 1656 ordered a force of its militia to march against the "many western and inland Indians" with "all the neighboring Indians . . . as being part of the articles of peace concluded with us." To protect the "nearby barbarians" who might help fend off the "outer barbarians," Carolina in 1680 created a zone extending 200 miles from Charleston where Indians were not to be enslaved or massacred.

Sometimes in America, as in Rome and China, the strategy did not work, but it was used often. As the whites advanced into Indian territory, there were always new "nearer" Indians. Consequently, practically no colonial "armies" had a majority of whites. For example, in 1712 Colonel John Barnwell led a force of thirty whites and nearly 500 Indians against the Tuscaroras in Carolina; and the next year Colonel James Moore commanded an even larger force of 850 Indians but only thirty-three whites. One Indian society after another was beaten into submission, caught between the advancing whites and the still more remote Indians.

The constant objective of the whites was simply land—land without people. Land was what the colonists had come to the New World to get. Each Indian society had to discover this for itself, but as soon as settlers began to construct houses and plant crops, Indians perceived a mortal danger. In 1607, from Jamestown, the British government was informed that the Indians "used our men well untill they found they begann to plant and fortefye." That was a message often repeated but rarely heeded as the whites moved inland. In desperation, the Indians sometimes struck at the settlers, trying to terrorize them into leaving. The result was always the same: they drew down upon themselves the armed might of colonial governments.

Such "search and destroy" missions, as George Percy related in 1610, do not figure among the dozens of "wars" but were much more frequent. Guided by an Indian captive in irons whom they beat and threatened to kill, Percy's men surrounded a "dissident" village and massacred all the inhabitants except for the "queen," her children, and one man. They immediately beheaded the man, burned down the houses, and hacked down the corn crop. When they got back to their boats,

> A Cowncell beinge called itt was Agreed upon to putt the Children to deathe the w^{c}h was effected by Throweinge them overboard and shoteinge owtt their Braynes in the water yet for all this Crewellty the Sowldiers weare nott well pleased And I had mutche to doe To save the quenes lyfe for that Tyme.

When they got back to Jamestown, Governor Lord de la Warr "thowghte beste to Burne her at the stake, but, mercifully, the soldiers merely "putt her to the Sworde."

By 1622, after repeated episodes like this, the Virginia Indians had become convinced that the whites would kill them all. The murder of a prominent Indian warrior triggered their first major counterattack. Led by their war chief, Opechancanough, they determined to kill or drive out the whites. They nearly succeeded. On the appointed day, unarmed Indians infiltrated the settlements, bringing gifts of food and helping in the cornfields; then they grabbed the colonists' own weapons and fell upon them, killing 347 men, women, and children and burning down their new settlements.

The Powhatan Uprising of 1622 was the first of a series of wars between Indians and whites—wars that would rumble across the continent for more than 250 years. Following it, the mask of fellow feeling dropped, never to be replaced. The governor of Virginia, Sir Francis Wyatt, proclaimed in 1624, that "our first worke is expulsion of the Savages to gain the free range of the countrey for it is infinitely better to have no heathen among us, who at best were but as thornes in our sides, then [sic] to be at peace and league with them." In a declaration of policy, the emphasis on taking over Indian lands was made explicit:

> Our hands, which before were tied with gentleness and fair usage, are now set at liberty by the treacherous violence of the savages, not untying the knot, but cutting it. So that we, who hitherto have had possession of no more ground than their waste . . . shall enjoy their cultivated places, turning the laborious mattock into the victorious sword (wherein there is more both ease, benefit, and glory) and possessing the fruits of others' labors. Now their cleared grounds in all their villages (which are situated in the most fruitful places of the land) shall be inhabited by us, whereas heretofore the grubbing of woods was the greatest labor.

To speed up the process of dispossession, the settlers at Jamestown summoned the Indians to a peace conference. There, after speeches affirming brotherhood, one of the leaders of Jamestown proposed that they all drink to good relations and brought out a barrel of wine. The barrel was laced with toxin "to poysen them." To reassure the Indian leader, the whites pretended to taste it; when the others drank "some two hundred" were poisoned and then the Jamestowners killed "some 50 more" and "brought home part of their heads"—that is, scalps.

To ensure the extirpation of the natives, the governor declared a white-only zone in a triangle formed by the York and James rivers and the Virginia fall line. Indians could enter this zone only by official permission. Two years later, each settler was ordered not to talk to or trade with the Indians without special permission. Twenty years later, after the second war between whites and Indians, Virginia ruled that no Indians could enter its white-only zone "upon paine of death." If any unauthorized Indian wandered in, it was lawful "for any person to kill" him. Also, it was a capital offense for any colonist to entertain or conceal an Indian.

To keep the Indians under observation, Virginia adopted the system then being used by the British in Ireland: reservations on which Indians would be confined. Reservations would typify whites' policy regarding Indians for the next four centuries. But, since powerful groups still remained outside the reservations, warfare continued. When they could, the Indians resisted. When they could not, they retreated. As the Powhatan told the resident secretary of the Virginia Company, Ralph Hamor, "I am now olde, and would gladly end my daies in peace, so as if the English offer me injury, my country is large enough, I will remove my selfe farther from you." To retreat beyond the next hill, mountain, or river seemed a hopeful strategy: the Indians thought that surely the whites would finally get enough land and be satiated. Well into the eighteenth century, Indian societies tried to buy peace with land. But of course that strategy did not work. Ship after ship arrived with colonists hungry for land. So, beginning in Jamestown shortly after the colonists arrived, war after dreadful war punctuated the calendar of early American history. Indians, driven to the wall, tried to fight for their land; the colonists retaliated in growing fury.

When the whites realized that they could not fight the Indians on their own terrain, the forests, they adopted the practice of annihilation: they would starve the Indians by burning their corn fields, carrying away or destroying their stored grain, preventing them from fishing in the rivers, burning down their villages and driving away or massacring their women and children. These tactics, already begun in Jamestown as early as 1610, soon spread to other colonies and became standard in the long series of confrontations that followed.

In 1636, the newly formed militias in the New England colonies, known as "trainbands" and organized into regiments of several towns to give them overwhelming force, stormed the main Pequot town and massacred the inhabitants, burning some alive. They then hunted down and killed or sold as slaves all the Pequot they could find. Metacomet (whom the New Englanders called King Philip), the leader of the Wampanoag of southern New England, is reported as saying in 1675: "Brothers, these people from the unknown world will cut down our groves, spoil our hunting and planting grounds, and drive us and our children from the graves of our fathers and our council fires, and enslave our women and children." It did not take long for his prophecy to be borne out, in one of the bloodiest wars in American history, King Philip's War.

King Philip's War was virtually inevitable from the day the settlers first landed. Each Indian society had to learn for itself the lesson of this war. None learned until much too late from what happened to their neighbors. A few tried to resist, but the odds were always against them. The incoming whites, generation after generation, not only wanted the Indians' land but—regarding Indians as "little better than wild beasts, perhaps even children of the Devil"; "poor, brutish barbarians . . . not many degrees above beasts"; "barbarous scum and offscourings of mankinde"; "wilde and savage people, that live . . . like heards of Deare in a Forrest"—violated every truce and treaty. Philip's people, the Wampanoag, had entered into an alliance to help defend the Pilgrims against their Narragansett enemies. Later, they stood aside while the Puritans massacred the Pequots. Their own turn came next. On behalf of the whites, the Mohawks and Mohegans attacked them.

Farther south, the Tuscaroras first tried to leave their ancestral lands and migrate northward to join their distant relatives, the Iroquois. They begged for asylum in Pennsylvania but were refused. Finally, driven to desperation by raids in which Carolinian whites kidnapped Indian women and children for slaves, they retaliated in the fall of 1711 while the colony was in the midst of what amounted to a minor civil war (Cary's Rebellion) and killed 130 colonists in just a few hours.

The whites quickly counterattacked. They killed thousands of Indians; burned villages; and destroyed crops, food stores, and orchards—the work of years. Peace, often "a desolation"—as the great Roman historian Tacitus described the aftermath of similar Roman search-and-destroy operations—would be imposed on the defeated Indians, with the price of the cession of more land.

In area after area, a few years of relative calm would follow a major conflict. Then, as settlers pushed forward into newer areas, the cycle would begin again in mutual recrimination and freshly invigorated hatred—leading always to seizure of Indian land.

The Indians appeared to the colonists, as Governor William Bradford of Plymouth said, "skulking about them." We no longer think of Indians as "skulking" (although we realize that they had good reason to be cautious); but to us as well, they seem just beyond the range of our knowledge. We cannot hear their voices—even from that best-known of all the early Indians, Pocahontas, we do not have a single word—and we can seldom identify individuals. One historian who tried to break through this veil of silence is Carl Bridenbaugh.

Bridenbaugh tells the story of a "very tall young man" whom the Spanish general Pedro Menéndez de Avilés acquired in 1561 from Chesapeake Indians. We know the beginning of the story and, perhaps, the end. But the connection between the two is speculative. As Bridenbaugh writes, Menéndez took the young Indian back to Spain and informed the king that he was a noble Indian of great potential value. The king agreed and ordered that he be turned over to the Dominicans and trained to return to America to help establish Catholic missions. This was not an unusual event: the Portuguese did it in Africa and the English would later do it in Virginia.

Sent to New Spain, young Luis, as he became known, had a close view of what colonialism meant to the Mexican Indians. To induce the Spaniards to send him home, he pretended to be their willing tool and a sincere Catholic. This pretense worked. The Spaniards finally sent him home to the Chesapeake in 1568. There he was warmly received by his brothers, including, according to Bridenbaugh, "almost certainly" Powhatan. The priests who accompanied Luis, and presumably Luis himself, were astonished to find that the diseases which would soon ravage all Native Americans had already struck the people in Virginia. Possibly spurred by what he saw, Luis threw off his

Western clothes and customs and "went native." Most horrifying to the priests, he took several wives, and they publicly damned him for his "sins." Furious and humiliated, Luis cut off contact with them. Finally, after they tracked him to one of the villages, he led a group of Indians to kill all the Spaniards—with the exception of a young servant boy who survived to tell the tale. When the Spaniards heard the boy's account, they sent a punitive expedition in 1572. The members of this force killed forty Indians and kidnapped others (whom they subsequently hanged). Luis was not among these.

Up to this point the story is documented, but Bridenbaugh speculates that Luis was the war leader of the Virginia Indians, known to later English colonists under his Indian name, Opechancanough. It certainly was Opechancanough who led the uprisings against the English colonists in 1622, nearly wiping out the colony; and in 1644, making his last attempt "to roote out all the English."

If Bridenbaugh is right that "Luis" and Opechancanough were the same person, he lived an extraordinarily long life, but even if he acted only in the events of 1622, his story would be a remarkable confirmation of the overlap and interplay of the Spanish and English in North America. It is an even more striking example of our ignorance of the Indians: we have not one word from any English source about the experience of Opechancanough in Spain or Mexico. Perhaps this leader was simply being wise: to let the English know much about his background would have invited assassination as an opponent too dangerous to be left alive. Telling tales about his experience might also have caused his own people to doubt his sanity. That was what happened to a group of Indians from the Mississippi valley who, more than a century later, were taken to France, were presented at court, and met Louis XV. When they returned, full of tales of "vast cities, carriages, fountains and other exotica which so little accorded with what the Indians saw of Frenchmen among them," their fellows dismissed them as incorrigible liars. Opechancanough would have been wise to keep quiet.

Keeping quiet was natural for the Indians. Few of them visited royal courts, but many were sent off to slavery on Caribbean sugar plantations and none are known to have returned. Enslaving Indians and selling them to sugar planters was a profitable business. In enslavement, as in many other things, the English colonists learned from the Spaniards. They claimed that they were fighting a "just war" against the recalcitrant or rebellious, and that they were therefore right to enslave the prisoners. In 1671, the Virginian government proclaimed that those Indians who refused peace on the terms dictated at the end of any engagement with whites should be shipped away, usually to the Caribbean, as slaves. Engagements did not come often enough, so the policy was extended to get the Indians to enslave one another. It was nearly always possible to incite one Indian people to fight against another. Some Indian tribes, like some African societies, soon came to specialize in the slave trade. Slaving was a profession easy to begin.

In the South, the Chiscas and the Lower Creeks raided Apalachees; Yamasees ambushed the Timucuans; Cherokees drove out the Yamasees; the Savannahs kidnapped the Westos; and the then powerful and numerous Tuscaroras indiscriminately raided them all. From an Indian perspective, the years around 1700 were a vicious circle of depopulation, cultural breakdown, and enslavement. Queen Anne's War in 1702–1713 dramatically increased the tempo of slave raids by encouraging attacks by Spaniards and English colonists back and forth across the vague, disputed Florida frontier. The raids were devastating to both sides, but it was the Indians who paid the largest share of the price. The Spanish governor at Saint Augustine reported that at least 20,000 Indians had been carried off by America colonists. An expedition sent by Governor Moore of Carolina against the Apalachee was perhaps the greatest slave raid ever in America.

In the North, the Iroquois kept up a relentless campaign against other Indians. By 1680, when they raided the "Great Village of the Illinois," they killed or captured about one in ten Illinois. The Illinois themselves were engaged in slave raiding among the Pawnees and Sioux. Soon, the Illinois and other neighboring societies like the Miami found themselves caught in a vise: aggressive Iroquois attacked from the east, and vengeful Sioux from the west. According to French missionaries and traders, the Illinois were one of the largest Indian societies, with nearly 100,000 people living in sixty villages; but even so, the Illinois concluded that their only hope of salvation was an alliance with the French, which would bring them guns. Champlain's strategic insight (and French weaponry) had multiplied the influence of the French vastly beyond their small numbers. "You are one of the chief spirits because you use iron," the French were told by a leading Indian. "It is for you to rule and protect all men. Praised be the Sun [god], who has taught you and sent you to our country."

Protecting all men meant providing safe areas around French forts and missions where towns sprang up and quickly filled with refugees. A group of Shawnee was one of many: about 1,000 people were so terrified by Iroquois raids that they gave up their homeland on the Cumberland and settled at Fort St. Louis. There they were joined by the remnants of other societies, so that the ballooning villages became a new form of society, geographic neighborhoods without ethnic or linguistic homogeneity, settlements where people were both neighbors and strangers, and where everyone had lost the sense of tradition that had given him identity and his life meaning.

Meanwhile, rivalry between the British and the French over dominance of North America had spilled into the Ohio valley. There it was often fought by proxies on the British side—traders, settlers, and land speculators, some of whom were also British officials. On the French side, it was fought by Catholic missionaries and French regular troops. Indian societies, under intense pressure from one another and from the French and British and their colonial allies, reacted with a desperate attempt to sustain themselves. The action they took has come to be known as "mourning wars."

Mourning wars appeared to be a particularly Indian phenomenon, but in fact they were very similar to events then taking place in Atlantic Africa. I will first describe the Indian pattern and then draw a comparison with Dahomey.

One purpose of mourning wars was to make up for the large number of group members lost to disease, raids by other Indians, and wars with the colonies by incorporating men, women, and children captured from other Indian nations. The larger the losses, the more important it became to seize as many new tribesmen and tribeswomen as possible. Those to be incorporated were adopted by individual families. "A father who has lost his son adopts a young prisoner in his place. An orphan takes a father or mother; a widow a husband; one man takes a sister and another a brother." So wrote Philip Mazzei in 1788 of a long-standing practice. As thousands died, wars became almost constant.

Those Indians who were not considered candidates for absorption were usually killed. That was the second purpose of mourning wars, to "cover" the dead and assuage their kinsmen's sadness. Similarly in Dahomey, captives were divided into two groups. Those who were to be kept alive were added to the workforce of the kingdom; the others were massacred by the hundreds to "water the graves" of ancestors of the reigning king "with the blood of the victims and recommitting the nation to the care of ancestral spirits."

The Mohawk did one thing the Africans in Dahomey did not do: they often ate the captives. They probably believed that this had the effect of transferring the captives' inner spirit and bravery to the winners.

In both Africa and North America, of even greater long-term importance than the suffering of the victims was the effect on the societies from which these people had been kidnapped: families were devastated, links between generations were broken, the cultural heritage was lost, and the terrorized survivors often fled their homelands to attempt to reconstitute themselves elsewhere. Among the Indians, this created "detribalized" clusters living alongside other, quite alien, peoples in refugee villages, and dependent on the protection, charity, and guidance of Europeans—usually the French. Such a community came into existence after Antoine Cadillac founded Detroit in 1701.

Detroit became a haven for collections of Sinagos, Kiskakons, Sables, Huron-Petuns, Miamis, and others who, within a few years, together numbered nearly 6,000. Over time, such aggregations acquired a new economic life of their own; the new towns began to attract large numbers of French merchants who settled there; one town had at least 150 French residents who mingled with the various Indian communities, acquired Indian "wives," and—as the missionaries furiously complained—went native. As a degree of security was established, when after 1700 the Iroquois lost their edge of military superiority, these new composite societies gradually found common interest. Some even began to relocate from military outposts to places more accessible to European trade.

The French led the way, but they were not the only Europeans who traded with the Indians. The Iroquois confederation had long relied upon the Dutch for tools and weapons; the Mohawks called the Dutch *kristoni,* metalworkers. When New Netherland was taken over by the English and became New York in 1664, the earlier relationship expanded into what became known as the "Covenant Chain," an alliance meant to balance the relationship that existed between the French and their Indian allies. This relationship was so successful that the Iroquois, with the active support of the colonial government of Pennsylvania, tried to impose it upon other Indian groups. To achieve these alliances, major efforts of diplomacy were required.

As practiced between the various Indian societies and the British or French, alliances were seen as an extension of kinship. The dominant party was called the father or uncle; the subordinate party was called the son or nephew. Accordingly, the Algonquians, the Illinois, and other groups around the Great Lakes referred to the French governor general or the French king as their father. From the name of the first governor, de Montmagy, the governor of New France was thereafter known as "Great Mountain" (Onontio). Onontio was father to the Ottawas, and they were his eldest sons as senior members of a confederation of Indian societies under the overall protection of the French. The Iroquois and other British-oriented peoples similarly referred to the senior British official as their father or uncle. Each relationship was conceived of not only formally but effectively in these kinship terms: the "father" should be generous, wise, and supportive; the "son" owed respect and gratitude but was expected to be a full person and not a minion. To assert "brotherhood" was to declare independence.

Insofar as the Indians practiced it, diplomacy aimed to create or protect autonomy from the dominant European powers, to ward off invasion by the opposite European power, and to gain better trading relationships.

Since Indian society was less hierarchical and less disciplined than European society, a crucial function of diplomacy was to ventilate and so dissipate hostility. To work, it had to be open to all. Often hundreds of warriors and dozens of leading men, orators, and shamans participated in convocations at which policy was discussed. A survey of some 400 such encounters with the Dutch, French, and English from 1663 to roughly 1730 reveals over 500 active participants, men we might call the diplomats or negotiators. Decisions were not binding, so that if a meeting failed to produce unanimity, the disgruntled faction carried on with its own policy; to avoid this potentially dangerous outcome, every effort was made to achieve accord. Feasts were held, presents were exchanged, speeches were made, histories were recounted, the long-stemmed calumet pipe was smoked, and other ceremonies were performed in a process that might last for weeks.

Dealing with the Indians, the French had the advantage of frightening them less than the English did. By 1700, only about 15,000 Frenchmen were in Canada and another 10,000 or so in the Mississippi River system, while colonists in British America numbered ten times that many. But the English had the advantage of being able to offer cheaper and better goods in exchange for furs. Sometimes, when British or American colonists interdicted the sea lanes, Frenchmen had no goods at all to offer the Indians. Such was the result of the capture of Louisbourg by colonial troops from New England (during "King George's War") in 1745. Thus, although the Miamis and other western Indians recognized the value of French diplomatic and military support, they were lured into the British trading network.

Beyond commercial relationships was the painful issue of cultural values. The three European powers differed in their attitude toward native customs. Spaniards focused their attention on religion and set out to destroy native culture by regrouping Indians in missions (*reducciones*). The French were far more tolerant of native customs, less vigorous in proselytizing, and more inclined to seek commercial and military ties with Indian societies, but they also grouped potential converts in *reserves.* Anglo-American Puritans shared some of the attitudes of both the Spaniards and the French. They too sought to make "their" Indians live in "praying towns" where the Indians could gain "*Civilitie* for their bodies" as the first step toward "*Christianitie* for their soules." They forced converts to adopt English names, cut their hair in an English style (imposing fines on those who wore long hair), put on English clothes, and marry one another by English law. They even regulated women's conduct during menstruation. However, unlike the Spaniards and the French, most Anglo-Americans were not much interested in Indian culture; what they wanted was Indian land after it was cleared of Indians. Justifying themselves by reference to the

Old Testament, they often branded Indian captives and sold these captives into slavery.

Before Indians could decide how to react to these major assaults on their lives, they were caught up in the great war between the French and the English.

PART III

Breakdown of the Imperial System

CHAPTER 12

The French and Indian War

The Seven Years' War (1756–1763), with operations stretching from Canada across Europe to India, has been called the first true world war. The part that was fought in the New World came to be known as the French and Indian War. That name obscures its complex nature: it was actually a many-sided conflict. The major European powers were France and Britain with Spain playing a relatively minor role. The Native Americans were tactically split between French and British, with some groups fighting on each side, but strategically they were opposed to both. Important parts were also played by the American colonists, some 12,000 of whom fought for the British against the French and their Indian allies. The colonists' involvement set a precedent that would be followed down the generations through the American Revolution to encompass some 200 wars Americans would fight as citizens of the United States.

Britain never placed much of its small army in America. Throughout the seventeenth century, from the imperial perspective, the American colonies were neither a valuable asset nor a strategic threat. Much more important were the flourishing trading outposts in the Ottoman and Mughal empires and the sugar-producing islands of the Caribbean. Consequently, modesty was the hallmark of imperial policy toward America. Its keynote was sounded by the philosopher John Locke at the time of the Glorious Revolution: he advised King William and Queen Mary that it was not necessary for Britain to expend its treasure to defend the American colonies, since they could be trusted to defend themselves with their own militias.

Locke's recommendation was more sound than he knew; or, rather, it was sound for reasons that he did not consider. All European armies were organized for the sort of landscape Europe had developed after centuries of farming. What generals did was to align troops in rigid formations, with each soldier's shoulder next to the shoulders of two fellows, facing a similar line of the enemy. The troops then marched deliberately in lockstep toward one another, and when they were only about twenty paces apart, they fired the land equivalent of a naval broadside. Their smoothbore muskets could not be aimed accurately, so the sheer weight of lead was relied upon to shock and disorganize the opponent. Neither side could reload, so after firing, the soldiers were expected to charge forward using bayonets 2 feet long fixed on muskets 4 feet long as virtual spears. In short, their tactics resembled those of the ancient Greek phalanx.

But in eighteenth-century North America, such tactics could not often be used, because forests had not been cut and swamps had not been drained; even in relatively open areas, gullies, rivers, and other natural obstacles made the formation of long lines of soldiers difficult or impossible. The problem of terrain was compounded by the tactics of the Indians. They wisely refused to stand in lines to fight; rather, they fought as guerrillas, ambushing when an ambush was opportune and disappearing when it was not. Thus the European order of battle was usually ineffective. While some European armies were beginning to deploy what were called "light troops" or jaegers (after the German word for hunters) who could move rapidly in uneven terrain, the English were slow in using such troops. British soldiers moved ploddingly because they were weighed down with forty or fifty pounds of equipment; wisely, the French had their soldiers pack only twelve pounds of supplies, and the Indians carried practically nothing. Except for the Indians, who lived frugally, supply was a nightmare in America. Roads hardly existed; and farms—upon which, in Europe, the poorly provisioned armies depended for food—were widely scattered and usually had little for soldiers to loot.

Although the French forces were always small, they were more adept than the British at using Indian auxiliaries. It was the British generals' failure to use Native Americans, who did not seem sufficiently "military," that led to the massacre of General Edward Braddock and his redcoat regulars at the start of the French and Indian war in 1755. Braddock's men, who outnumbered their French and Indian opponents and were better armed, were mowed down in an Indian-style ambush for which their European training had not prepared them. Of his 1,100 soldiers, 714 were killed; and 63 of his 86 officers were either wounded or killed. After abandoning their artillery, food, equipment, and transport, those who could still walk retreated all the way back to Philadelphia. There they were a sorry sight. As Benjamin Franklin wryly commented, "The whole transaction gave us Americans the first suspicion that our exalted ideas of the prowess of British regulars had not been well founded." That episode may be taken as the beginning of Americans' belief that they did not need the British army.

Although Franklin said, one suspects with tongue in cheek, that he made no pretense at military expertise, he reported that he had warned General Braddock in advance of this lesson of American colonial wars. "The general," Franklin wrote in his autobiography, "was, I think, a brave man, and might probably have made a figure as a good officer in some European war. But he had too much self-confidence, too high an opinion of the validity of regular troops, and too mean a one of both Americans and Indians." Warned of the danger of an ambush on his exposed long supply train, "he smil'd at my ignorance, and reply'd, 'These savages may, indeed, be a formidable enemy to your raw American militia, but upon the king's regular and disciplin'd troops, sir, it is impossible they should make any impression.' " As he lay dying, General Braddock is reported to have said, "We shall better know how to deal with them another time." When "another time" came, fifteen years later, it showed only that British generals learned slowly. Then Braddock's failure was to be echoed in the disaster that befell General John Burgoyne during the Revolutionary War.

The generals should have learned that fighting better meant doing it the French way—that is, getting help from the Indians who knew how to fight in the wilderness. The Indians' contribution was laid out by the French general Louis-Antoine de Bougainville. Like Braddock, he was often annoyed by the Indians; but, as he commented, "In the midst of the woods of America one can no more do without them than without cavalry in open country." Being great trackers, they were the only possible scouts and spies:

> They see in the tracks the number that have passed, whether they are Indians or Europeans, if the tracks are fresh or old, they are of healthy or sick people, dragging feet or hurrying ones, marks of sticks used as supports. It is rarely that they are deceived or mistaken.

One British general, the Scottish earl of Loudoun, thought he had found a method of combining the best of English disciplined warfare with the woodland skills of the Indians. He wrote in 1756, "It is impossible for an Army to Act in this Country without Rangers." Rangers—that is, men who "ranged" between distant outposts—were colonists who knew the terrain and had learned the Indian methods of fighting. Unknowingly, Loudoun was also echoing Captain John Smith, who had taught the first English colonists to crouch behind trees while they fired their muskets at massed Indians in ceremonial costume and parade. The colonists taught guerrilla warfare to the Indians; now the British had to learn it from them.

In addition to failing to appreciate the Indians, the British held the American colonists in contempt, treating them like London "poore folke." British generals sent military task forces into American towns to "impress" citizens into the army. In one raid in New York City in the spring of 1757, the press gang was made up of three battalions—a large part of the British forces in North America—who rounded up about one of every four adult males they encountered. Poorly armed and unorganized, the colonists were unable to stand against such displays of military might. Soon, however, they hit upon what would later become one of their most potent weapons: taking the officers and men of the army to American civil courts, where they charged them with trespass and other crimes. Whether confronted in the streets with bricks and bullets or in the courts with writs and hostile juries, the British found that their recruitment of Americans virtually dried up: from 1755, when enthusiastic militiamen had flocked to the colors, enlistment fell to only 670 in 1762. Faced with this failure to win colonial support, Britain was forced to send regulars out from England.

One man who actively supported the British was the then twenty-three-year-old George Washington who commanded the 700 Virginia militiamen. He was furious, however, when he discovered what he regarded as an insult—that colonial officers of senior rank were regarded as inferior to

British officers of junior rank and that those of equal rank were paid less. Forthwith, he resigned his militia commission. Virginia's royal governor, Robert Dinwiddie, lured him back in the time-honored way, by giving him a promotion. Perhaps grudgingly, Washington returned, because he apparently hoped that General Braddock, with whom he thought he had a supportive relationship, would intercede to get him a commission in the regular British army. (How ironic for the course of American history that "career decision" would have been!) A vain man for whom honor was supremely important, Washington never got over the British affront to his dignity. Nor was his experience unusual. During the same war, the officers of the Massachusetts provincial forces recorded daily insults in their diaries and letters. Veterans and relatives of the citizen-soldiers whom Braddock had despised would, a decade later, evolve into the American army.

The Indians also were affronted by British hauteur. When Braddock's would-be Indian allies asked what his intentions were after he had driven out the French, "Braddock summoned all his considerable reserves of arrogance, and replied, 'that the English Shou[l]d Inhabit and Inherit the Land. . . . No Savage Should Inherit the Land.' " Hoping to change his mind, the Indians arranged another meeting at which they warned that "if they might not have Liberty To Live on the Land they would not Fight for it, To which Genl Braddock answered that he did not need their Help and had no doubt of driveing the French and their Indians away." The result of this exchange was soon evident: in the next engagement, the French were assisted by 637 Indians while Braddock could muster only eight.

Despite having competent agents who worked among the Indians, the British were rarely able to win their support. Perhaps the British were inhibited by their recognition that it made no sense for the Indians to help them. As Samuel Johnson pithily commented, the British and the French had been "endeavouring the destruction of the other by the help of the *Indians,* whose interest it is that both should be destroyed." The accepted British strategy was what their agent among the northern Indians, Sir William Johnson, proposed: to "raise up Jealousies of each other and kindle those Suspicions So natural to every Indian, and which it's now our Business to encourage, and foment as much as possible." In short, divide and conquer.

A policy of destroying the Indians pleased the colonists, but they would have been delighted as well to see the British army mauled. Their only regret was that the British, French, and Indians could not all be defeated. The independent spirit that led to the Revolution was already beginning to find expression. As the colonists watched the British army at closer range than ever before, they became increasingly hostile to the very idea of standing armies. So visceral was this feeling eventually to become that, even in the darkest days of the Revolution and against all logic, they were to hamper the growth of their own "regular" army, the Continental Line. For them, the militia, the very group the British so despised, was the only acceptable form of military organization. It was the "nation in arms."

On behalf of the colonists, during the decade following Braddock's defeat, Benjamin Franklin urged on the British government Locke's proposal for reliance on the militia. His aim was political, to avoid antagonizing the public, but he couched it in military terms: European armies simply did not work in America. His opinion was not widely shared. George Washington, for whom Braddock's defeat was a particularly painful memory, drew exactly the opposite lesson from it: Washington concluded that American soldiers must dress, drill, and fight in the European manner. For him, the British army, which he had wanted to join, was the pinnacle of the military profession. Later, during the Revolutionary War, when he had time to devote to the internal management of the fledgling American army, particularly in the winter encampment at Valley Forge, he attempted to recast it in the British mold. Using the British Army's code and assisted by his German drillmaster, General Friedrich von Steuben, he tried, literally, to whip American farmers into British regulars. The popular myth that in the Revolution sturdy American farmer-soldiers, the militia, took down their trusty long rifles and fought in the Indian manner against the foolish redcoats is just that, a myth.

It was a myth George Washington did not relish. In his letters, Washington excoriated the "nation in arms," the militia. He almost echoed the British officers whose company he wished to join. One of Braddock's colleagues, General James Wolfe, the conqueror of Quebec, had said that American soldiers were "the dirtiest most contemptible cowardly dogs that you can conceive. There is no depending on them in action. They fall down dead in their own dirt and desert by battalions, officers and all." Another British officer, General James Murray, thought he knew why. As he put it, "the native American is a very effeminate thing, very unfit for and very impatient of war."

Impatient of war the part-time American soldiers certainly were: they wanted to go home to tend their new farms or work at their trades. In part, this was because they or their parents had already won a "war." Having killed off or pushed the Indians to the west, they mostly lived in areas where shots were rarely fired in anger and never at an alien. The frontier, which in earlier years had occasioned the formation of the militias and fostered the martial arts, had moved far from the population centers along the coast. For most Americans by 1760, firearms were a distant memory.

Firearms were also expensive. A used but reasonably serviceable musket cost roughly two weeks' wages for a laborer, or the equivalent of enough wood or coal to heat a house for the winter. The choice was easy: it was better to be warm. In all the colonies, there were only three arms factories. By the generation before the American Revolution, most Americans no longer owned or even knew how to use guns. In the French and Indian War, the British actually had to teach militiamen to fire muskets. Locke's argument in favor of a colonial militia had been right in its time, but that time had long since passed. The ability to make war and the spirit to do it, those British officers who knew the Americans best were convinced, no longer existed. Americans wanted to make tobacco, not war. As the Virginian patriot Arthur Lee would lament in the years just before the Revolution, the colonials were "Unarm'd already and undisciplined, with our Militia laws contemned, neglected or perverted."

Yet the British enlisted about 11,000 Americans in the French and Indian War. This was because, relative to the cost of hiring troops from one of the petty German principalities or bringing over English soldiers, Americans were cheap. Cheapness was important. Training men to maneuver in the elaborate drill used by the British was nearly a lifetime task and was so expensive that their army was always small. It rarely numbered as many as 40,000 men. In Europe, wherever feasible, generals avoided combat; and when combat was unavoidable they did everything possible to avoid casualties. Even if he won his battle, a commander who suffered significant casualties had blighted his career. "Avoiding loss" was the fundamental definition of success.

Success in operations of the army often depended upon the fleet. The Royal Navy was, in the British expression, the "senior service." Its ability to move regular army units from place to place enabled even small forces to achieve "theater superiority" and so to win battles against enemies with vastly superior total forces. But the fleet was ruinously expensive. About 1750, including service on military-related debt, it took up over 75 percent of government expenditure. In 1763, the British fleet was composed of some 270 ships of all classes, manned by some 76,000 men; during the period of the Revolutionary War, the numbers would rise to 430 ships with 107,000 men. That amounted to a "floating population" larger than any English city except London. During the Seven Years' War, a total of nearly 150,000 died or were listed as missing. Such casualties were a severe drain on the English population and economy.

Casualties were not the only drain. Standing behind the men at sea were upwards of 3,000 shipwrights who, collectively, constituted England's "military-industrial complex." Using about 22,000 tons of oak and elm each year, Navy dockyards were deforesting England. So highly regarded were the fast, seaworthy, inexpensive American ships that particularly as Britain began to run short of native timber, it imported them in large quantities. In the decade before the Revolution, the American colonies produced at least 40,000 tons of shipping yearly, of which nearly half was exported to England.

Britain needed both American ships and American shipping. During the years before the Revolution, in the quest for individual security, American merchant ships had become a formidable force. While they did not include any "ships of the line," comparable to the floating fortresses of the European navies, they were numerous. By 1775, some 2,000 privately owned ships, one-fifth as many as Britain then had, carrying about 18,000 cannons, were being sailed by some 70,000 Americans on the Atlantic and were finding their way around the globe. Although they were essentially merchants or fishermen, they did not limit themselves to peaceful tasks. When one encountered a foreign vessel, the American often tried to take it. From such profitable and exciting experience grew the seaborne counterpart of the land-based militia.

The British found this seafaring tradition useful; with British sanction, American privateers took part in the British attacks on French-held Quebec in 1690 and 1745. Colonists' ships, Benjamin Franklin commented, made up "A Naval Force, equal (some say) to that of the Crown of *Great Britain* in the Time of Queen *Elizabeth*." During the French and Indian War, hundreds of letters of marque, which turned part-time pirates into privateers, created a second, "virtual," Royal Navy. It was the training ground for the men who in the Revolution would harass the British merchant marine and form the nucleus of the navies of the separate colonies and the Continental Congress.

Ominously, this numerous collection of ships also held an assemblage of sailors who could be "impressed" or kidnapped, when the ships of the Royal Navy fell short of men. Impressment, which was not supposed to be practiced in American waters after 1713, occasioned several riots—in one of which the enraged citizens of Newport opened fire with cannon on a ship of the Royal Navy and released four impressed fishermen.

Meanwhile, Braddock's humiliating defeat had unleashed attacks by the Delawares, the Illinois, and other northwestern Indian societies. Even after the French were defeated, the Illinois, the Shawnees, and remnants of the older societies continued to resist. Occasionally encouraged by French officers disguised as Indians (as in October 1762 at Detroit), the Indians still hoped that the French would return to create a counterpoise to the British. But this did not happen. For the French, North America was an unprofitable sideshow—and without the French, the Indian strategy fell apart. Elated, the British commander in chief, General Jeffrey Amherst, echoed almost word for word General Braddock's opinion of the Indians' military irrelevance. No longer worried about the French, Amherst saw no reason to cater to Indians. Good auxiliaries they may have been, but absent the French, they were auxiliaries to nothing. So, to save money, he cut down on gift and tribute goods—arms, ammunition and clothing—which the Indians regarded not as gifts in the English sense but as payment for the land on which the British had built forts. The reduction in gifts, they felt, was unfair; worse was the British decision no longer to supply them with muskets, a decision the Indian leaders viewed as preparation for a "final solution" to the "Indian problem."

Fearing annihilation, the Indians, although scattered and mutually hostile, made one last, dramatic, but catastrophic effort to shake themselves free. That great upheaval is known as Pontiac's Rebellion but might better be described as an Indian "great awakening." In the East, against the British, it was inspired by the Delaware prophet Neolin, the Ottawa war chief Pontiac, and other Indian leaders. Not all Indians responded, but Ottawas, Potawatomies, Hurons, Chippewas, Shawnees, Delawares, and dozens of smaller groups did. Beginning with an attack on Detroit on May 9, 1763, the Indians captured ten of the eleven British forts that dotted the formerly French-held Ohio territory (today divided among New York, Pennsylvania, West Virginia, Kentucky, Indiana, and Illinois). They killed about 2,000 settlers, drove away thousands more, and killed some 400 British troops. Combat was literally "war to the knife." General Amherst ordered that all Indian prisoners "immediately be put to death." He also instructed his field commander to distribute smallpox-infected blankets, "as well as to try every other method that can serve to extirpate this execrable race." Indians sickened and died by the thousands. But genocide through biological warfare was insufficient to pacify the frontier. Jon Butler has commented, "It was only ironic, then, that achieving an imperial victory in the Seven Years' War quietly undermined the very relationship that the war had been designed to preserve." Having realized that they could not kill all the Indians, the British attempted to cut their losses with the Indians at the expense of the settlers. To that effort, I now turn.

CHAPTER 13

From International War to Colonial War

Allowing the Indians somewhere to live seemed to the British government merely prudent. Invaded, Indians would have to fight, and Britain would have to pay most of the cost of warfare. But in 1763, peace was not at hand. The treaty that had been signed five months earlier in Paris to end the war between England and France had not brought peace to the American frontier; rather, it had made the Indians desperate. Although their strength had declined, those living north of the Ohio River were still militarily significant. They would become formidable if they formed alliances with the 30,000 Creeks, Chickasaws, Choctaws, and Cherokees to the west and south and the numerous, warlike Iroquois to the north and east. Thus pacifying them was a major objective of British policy.

Finding a way to accomplish this seemed urgent to the British government. Britain had come out of the Seven Years' War with a vastly increased public debt and was determined to cut expenses, but it obviously could not reduce spending if it had to keep a large military establishment on the American frontier. So it reined in its aggressive military commanders, and, while it began to consider measures to increase revenue, it sought to prevent further incitement of the Indians. To this end a royal proclamation drew a "line" on the crest of the Allegheny mountains beyond which settlers and speculators would not be permitted. The British government went even further: after admitting that "great frauds and abuses" had been perpetrated on the Indians, it ordered squatters off Indian lands.

Underlying the proclamation were new attitudes on the part of the British and Americans toward one another. By 1763, the British "mother" had come not to trust, or even to like, her colonial daughters: they did little to help her while she spent much to protect them. Colonial and imperial interests, never fully compatible, had increasingly diverged; as they did, the Americans evaded British law, pursued their own objectives, and sometimes even collaborated with Britain's enemies. Now, it was evident that the Americans' unbridled lust for land was costing British lives and money. To maintain minimal order, the government had stationed about a third of its effective army, fifteen battalions, in North America at a cost of about 5 percent of its annual budget.

The colonists' view was rather different. They claimed that their own daring, labor, and money had created the American part of Britain's empire; Britain had done little to help them but had grown increasingly restrictive as they began to prosper. Now Britain proposed to prevent them from taking possession of what they had always regarded as prospectively theirs, the "empty" interior of America. To them, Britain's plan to deny them their rights was unforgivable.

Colonists neither forgave nor obeyed. Several thousand settlers were already, by some definitions even legally, established beyond the interdiction line of the proclamation. Well before Indian rights were "extinguished," Pennsylvania had authorized some private citizens to issue what amounted to titles, known as "Bulunston's licenses" for unregistered and unsurveyed lands. With or without such documents, hundreds of families were seeping across the frontier as inexorably as water flows downhill. Land was what they were after, and the only place to get it was in the forbidden zone. Unless they were driven away by British troops, as a few actually were, they kept on crossing the mountains, building their huts, cutting down trees, and taking land by "tomahawk right." They did not respect Indians' rights and had little or nothing with which to "buy" land. Furthermore, even if they had been willing and able to buy land, the Indians could not have sold it, since land ownership was communal. Even chiefs, who the settlers thought were like territorial monarchs, had no authority to dispose of tribal land. To get land, settlers had to drive off the Indians; to keep it, they had to defend it. They were free to use lethal force, since no white jury would condemn a white man for murdering an Indian.

Not only the homesteaders disobeyed the royal proclamation. Merchants, plantation owners, members of the assemblies, and even British officials were funding, inciting, and leading the charge onto Indian lands. Among the colonies, Virginia took the lead. After watching the process for nearly a decade, the British agent for Southern Indian Affairs, John Stuart, commented in 1770, "Virginians are insatiable. . . . In short they want all the Cherokee hunting grounds." That desire was not new—it was merely a westward extension of the process begun at Jamestown in 1607.

What was new, at least in emphasis, was the involvement of royal officials. After 1763, most of the royal governors of the colonies along with key British officials—such as Sir William Johnson, the superintendent of Indian affairs, and his deputy, George Croghan, the very men charged with defending the Indians—were getting rich by acquiring vast stretches of Indian lands. For personal gain, Johnson gave his government misleading intelligence appraisals, violated explicit orders, and actually redrew the lines laid down in the royal proclamation. And even the new commander in chief of the British forces, General Thomas Gage, turned his hand discreetly to illegal land-grabbing.

Illegal these actions were, but in those times it was expected that officials would use their position to build a personal fortune. The British system, from top to bottom, was a dynamic balance in which the pursuit of private gain was somehow made to serve public purposes and public service was expected to feather the nests of the ruling aristocracy. The king himself sold commissions in the army; and one of his appointees, the First Lord of the Admiralty, sold commissions in the navy. Decorations for meritorious service had to be paid for. At each stage, everyone had his hand in someone else's pocket. Grabbing Indian land hardly raised an eyebrow.

So it was that the Americans who had their eyes on the western lands just followed the parade. Members of the legislatures, merchants, explorers, Indian traders, surveyors, and militia leaders found common cause with British officials. Helped and emboldened by them, and aided by their agents in London, speculators enlarged their own circles to include members of the court, the Privy Council, and Parliament. All along this golden road to wealth, the common objective was land. The Creeks summed it up well: their name for the colonists they knew was *ecunnaunuxulgee*—"people greedy for land."

Greed for land had begun at Jamestown; active solicitation for land grants under a sort of partnership that was to grow into the Ohio Company of Virginia began in 1747, well before the French and Indian War. The lieutenant-governor of Virginia at this time, Sir William Gooch, reported that he had been "lately much solicited" by unnamed individuals who sought "grants for lands lying on the western side of the great mountains." (That land, claimed by Virginia, comprised substantially all of the modern states of West Virginia, Ohio, Indiana, Illinois, and Kentucky, with parts of what is now western Pennsylvania.) Already living there were roughly 20,800 settlers. Those settlers were not, however, the people for whom Gooch was being importuned. They were squatters—farmers without the means to buy land—whereas those who sought Gooch out wished to acquire land to sell. Gooch was on their side.

Gooch argued that for strategic reasons, Britain should open up this vast territory to settlement. The French, then still firmly planted in Canada, already had extensive connections with the native inhabitants through missionaries and traders and were beginning to build a string of forts to bridge the gap between their Canadian colony and their Mississippi plantation. Unless this French intrusion was halted, Gooch asserted, it would block British expansion.

Expansion would also be beneficial commercially, because of the "skin trade" with the Indians. Not only were furs valuable in themselves, but exchanging them for English products helped to stimulate English manufacturing and shipping. By encouraging settlement, English and colonial "adventurers" would attach this valuable area to the empire. What Gooch chose to overlook was that white settlement would destroy the fur trade, make forts more necessary, and probably force Britain to send expensive military forces to protect the frontier, since the besieged Indians would try to defend their lands. It might also precipitate a war with France. It did. It was a major cause of the French and Indian War.

When the English or colonists made some pretence at accommodating the Indians, transfer was often effected by fraud: a sale was disguised as a trading contract or was couched in "lawyers' language," which the Indians could not possibly have understood. Often governors simply assigned deeds to vaguely identified areas. During his tenure in Virginia, Robert

Dinwiddie gave grants of tens of thousands of acres to his friends. But title to such assignments could not be "proved" without the concurrence of the government in London. To get its agreement, Dinwiddie compiled a revealing roster of officials he had suborned. It included nearly the whole government. On March 16, 1749, his campaign paid off: the government approved the grant of a total of 500,000 acres, almost 800 square miles, of land south from the modern city of Pittsburgh, "within the colony of Virginia," to a group of mostly unnamed "associates."

The Indians living in the Ohio valley—the Mingoes, Shawnees, and Delawares—could not know what was happening in London or Richmond. But they did know that, already weakened by disease, liquor, and warfare, they lacked the force to stop the rush onto their lands. They concluded that their best strategy was diplomacy. Hoping that where the tomahawk had failed, the peace pipe might succeed, they sent a mission to the frontier town of Carlisle in September 1753 to meet with a Pennsylvanian delegation (in which Benjamin Franklin, then a member of the Pennsylvania Assembly, participated). The delegates ranged over the whole scope of Indian affairs—how many trading posts could be established, how much rum could be sold, and how gouging by Pennsylvania's traders might be restrained. These were all important matters, but Pennsylvania was not the leading actor on the white side; Virginia's aggressive governor, Robert Dinwiddie, played the major role, with young George Washington as his agent.

It is during a mission into the western "wilderness" in pursuit of the joint objectives of the colony and company that Washington first appears on the stage of history. In 1753, at age twenty-one, Major Washington was sent by Dinwiddie to meet "the Half-King [or viceroy] . . . Monaastoocha, and other Sachems [Algonquian, "chiefs"] of the Six Nations" and to urge the French to quit the Ohio River valley. The principal Indian with whom he met, Tanaghrisson, dismissed Washington as "a good-natured man [who] had no experience." The French also were unimpressed. They misread both Washington's character and his definition of diplomacy. As his diary testifies, diplomacy for him was another word for acquisition of land. When he reached the first French outpost, he commented, "We passed over much good land since we left Venango, and through several extensive and very rich meadows, one of which, I believe, was nearly four miles in length, and considerably wide in some places."

Excited by Washington's report, Governor Dinwiddie decided to send a militia force to protect the claim of the Ohio Company of Virginia against the French. Since he had to rely upon militia forces, Dinwiddie went before the Virginia House of Burgesses to paint a lurid (and inaccurate) picture of French-inspired Indian rapine: "You see the infant torn from the unavailing struggles of the distracted mother, the daughters ravished before the eyes of their wretched parents; and then, with cruelty and insult, butchered and scalped." In his purple prose, patriotism and humanity became the justifications for a push westward. More practically, those burgesses who agreed were promised portions of some "200,000 acres of his majesty's lands on the Ohio."

Leading the charge, Washington set out with a motley force of aged men and indigents, burdened by many women and children and lacking suitable provisions, blankets, and tents. When his little army got to within about 75 miles of an outpost, known later as Fort Duquesne and still later as Fort Pitt, he learned that the garrison had already surrendered to the French. After some delays and indecision, Washington moved ahead. Toward disaster. Following an initial skirmish on May 28, he too was forced to surrender. This engagement has been described as the first shot in the French and Indian War.

That war soon moved into a new phase: the British determined to use their regular army to drive the French out of the Ohio Company's claim and sent Major General Edward Braddock to do the job. As I recounted in Chapter 12, that job was beyond his capacity; but over the next several years the British finally prevailed, so in the peace treaty of 1763 the vast north-central hinterland became "British."

Under the British flag, the colonists planned to move ahead. Moving ahead into Indian lands would be a major interest of George Washington and his Virginian neighbors and relatives until the outbreak of the Revolution twenty years later. The organization they chose was by then a traditional model, a joint-stock company. Grouped under its legal umbrella, investors and speculators cast their eyes on, and set other men's feet along the way toward, the western wilderness. In principle, the British government should have welcomed their venture; after all, the government would find it easier to work with a coherent organization on an identified piece of land than to try to police the movements of scattered groups of poor, usually illiterate farmers. A large-scale organization presumably also could defend itself, whereas isolated clusters of settlers were likely to provoke an Indian attack. This argument was put forward by the promoters. In London, the government was not so sure. It shared the objective of taking Indian lands; but as the proclamation of 1763 made clear, it feared that a very aggressive invasion would meet with a costly response from the Indians. Argument on these issues, begun in that year, would run back and forth across the Atlantic for a decade. The last move came in 1774, on the eve of the Revolution, when the British government made a vain effort to void land grants in the West, thereby infuriating the growing population of frontiersmen and helping to precipitate the Revolutionary War.

In those years, failure to make peace was also expensive in terms of men and money. Policy, or the lack of it, was unpopular among the constituency that mattered to the British government—English landowners. They were tired of high wartime taxes, which went beyond the 10 percent they had lived with since 1692; and they demanded a cut in costs. But expense was not the only problem. The American colonists were not accustomed to British soldiers; even in the recent French and Indian War, action against the French usually took place far from the American colonial settlements. Opposition to a standing army, even one ostensibly sent to protect them, would rally those who were coming to see "mother England" as a tyrant. The year 1763 thus marked a turning point not only in the organization of Britain's North American empire and its policy toward Native Americans but also in the attitude of the Americans toward the British.

That year was, of course, just one stage in an ongoing process that was already in its second century. In the years before the outbreak of the Revolutionary War, thousands of men, women, and children trekked across the vaguely drawn frontier into Indian territory. Most were unrecorded individuals, but one large group emerges clearly enough to be described, the people who became known as the Regulators of the Carolinas.

By the middle of the eighteenth century, North Carolina, like Virginia, Maryland, and Pennsylvania, had become divided into groups with often conflicting interests. Along the coastal plain, a plantation economy was already well advanced. The families who dominated it were not so numerous as their counterparts in Maryland and Virginia, but they were prosperous, well-organized, and in control of the newly established legislature. Very different were the widely scattered families on the still sparsely settled western frontier. They were only beginning to recognize themselves as a distinct group in the late 1750s and to organize to protect their interests in the 1760s. Moving between the coastal elite and the frontier settlers was a small but powerful group of speculators.

Using their influence with the government in London, and using their money to bribe local officials, the speculators crossed the interests of the existing coastal establishment (which thought of itself as a sort of extended family) and the frontiersmen in their attempt to obtain vast stretches of land, which they wished to sell to settlers who were desperate for land but were unwilling (even when able) to pay unknown "foreign" speculators for it.

These different ambitions and perceptions opened wide the division between the coast and the interior during the middle of the eighteenth century. Increasingly, the people of the interior felt themselves to be deprived, exploited, and excluded; the more affluent people on the coast, both the establishment and the speculators, thought of the settlers as uncouth, unauthorized, and uncivilized—indeed, as "white Indians."

These "white Indians" had migrated to the interior of North Carolina from Pennsylvania, Virginia, Maryland, and various parts of Europe to hack out farms. In their eyes, their labor—chopping down trees, hoeing and plowing the land, planting and harvesting—and of course their fights with the Indians had turned their "tomahawk rights" into actual rights.

With their eyes fixed firmly on local concerns, the settlers were not much distracted by the French and Indian War. Left more or less alone during the six years of the war, they solidified their position; indeed, the war worked to their benefit by preventing speculators from contesting their assertions of ownership. Those who did affect—and infuriate—them were the sheriffs and other royal (or coastal government) officials who imposed fines, demanded high fees for services, and levied taxes in British currency, which was then almost unobtainable. When the settlers could not pay, in what today we call "hard" as distinct from "soft" currency, their newly built houses and newly plowed land were sold, often to friends or relatives of the sheriffs. The settlers' anger grew until, in 1759, a group armed themselves and seized a sheriff they regarded as a bandit.

Over the next few years, such incidents multiplied; but they produced no general reaction, because the settlers were few and widely scattered. Then, as the French and Indian War ended in 1763, hundreds of new "white Indian" families arrived. Rather suddenly, the frontier had a "society" that was prepared to defend itself. To families that were struggling to tame the land and get enough to eat, intrusion by government officials or speculators' agents was a burden too heavy to tolerate. And, growing more numerous, they were less willing to tolerate it. So when speculators began to arrive with papers from the government to assert ownership of land cleared by the settlers, the settlers chased them away. The speculators, who had organized, paid, and planned for years to get those lands, would not be put off. Both sides armed themselves.

A clash probably was inevitable, but it was delayed for a few years because North Carolina was the fastest-growing of the colonies. Not only was the population doubling each generation, but exports of agricultural products were rapidly rising. In the process, a group of relatively prosperous people who have been called "improving landlords" came into being on the the "nearer frontier." We can monitor their growth in the 1760s. Increasingly, they were finding their way into the legislature, the militia, and various local assemblies. Having become more established, more prosperous, and more convinced of their opportunities, they came to oppose the wilder, more recent settlers farther west. Between 1765 and 1771 prosperity converted many squatters into settlers. They began to think that perhaps government could find a means to redress their grievances. But, of course, not all benefited.

One of those who had not benefited was Hermon Husband, who in 1766 organized those who were still squatters into a loose association, known as the "Regulation." The Regulation, counted in terms of the North Carolinian society of that period, soon became virtually a mass movement. At its peak, it could rely on the support or active participation of perhaps as many as 6,000 of North Carolina's 8,000 taxpayers. Using vigorous propaganda, Husband and his colleagues campaigned for control of the state assembly. They got thirty-eight of their members elected but were thwarted by rigged elections, so they never managed to elect more than a minority of the eighty-four delegates. Still, encouraged by the results, they began to bombard the royal governor with petitions demanding legislative remedies for their grievances. Some relatively prosperous settlers joined the movement, which became more conservative as many preferred to work within the system rather than try to destroy it. Getting elected to the colonial assembly was the preferred route.

Exposed in the assembly to currents of thought from other colonies, the new members merged the demands of the Regulators into the more general colonial opposition to such British policies as the Stamp Act. Thus North Carolina's legislature passed essentially the same act as the better-known Virginia House of Burgesses against British impositions. This vote caused the royal governor to dissolve the assembly. Infuriated, the Regulators decided to harden their resistance and refuse to pay taxes. As negotiations broke down, the governor had the Regulators' leader, Hermon Husband, arrested, tried, and imprisoned. These moves were taken as virtually an act of war. Some 700 people, of whom only a part were Regulators, broke into the jail and released the prisoners. The governor, William Tryon, could not ignore this action, but he at first tried to calm the situation. Tryon—an able and experienced man who would later become governor of New York and then a British general in the Revolution—proceeded with discretion. He promised to investigate the Regulators' grievances if they disbanded. Privately, he admitted the justice of their complaint, saying, "The sheriffs have embezzled more than one-half of the public money ordered to be raised and collected by them." But, as he wrote to his superiors in London—and this was still six years before the Revolution—he discerned "a republican spirit of Independency rising in this Colony."

He was right. On August 12, 1770, 500 men gathered, giving the governor's undercover agents the impression that they would march on the capital and, if the governor did not meet their demands, burn it. Acting on this intelligence, the governor called up the militia. He also persuaded the legislature on January 15, 1771, to pass a "state of siege" act which, for the first time in American history, authorized an attorney general to proscribe anyone who refused to appear in court.

By that time, the frontiersmen's community of interest had cracked; not only the governor but also the increasingly prosperous men of the inner frontier had become frightened by the breakdown of law and order. Moreover, they found that they could settle their complaints legally. Issues of ownership began to be resolved in a sort of tide running westward. As each area found its grievances at least partly settled, it became concerned that further unrest would damage local interests; then man after man, group after group, shifted away from the Regulation.

So when the governor called out the colony's militia, prudently encouraging loyalty by offering a bounty of 2 pounds—then about the cost of 100 acres of land—over 1,000 militiamen appeared. With them, Tryon marched to confront some 2,000 Regulators at Alamance Creek in western North Carolina. By colonial standards, or even by the standards of the Revolutionary War, the ensuing battle was massive. It was the greatest battle ever to take place solely among colonial Americans. But its outcome was never in doubt: the Regulators had few muskets, no discipline, no commander, and—perhaps most important—no artillery.

With his overwhelming advantage, the governor offered the Regulators a stark choice: disperse or be shot down. The Regulators refused to give up. Disorganized as they were, they still fought desperately, but in two hours they were defeated. Most of them then fled, but Tryon's forces managed to seize about fifteen men, of whom he immediately executed one and placed the others on trial. An additional six were then executed. With that example established, he then offered pardons to all who would submit. Eventually, some 6,000 signed oaths "Never to bear arms against the King, but to take up arms for him, if called upon." All but Hermon Husband, who had led the movement toward independence, were forgiven. Four years later, many of the men who had fought at Alamance took up arms again in the larger cause. Thus, two wars, the French and Indian War and the War of the Regulation, helped to set the stage for the Revolutionary War.

CHAPTER 14

Production and Commerce

How the colonists earned a living; what resources they learned to exploit; how they organized themselves to manufacture what they needed; and how they interacted with, were thwarted by, and learned to evade the restrictions of the mother country—all these factors shaped the America that made the Revolution.

When attempts at colonization began during the reign of Elizabeth I, would-be colonists were tempted by Richard Hakluyt, the patron saint of English colonization, with reports that the New World rivaled the garden of Eden: trees dripping with luscious "fruites very excellent good"; woods filled with edible animals; waters teeming with fish, "the best of the world." "The soile is the most plentifull, sweete, fruitfull and wholsome of all the worlde. . . . ourselves prooved the soile, and put some of Pease in the ground, and in tenne dayes they were of fourteene ynches high." In this Eden, with little effort a family could presumably lead a life beyond the dreams of any but the most wealthy inhabitants of the British Isles. Disillusionment came soon and painfully.

During the first few years of their residence in Virginia, the original settlers struggled desperately just to stay alive. With little support from their patrons, they had to cope or die. Coping was not easy in Virginia. After most of the original group died of starvation, those still alive were saved by two revolutionary changes.

The first change was in the relationship of labor to land. The settlers began with collective farming, but that obviously was not working, so when a new governor arrived in "James Towne" in 1613 he began assigning each settler a small plot of private land on which he could grow at least his own food. This incentive produced such a dramatic improvement not only in production but in morale that a program of granting 50-acre private plots was begun. It too worked well; and despite their opposition to such "particularism," even the Puritans of New England were forced to adopt it. Shortly thereafter it grew into the "head right" system that set the relationship between immigration, labor, and land in most of the colonies, including Dutch New Netherland, for the next century.

The second revolutionary change was the adoption and commercial refinement of one of the Indians' crops, tobacco. In 1612, John Rolfe, best known as the husband of Pocahontas, learned by reports from Barbados how to cure tobacco in a style that made it salable to Europeans and so gave the little colony the basis for its growth and enrichment. From that time onward, the whole of the Chesapeake region realized the dream of all developing economies: having an export crop that earned hard currency. Tobacco was the salvation of Virginia.

No such salvation was in sight for the colonists sent north by the Virginia Company's sister, the Plymouth Company. It tried to establish a base on the Kennebec River in what is now Maine. The original intent was to establish not a "colony" on the Irish or Virginian model but a "factory" on the model of the contemporary and highly profitable Levant and East India companies, wherein factors (merchants) could trade with natives. That proved impractical. The local Indians had little of value except furs, which did not become commercially significant in England until some years later, in the 1630s. The company's hopes withered. To encourage English investors to keep pumping in funds, some new incentive was required. So, in 1619, a "Councill, [was] established at Plymouth, in the County of Devon, for the planting, ruling, ordering, and governing of New England, in America" and was given a patent for all the lands between the site of modern Philadelphia and the Gaspé Peninsula and from "sea to sea." To sweeten the grant, the company was awarded a monopoly of offshore fishing.

Fishing seemed promising, since Francis Higginson in 1630 found "The abundance of Sea-Fish are almost beyond beleeuing." Lobsters grew to 20 or more pounds and were so numerous that an Indian was observed to have caught thirty in an hour and a half. Oysters were to be found in profusion and were so large that they had to be cut in several pieces before they could be swallowed.

On land, the small group of "New Englanders" faced a much less inviting prospect. As has been wryly pointed out, "the fields of New England produced grain and stones, the latter more easily and abundantly than the former." Not only was the land stony, but the growing season was short, the winters were harsh, and there was little market in England for the only suitable crop—grain—because England also was a major grain producer and was seeking markets rather than sources. It regarded colonial ventures in grain production as dangerous competition. So at least one adult male colonist in ten worked in some capacity at fishing.

Fishing could hardly then be called an "industry," but it did promote a variety of activities. Fish had to be processed, salted, and sold, and first they had to be caught. To catch fish, men needed boats; so from hammering together small skiffs, craftsmen progressively graduated to more sophisticated designs like the "gondolos" and galleys that acted as peddlers' carts, taxis, and ferries on the waterways. Step by step, craftsmen, merchants, and sailors became more ambitious, building shallops and coastal sloops, then copying British designs for larger ships, and finally designing their own ships. Their progress can be measured by the few ships for which records remain: from the primitive *Virginia* in 1607, a more sophisticated 30-tonner, *Blessing of the Bay,* was built at Medford in 1631; in 1636, craftsmen in Medford essayed the 120-ton *Desire.* So successful was it that in 1641, carpenters at Salem launched the 300-ton *Mary Ann.* Two years later, Bostonian shipwrights began to turn out a small fleet. According to local tradition, the first schooner built in North America came off the rails in 1713 at Gloucester. It was followed by so many more that by the time of the Revolution, travelers mentioned seeing "forests" of masts of American-built ships in the harbors of Boston, New York, and Philadelphia. During the middle of the eighteenth century, New England yards alone were producing about 40,000 tons of ships, worth the huge sum of 300,000 pounds. Nearly half of these ships were sold abroad, mostly to England. Fishing was the stimulus for shipbuilding and was the forerunner of for seaborne commerce, whaling, piracy, and slaving.

Whaling gives us a precise date and a clear motive for the transformation. The Boston *News Letter* of March 20, 1727, reported:

> We hear from towns on the Cape that the whole fishery among them has failed much this winter, as it has done for several winters past [since whales no longer come into harbors], but having found out the way of going to sea upon that business . . . they are now fitting out several vessels to sail with all expedition upon that dangerous design this spring.

Piracy was even more alluring to New England sailors. One captain, William Phips, lived a typical success story, the prototype for what later Americans would see as a Horatio Alger. He began his career as a ship's carpenter. Then, using the capital he accumulated by marrying a rich widow, he bought his first boat. In it, he attacked a Spanish galleon filled with treasure with which he was able to buy a knighthood from James II. Subsequently he became the first royal governor of Massachusetts.

What ventures at sea offered to New England, and tobacco provided to the Chesapeake, the Carolinas reaped from rice. *Oryza sativa,* known as "Carolina gold," was imported from Madagascar; *Oryza glaberrima,* a red rice, was probably imported from West Africa, along with black slaves who knew how to grow it efficiently. Rice, marvelously prolific, proved an ideal crop for the steamy, wet coastal lowlands. A unit of 4 acres, the area that could be tended by one black slave, could produce yearly as much as 1 ton of rice, which in a good market was worth about 41 pounds sterling, just about the purchase price of a slave. Thus rice made slavery extremely profitable. But the British government discouraged exports by requiring ships to transit England to sell to Europe. Producers always had to juggle their prices and costs carefully while they lobbied to get a change in government policy or smuggled their crop abroad. Not until 1730 were they permitted to sell directly to the Mediterranean countries; and for the bulk of their trade, which was in northern Europe, they were unable to shake loose from British restrictions.

Very different was the British attitude toward the other main crop produced in the southern colonies: the indigo plant, from which dye was made. When it was introduced into the Carolinas and Georgia from Antigua in the middle of the eighteenth century, the British government encouraged production, since indigo was useful to the British woolen industry and the Anglo-Irish linen industry. Awarded a subsidy, it was said to yield roughly 25 percent return on investment, and it flourished until the outbreak of the Revolution.

Tobacco, rice, and indigo required vast amounts of labor—which, despite putting all family members including small children to work, the original colonists did not have. So they encouraged the immigration of laborers from Europe. Some of the incoming laborers were free people who had paid their own way across the Atlantic; but the colonists—except in Puritan Massachusetts, which was very restrictive—took anyone they could get, including prisoners of war and felons who had been "transported" instead of being incarcerated or hanged. For these people the colonists used an adaptation of the traditional system of apprenticeship: indenture. The indentured laborer entered into an agreement to serve a patron for a given number of years. The servant gained by having the trip across the Atlantic paid for, and by receiving lodging, food, a small allowance, and often a sort of severance pay at the end of the contract. Many of these immigrants served in individual households; but, particularly in the seventeenth century, others did gang labor with axes, hoes, and sickles on plantations.

The colonists also forced Native Americans to work for them. When the colonists were at war, which was often, they captured Indians whom they either employed as slaves or sold to planters in the West Indies; when at peace, they often encouraged "inner" Indians to kidnap "outer" Indians for them. The whites did not regard this source of labor highly, because the Indians frequently ran away and were unaccustomed to the sort of labor associated with the plantation system. So the colonists turned eagerly to a third source, black slaves.

As I recounted in chapter 10, the first black slaves were brought to Jamestown by a Dutch ship in 1619. Thereafter, increasingly large numbers of blacks were brought from the West Indies or directly from Africa. The colonists found blacks more useful than the Indians because blacks could not so easily run away, were accustomed to farm labor in Atlantic Africa, and had more immunities to disease than the Indians. Soon blacks were the "motor" of the growing plantation economies. Also, some, particularly in the rice belt, were technicians, introducing irrigation practices that were common in the Bight of Biafra region of Africa. The white colonists had not known of these practices, on which their economy came to depend.

The colonial economies were growing so rapidly from about 1650 to 1750 that, despite the organization and exploitation of indentured servants, Indian and black slaves, and free settlers who arrived in increasing numbers, the price of labor rose steadily. By the 1770s, it reached about four times the going rate in England. What fueled this rise was not only agriculture but also the mutually reinforcing commercial and industrial enterprises based on tobacco and fish.

To survive, the colonists had to import from Britain or other parts of Europe what they could not make. Initially, this was almost everything. And to pay for these imports, they had to take from the land commodities that Europeans would buy. At first, all the exports were what are today called "primary products," particularly tobacco, rice, timber, fish, and furs. Also at first, these products were carried in British ships, which returned with manufactured goods such as cloth, shoes, tools, weapons, and various iron fittings.

Relying on Britain was expensive. To ship 1 ton of tobacco across the Atlantic cost roughly 18 to 20 percent of its gross sale price. Factors in London took large commissions, so that producers got only about one-third of the sale price. Worse, the goods the American colonists got in exchange for their tobacco were often what England could not sell on the open market. Iron implements, for example, were sometimes of such poor quality that they were virtually unusable.

As new immigrants with skills and money arrived, and as the earlier settlers also accumulated capital and proficiency, some craftsmen began to create what economists call an "import substitution" industry. At first it was on a small scale. One man, about whom we know because he became rich, got his start making buckles and buttons for his neighbors. Others built sheds, huts, and houses for which timber had to be cut and sawed, and rocks hauled and shaped. Still others hammered scrap iron into cooking pots, mended clothing or made clothes from imported cloth, and repaired tools and weapons. Piece by piece, new items were added—brick for fireplaces; glass for windows; paper for broadsheets; and—for release, warmth, and trade—rum. The basis of rum was molasses imported from the West Indies; to pay for it, the colonists had to find, produce, or fashion things that the molasses producers needed but could not economically make; collect those things in commercial quantities; convey them to the West Indies; and trade them. This process involved about two-thirds of New England's exports in the late eighteenth century and constituted America's original "business school."

At first, colonists did not have much to work with—just fish, grains, and wood products—but each step they took led to the next. For example, it was more profitable for them and more attractive to the West Indian molasses producers to transport barrel staves than raw timber, so some craftsmen found their way into carpentry and barrel making. To cite another example, this time dealing with the import side, they wanted to turn raw molasses into rum. That led to the building of distilleries. Distilleries required energy, for which the only available source was timber. One distillery in Alexandria, Virginia, burned enough wood each year to form a stack 4 feet high, 4 feet wide, and about half a mile long. Multiply that figure by seventy-three, the number of distilleries operating in the North American colonies by the middle of the eighteenth century—or by about 155, the number operating by 1770—to see what economists call the "multiplier effect" of subsidiary occupations required to produce rum.

Producing rum, of course, was only the first step. Next, it had to be packaged, carried, and sold or traded in identified markets. Much rum was used in the African slave trade and in the trade for furs among the Indians; disastrous as the effects were on the Indians and the blacks, they enormously stimulated the growth of commercial skills, organization, and wealth among the whites.

The British government was not pleased by the colonists' move into commerce and industry. Reflecting on policy toward colonialism in the sixteenth, seventeenth, and eighteenth centuries, the English economist and philosopher Adam Smith wrote, "Every European nation has endeavoured more or less to monopolize to itself the commerce of its colonies." In area after area, the British sought to push the colonists back to the production of only primary products—what Adam Smith called goods in the "rude state." As he pointed out,

> The most advanced or more refined manufactures even of the colony produce, the merchants and manufactures of Great Britain chuse to

> reserve to themselves, and have prevailed upon the legislature to prevent their establishment in the colonies, sometimes by high duties, and sometimes by absolute prohibitions.

American colonists soon began to struggle against these restraints. Assisted by the disruptions that affected Britain during the seventeenth century (the civil war, interregnum, Restoration, and Glorious Revolution) as well as by the primitive state of communications and the lack of sufficient military capacity, they had a good deal of success. The American historian Bernard Bailyn believes that the year 1643, when the British government was immobilized by the civil war, may be taken as the birth of independent commerce: in that year Americans sent five ships filled with cargoes for non-British destinations. Soon ships not only were making regular runs to the West Indies but also began to appear in ports all along the Atlantic and the Mediterranean. Following the routes that led to the products they wanted to buy and sell, the merchants and ships' crews fell naturally into a pattern that has become known as the "triangular trade."

At its simplest, the American version of the triangle was formed by the shipment of molasses and sugar from the West Indies to New England, where it was made into rum that was then shipped to Africa to buy slaves who were sold in the West Indies to buy molasses—at which point the pattern began again. The *Rainbow,* out of Boston, was probably the first American ship to engage in this trade; it sailed in 1644. Different versions typified the British and French triangular trade. The British "triangle" was estimated to take a sailing ship about six months to complete. The English brought gin, guns, and cloth to trade for slaves, gold, and ivory. They sold the slaves to English sugar producers in the Caribbean and then sailed to England with sugar and what remained of the African cargo. Often, however, they were unable to secure profitable cargoes in the Caribbean and had to return to England in ballast. The French had the most difficult triangle. They did not produce most of what they shipped to Africa but gathered cloth, brandy, iron bars, and guns from all over Europe and cowrie shells (the Atlantic African currency) from the Maldive Islands in the Indian Ocean. They took these goods to Africa to trade for slaves; then they sailed to the French Caribbean, where they traded with the Africans for

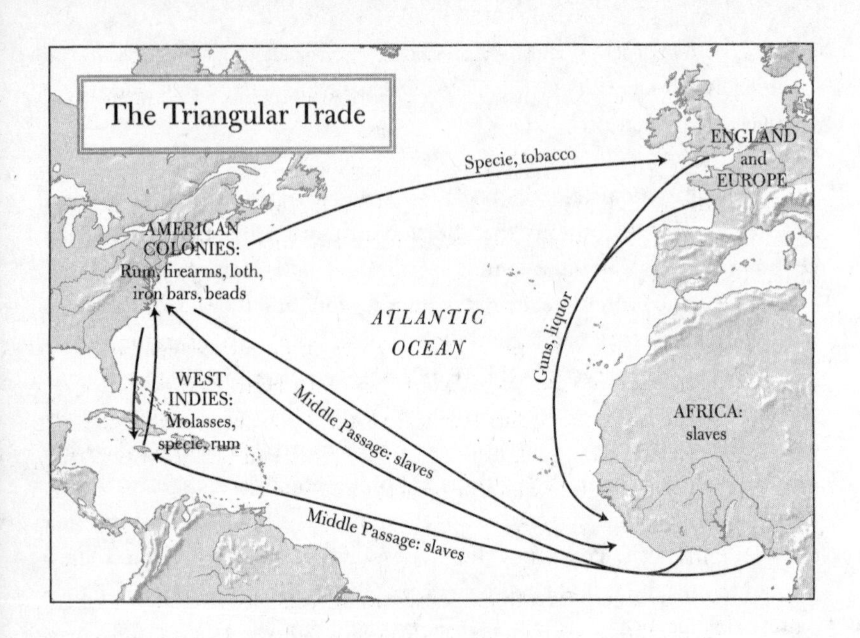

sugar, cotton, and indigo, which they returned to France. That trip might take almost three times as long as the American and British triangles—often nineteen months. This made it far less profitable than the British or American commerce.

The enterprising American sailors elaborated the triangle to enable them to adjust their activities to fit changes in prices and markets, and so they often came out better than either the French or the British. As opportunities arose, the shippers acted like captains of twentieth-century tramp steamers, buying, swapping, and selling a variety of both American and locally produced goods along their routes. Controlling, or even stifling, this trade became a preoccupation of the British. As the British government recovered from the distractions of the revolutionary seventeenth century and began to develop a more sophisticated understanding of how its overseas commerce affected its domestic economy, it clashed more often with the colonists. Under King Charles II, Parliament "enumerated" a list of products "confined to the market of the mother country." Despite Britain's attempts to confine them, colonists were smuggling goods abroad by the beginning of the eighteenth century, if not earlier, even to countries with which Britain was at war.

Not satisfied with "enumerated" restraint on the colonies' overseas trade, Parliament under King George II prohibited even trade between colonies within North America. As a merchant in Philadelphia wrote to Benjamin Franklin,

> No one Colony can supply another with wool, or any woollen goods manufactured in it. The number of hatters must be restrained, so that they cannot work up the furs they take at their doors; nay a hat, though manufactured in England, cannot be sent for sale out of the Province, much less shipped to any foreign market.

The colonists found such prohibitions particularly galling.

While Parliament could pass restraining acts, it had few means to enforce them. Other than the long distances and the lack of convenient roads, no significant barriers hindered movement of goods among the colonies. Goods that the different climates and soils of the colonies made necessary moved relatively easily along the coast and on inland waterways. Sugar producers in the South, for example, needed grain from the North. And by gathering local products for trade among the colonies, merchants learned how to provide for overseas commerce.

Overseas commerce put them athwart the British, but the colonists soon learned how to evade controls. One way they found was to profit from the European wars by pretending to be engaged only in returning prisoners of war under "flags of truce." Since wars were sporadic and prisoners were often hard to find, they were sometimes "sold" by privateers for as much as 500 pounds each. To avoid that cost, the American merchants often neglected to take the prisoners along with the contraband they were selling to Britain's enemies. When, occasionally, a smuggler was caught, he usually had little to worry about. Prudently, he had probably made a partner of the royal customs officer in his home port. There he could purchase a "flag of truce" from corrupt British officials for as little as 20 pounds. If he had neglected this inexpensive formality and a post hoc bribe did not work, the captain could be fairly certain that if he were taken to court, he was unlikely to be convicted by a jury of fellow Americans. John Hancock, "who signed the Declaration of Independence in large bold letters, so that the king of England might read it without spectacles," then had some 500 indictments against him for smuggling. It was Britain's attempt to stop his smuggling of tea, John Adams said, "that made profiteer John Hancock a patriot."

While the British sought to restrain economic growth, the burgeoning colonial governments actively managed their economies to promote it. Americans today assert their belief in free enterprise as a basic American tradition, but the nation's economic roots were in a much more managed economy. During the first years in Jamestown, there was no private property and all labor was performed by regimented gangs. That system quickly broke down, but elsewhere during the next century government tried to regulate prices and wages, attempted to balance "ingate" (imports) with "outgate" (exports), endeavored to prevent overproduction and strived to control quality. Even when public authorities did not undertake "internal improvements" (roads, bridges, and port facilities), they directed and subsidized those who did. Colonial authorities established towns, often determined who could live in them, and prevented those already settled from leaving.

Money was, of course, central to governmental intervention in the economy, or—to put it more accurately, the lack of money was central to intervention by colonial governments. From the very early days, money was in short supply. Applying an economic system known as mercantilism, Britain bled the colonies of their specie. Among themselves, colonists commonly substituted goods for money, bartering or paying debts and taxes with produce. As early as 1640, the Massachusetts General Court ruled that debts could be paid in "corne, cattle, fish, or other commodities, at such rates as this Courte shall set downe from time to time." Initially, price fixing set corn at 4 shillings a bushel, rye at 5 shillings, and wheat at 6 shillings. Money then became even tighter, so by 1682 Maryland ruled that produce could be used as "currency" for private transactions. A further refinement turned even receipts for deposit of goods in official warehouses into "money."

None of these substitutes for money worked satisfactorily, so the colonies competed with one another to attract silver and gold coined currency. They did this by "overvaluing," or paying a premium on, coins in terms of the paper currency—which, following the example of Massachusetts, almost all the colonies were issuing by the middle of the eighteenth century. These actions greatly disturbed the British, who attempted to put a stop to the colonies' dealings in currency. One attempt, in 1741, was even more unpopular in Massachusetts, according to John Adams, than the later Stamp Act. In 1751, at the behest of English merchants, who feared that the paper currency they received for their goods would be worthless, Parliament forbade the New England colonies to issue paper money. And on April 19, 1764, the Paper Currency Act made all colonial paper money no longer legal tender. Like restrictions on manufacturing and trade, however, currency control could not be enforced. The colonists had to have a medium of exchange.

Money was a constant source of annoyance, but the endeavor that best illustrates all the elements of the move toward colonial development was the iron industry. It demonstrates the inherent fatal flaw of colonialism: that as colonies grow in wealth and skill, their initial shared interests begin to diverge, and cooperation gives way to competition. When the American colonists began producing iron in the seventeenth century, they lacked the technical skill to turn iron into steel. (By contrast, as I pointed out in chapter 6, some of the Atlantic African societies had been making steel for a long time.) Consequently, the colonists were willing to sell iron to Britain in its "rude state." This suited British interests because, as Adam Smith noted in 1776, Britain wanted its own manufacturers to refine iron into steel and steel products. American exports of crude iron to England reached impressive figures by 1750, when New England shipped 21 tons; New York, 75 tons; Pennsylvania, 318 tons; and Maryland and Virginia combined, 2,508 tons. So important was the growth of an iron-based industry to the birth of America that we must examine it carefully; three aspects demand our attention: first, the British attempt to prevent Americans from processing iron into steel, which fostered an American belief that Britain, no longer a benign "mother," had become a competitor and was "cramping" colonial growth; second, the ways in which the Americans organized their efforts to produce initially iron and subsequently steel; and, third, the effects on American society of this process.

British rivalry took the form of a law designated as 23 Geo. II., c. 29, which was widely publicized by John Dickinson in his influential *Letters from a Farmer in Pennsylvania.* The law specified that "from and after the twenty-fourth day of June 1750, no mill, or other engine, for slitting and rolling of iron, or any plating forge, to work with a tilt hammer, or any furnace for making steel, shall be erected, or, after such erection, continued in any of his Majesty's colonies in America." Dickinson commented that, given the shortage of hard currency, that is sterling as distinct from the paper money minted in the colonies, colonists "will not be able, in a short time, to get an ax for cutting their firewood, nor a plough for raising their food." His comment was echoed by a merchant in Philadelphia who wrote bitterly to Benjamin Franklin, "The iron we dig from our mountains, we have just the liberty to make into bars, but farther we must not go: we must neither slit it nor plate it, nor must we convert it to steel, though 'tis a truth well known, that we cannot have steel from England fit for use." Getting steel "fit for use" was critical, of course, if the colonies were to grow and prosper. They knew how to turn iron into steel and they had the raw material, iron, but they could not legally refine it.

Extraction of iron from ore, given the technology of the times, was a difficult and labor-intensive process. It was made possible in the colonies by the availability of vast amounts of timber, which could be converted into charcoal. The process was inefficient and environmentally destructive. Producing one ton of pig iron required the burning of a pile of timber 4 feet high by 4 feet wide by 100 feet long, that is about fourteen cords. As annual production increased to reach roughly 30,000 tons shortly before the outbreak of the Revolution, about thirty square miles had to be deforested every year. From our perspective, concerned as we are with the destruction of forest lands, this was reprehensible at best, rather as we view activities in Brazil, Burma, and Thailand, but ours is almost exactly the reverse of the eighteenth-century view. As Adam Smith summed it up, "There is no manufacture which occasions so great a consumption of wood as a furnace, or which can contribute so much to the clearing of a country over-grown with it." As areas were swept clean of their usable timber, producers had to move to virgin land where they could start all over again. Astonishing as it may seem, iron production was a nomadic industry.

John Dickinson

At that time, trees had to be cut with the very inefficient straight-pole ax whose iron head had only the edge "steeled." The edge dulled quickly and was easily shattered by the force of a blow on hardwood. Once a tree was down, the trunk had to be shorn of its branches and then split and cut into transportable logs. If the timber grew near a stream, the logs could be floated to the kiln. However, even if a stream was handy when work began, the supply of nearby trees would soon be exhausted, and it would be necessary to bring logs from farther and farther away. Then, teams of oxen or horses would drag the logs or haul them on carts. Upon arrival at the kiln, the logs had to be manhandled into stacks for the burning that would reduce them to charcoal. Each step in this process required prodigious amounts of human labor. Much of that labor could be performed by semi-skilled men, although the various tasks were hardly simple. Wielding a fragile ax, loading logs so as not to break the axles of the cart, hitching a team of draft animals, and driving the team required not only muscle but experience. Men with more advanced skills as well as muscle were needed at the next stages. The newly cut logs had to be stacked in pyramids that would smolder evenly, not burning too fast and not ceasing to burn despite wind and rain.

Meanwhile, ore was dug with picks and shovels out of open pits or shallow shafts and hauled to the site of a furnace. Then, when the charcoal had been prepared, it had to be fanned into fires sufficiently hot to melt the ore so that impurities could be drawn off. This was the most dangerous of the tasks, and it required skills that were always rare in colonial America. Workers with these skills were among the most prized immigrants from the British Isles and Germany in the early eighteenth century. To see the process in action, consider one operation in New Jersey.

In 1740, the Oxford furnace was established, and an iron ore mine was dug near the Delaware River. The furnace and mine were to continue in operation for nearly two centuries, the longest of any American endeavors in this field, and ultimately gave birth to the city of Scranton, Pennsylvania. Oxford centered on a pyramid built with timber and stone 38 feet high. Into this building the workers dumped a "charge," that is, a mix of six baskets of charcoal, each weighing 70 pounds; five similar boxes of iron ore, broken down into pellets each about the size of a walnut; and a box of limestone weighing about 100 pounds.

The labor force at the furnace consisted of four teams. Each team included a "filler" whose job it was to keep the furnace full of ore brought from the mine shaft. He would dump his wheelbarrow loads of ore in front of two men who smashed it into pellets; then, under the direction of a "keeper," another man would dump the pellets into the furnace. Massive bellows, driven by waterpower from a mill on the adjacent stream, raised the temperature in the furnace to the required level. Then, about every eight hours, the molten iron was ladled out into molds to form pig iron.

The Oxford furnace produced between 13 and 17 tons of pig iron weekly and consumed about 350 bushels of charcoal for each ton of iron it produced. Thus over the course of a year, it required about 275,000 bushels of charcoal. With the technology of the time, one unit of hardwood produced about half a unit to six-tenths of a unit of charcoal. So enough trees had to be cut each year to yield a stack of wood 4 feet wide by 4 feet high by more than 4 miles long. Cutting that much timber forced the original investors to expand their holdings from less than 2 square miles to more than 10. Cutting, trimming, and transporting the timber to the furnace required a labor force of twenty or so men.

The labor force was diverse. In 1754, several of the men at the Oxford furnace were black slaves. Slave workers at iron furnaces were apparently common in New Jersey. Other workers, some bought as indentured laborers off ships, came from Europe. Among those listed in 1754 were Welshmen, Germans, Irishmen, Englishmen, Danes, Hungarians, and one man from the island of Elba. This diversity was probably typical of the eighty or so furnaces in operation shortly before the Revolution. Their workers formed part of a labor force of about 30,000—more than the total population of the five largest American cities at the time—who produced about one-seventh of the world's iron production.

Twenty years before the Revolution, the owners of the Oxford furnace decided to go "downstream" by adding to the production of pig iron an illegal forge. To underwrite this transition, they sold a quarter of Oxford in 1754 for 1,500 pounds. Making iron into steel required a change in both the size and the skills of the labor force. The best sources of workers were the little states of central and southwestern Germany; there, men who today would be called headhunters recruited hundreds of specialists and sent them to the New Jersey highlands. Their skills had a ready market.

Iron had many uses. Northeastern America was cold, and houses were heated by open fireplaces. To maximize the output of a fireplace and radiate heat even after the fire had burned low or gone out, it became customary about the middle of the eighteenth century to place an iron panel at the back. Shortly thereafter, Benjamin Franklin designed his famous "Pennsylvania fireplace," which became the first large-scale, "durable" product of American industry.

Nails, iron bars, and a variety of tools and fittings flowed from the Oxford forge and its adjoining smithies. During the years of the Revolution it would also produce cannonballs and the steel from which a member of the family that had set up the Oxford furnace, William Henry, fashioned his famous muskets.

Without the development of the iron industry in the half century before 1775, Americans could hardly have fought the Revolutionary War. Although they acquired most of their ordnance from France and by seizing British supplies, they were dependent upon local resources for many of the implements that made possible the feeding of troops and civilians, moving men and supplies, and manufacturing clothing. Politically, the "cramp" of the iron industry by British policy was one of the issues which convinced Americans that Britain was less a "mother" than an enemy.

Perhaps most important—as those who struggled with development projects in the twentieth century learned—it is not so much "things" such as roads, bridges, factories, dams, and canals that matter, but rather the skills required to build, maintain, and use them that constitutes "development." Thus, not only was the physical plant, typified by the Oxford Furnace, essential but experience, skill and self-assurance Americans gained from building and managing iron works was an essential precondition to independence. In facing the challenges of mobilizing, equipping, feeding, and moving the armies of the Revolution, one of the key men was a "graduate" of the iron industry, William Henry. Having learned steel technology, Henry put his organizational experience and skills to work to procure and supply food, clothing, and armaments that sustained the Revolutionary army in the crucial Pennsylvania-New Jersey area.

In addition to the growth of skills and organization, the people of the American colonies on the eve of the Revolution had acquired capital, which has been estimated at 110 million pounds—or, in terms of the American currency as of 2000, more than 10 billion dollars. From the tenuous beginnings at Jamestown, America had become a prosperous nation, economically ready, or so many of its people then thought, to stand alone. What stood in its way was Britain. Thus in commerce and industry, as in military affairs, the imperial system was breaking down.

CHAPTER 15

Representation and Taxation

The British viewed America much the same as the Americans viewed it: it was rich and growing richer. But whereas the Americans were proud of their accomplishment, the British government thought that they were inappropriately rich. Britain had to consider what it, as an imperial power, should do with its now flourishing but uncooperative colonies. The man who stood at the center of British decision making was George Grenville.

Grenville, one of the half dozen significant British statesmen of the eighteenth century, had a remarkable career. He served as president of the Board of Trade, as secretary at war, and as prime minister (then called the first lord of the treasury) from 1763 to 1765. A decade later, he played a major role in the events leading to the Revolution. He is not a figure treated with much sympathy. King George III found him insufferably boring, and historians have always put the worst construction on his motives. Yet he knew more about the British economy than anyone else in the government. Unswerving in his focus on fiscal problems, he had come to fear that Britain was on the brink of catastrophe. He had reason to think so. Emerging from the Seven Years' War (in America, the French and Indian War), Britain had a large public debt. No one knew for sure how large it was. No satisfactory accounting existed, and such records as existed were said to be twenty years in arrears. Grenville estimated the debt at 146 million pounds—about eleven times the government's yearly revenues. What he knew for certain was that, as the debt grew and lenders became apprehensive, the cost of borrowing money had almost doubled. Servicing the debt cost the government over half of its yearly revenues. The rest of the revenues covered only one-third of ordinary expenses; so each year the government was going further into debt. In 1760, total net government revenues were 9,207,000 pounds and total net expenditures were 17,993,000 pounds. Since paying for the minimal activities that the government then undertook was virtually inflexible, it had been forced to double the rate of taxes on the major landowners—that is, on its political supporters.

The debt, the cost of servicing it, and the level of taxation seemed to contemporaries to add up to an unacceptable burden, as Grenville warned the House of Commons on March 9, 1764. The political danger to the ruling party was great, but even greater was the national danger. Not only Grenville but all the men around the king and the king himself believed that taxes could not be raised further because doing so would make British goods uncompetitive abroad. If Britain lost its overseas markets, they reasoned, it would become poor indeed. Therefore, expenditures must be trimmed. The fiscal ax fell first on the Royal Navy, the most expensive part of the government; it was cut drastically to only twenty-six ships with 3,290 sailors, from a wartime high of 270 ships and 76,000 men. The army withered to almost nothing. But after those cuts, there was not much more "fat" in the budget.

New sources of revenue had to be found, and the colonies were the obvious source. Americans had emerged from the French and Indian War relatively rich: they could afford to import at least twice as much from England as they had imported before the war; they were illegally engaged in contraband trade worth perhaps 500,000 pounds; and their per capita income was probably at least double that of England. Yet, as Adam Smith pointed out, "The English colonists have never yet contributed any thing towards the defence of the mother country, or towards the support of its civil government. They themselves, on the contrary, have hitherto been defended almost entirely at the expence of the mother country." That expense had risen dramatically. In 1748, before the French and Indian War, the colonies cost Britain 70,000 pounds a year; by the end of the war in 1763, the cost had risen five times, to 350,000 pounds.

To Grenville, it seemed fair that the North American colonies, which had benefited from the expulsion of the French and the defeat of their Indian allies, should foot some of the bill. His logic was undeniable:

Americans had benefited, so why should they not help Britain cover the cost? The answer lies, of course, not in logic but in politics. The colonials sounded the clarion cry that their own legislatures should decide when, how, and indeed whether to raise such money as Britain requested. In essence, they were asserting the revolutionary idea that their representative bodies in the colonies were the equivalent of Parliament in England.

The slogan "No taxation without representation" was popular, but it was not new. A Puritan clergyman in Watertown, Massachusetts, had voiced it in 1632, not against Britain but against the Massachusetts Bay government. More recently, after their own parliament had been hobbled by the Declaratory Act of 1720, Irish patriots had taken up the call. When they were suppressed, American colonists, who were well informed about the struggles for "liberty" on the other side of the Atlantic took up the refrain. The burgeoning American press was filled with accounts of the "sons of liberty," the "freedom fighters" of their day. The colonies' growing Irish population was in contact with relatives in Ireland, and a considerable trade was developing between the colonies and Ireland. Many colonial Americans saw in England's exploitation of Ireland, and in the desperate attempts of the Irish at resistance, a warning about their own future. In short, the issue of taxation was not just about money: it was also about liberty, representation, and sovereignty.

By the end of the seventeenth century, Americans had already contested the authority of Parliament. At first, their economies and societies were too weak to prevail, and so after each challenge they drew back. But their capacities grew dramatically during the eighteenth century. Then, more numerous and more confident, they advanced two arguments against imperial policy.

First, they proclaimed that they had been granted special privileges in their charters that limited the right of the mother country to interfere with their local governments. No one in Britain knew if that was true. Archives were not so carefully kept then as today, so Grenville put his staff to work collecting all the materials showing the wording of the charters and the circumstances of their issuance. From that study, he concluded that no such exemptions had been granted and that even if they had been, they would have been unconstitutional.

Second, the colonists also argued that there was a "general right of mankind not to be taxed except by their representatives." Yet they admitted that they were bound by parliamentary laws and regulations affecting overseas commerce. If they were so bound, Grenville countered, then they were bound not only in "external" affairs (that is, foreign trade) but also in "internal" matters (that is, taxation). Again, logic was on Grenville's side, but politics was not. During the 1760s, it began to be clear that what the colonists sought, although they were not yet using the exact words, was not to be bound at all.

What had happened was not only that the colonies were growing richer and more populated but that for a generation they had been left more or less alone to manage their own affairs. That period of neglect lasted during the government of Sir Robert Walpole (1721–1742) through the French and Indian War (ending in 1763). During that time, as a former customs controller at Boston, Nathaniel Ware, wrote in 1762, "seeds of a future independency" had taken root. Observing this, Grenville felt that America was not carrying its fair share of imperial burdens and was also becoming a "distinct kingdom." He was determined to stop this drift: "All colonies," he observed, "are subject to the dominion of the Mother Country, whether they are a colony of the freest or the most absolute government." So levying a tax seemed important not only for fiscal purposes but also to reaffirm the authority of Parliament. The king agreed.

What the Americans and Irish were complaining about was not just taxation but the imperial system as a whole. Indeed, they perceived a blatant rigging of representation in Parliament. At that time, although all Englishmen with any significant income were taxed, only about one in twenty was actually represented in Parliament. Only landholders with a yearly income of at least 2 pounds could vote; and since balloting was not secret, even a prosperous landholder had to vote as his lord advised—so that his vote was of little importance except in affirming the status quo. Also, proportionally fewer Englishmen qualified to vote than had qualified a century earlier. The trend was away from representative government. Like the Americans and Irish, the English themselves began to agitate.

That agitation echoed back and forth across the Atlantic as Americans listened to the English "public"—that is, the relatively wealthy and educated—

who had begun to find a voice in the popular press, in broadsheets and pamphlets, and in conversations at taverns and inns. England was far beyond America in each of these areas: nearly 9.5 million newspapers were sold in England in 1760, and by 1775 the number rose to more than 12.5 million. In 1760, London had eighty-nine newspapers, four of which were dailies; and the English provinces had thirty-five or so. Pamphlets poured forth in profusion, with editions averaging about 500 copies. Some, like a pamphlet on taxation and representation by an American expatriate, Arthur Lee, masquerading as a former member of Parliament, sold thousands of copies in both England and America. Of eighty-four known pamphlets published in the New World on relations between England and the colonies, thirty-seven were reprinted in England; and many pamphlets published in England were reprinted in the American colonies. In addition to printed materials, representatives of merchant houses wrote to their home offices and families, and their letters were passed from hand to hand. So what was printed or even just discussed in London quickly reverberated throughout the land and across the Atlantic.

Thus the grievances felt by the new professional and merchant class in England—and, less vocally but more violently, by the laboring class—created a receptive climate for the radical leader John Wilkes. Wilkes was a master propagandist, writing and speaking in appealing terms, full of humor, satire, and appositeness. When the government tried to silence him, he became a hero to the Americans as well as to the English. "Wilkes and Liberty" became the cry all over England. When the king's carriage was pelted with stones and horse manure while crowds shouted "No Wilkes, no king," it seemed to many in America that a revolution was in the making.

It was in this atmosphere that Grenville began to feel his way, cautiously and slowly, toward the means to levy a tax on the colonies. He first tentatively suggested, but did not then actually propose to Parliament, that it might be necessary to extend the stamp tax, (which had been applied in England since 1694) to the colonies. On the face of it, a stamp tax was not particularly onerous. Moreover, it was familiar to the colonists. Their own legislatures had recently imposed stamp taxes. Both Massachusetts (in 1755) and New York (in 1756) had passed acts authorizing a stamp tax for colonial usage. The later revolutionary leader John Adams, who supported such a tax, pointed out that it was "not as onerous as the excise tax on wine and liquor" and that it did not occasion customs raids on private establishments. Grenville understandably concluded that it ought to be acceptable to the colonists; moreover, it was easy and cheap to levy. Unlike other taxes, it was self-collecting: everyone who wanted to be sure that his documents would hold up in court would gladly pay a small fee to make them legal. (Parenthetically, I should point out that although stamp taxes are demonized in American history, they are still used to validate American documents.) Therefore, as Grenville told Parliament, he did not think that the proposed Stamp Act would be particularly unpopular; and, relative to other forms of tax, it would yield considerable and sorely needed revenue. He had been told that a stamp tax in America might net about 60,000 pounds yearly. In contrast, the cost of collecting customs duties there was larger than the duties themselves. Then, both to seek advice and, presumably, to begin to build a colonial constituency for his program, he turned to the only group of "experts" available, the colonial agents.

In a practice that began in 1624, when Virginia sent John Pountis to London "to solicite the generall cause of the countrey to the King and the Counciel," most of the colonies maintained representatives in London to act as their agents. Occupying a position somewhere between an ambassador and a modern lobbyist, the agent was both official and informal: official in the sense that he was appointed and paid by a provincial legislature and his status was recognized by the British government; informal in that he probably also had another job. He was expected not only to assert his sponsor's position on pending legislation or administrative action to the senior members of the British government but also to write for the press, form associations with merchants, and generally "win friends and influence people." Benjamin Franklin was the best known of these men and was a master at each aspect of the task, but he was just one of many. Thirty-three agents are recorded in the twenty years from 1756 to the outbreak of the Revolutionary war.

The British government often found the agents useful sources of information against which to check the reports it received from its own officials in America. It also occasionally "leaked" to them information that it wanted to be passed back to the colonies' legislatures. But the government was also annoyed by the agents' attempts to sway English opinion, and so it mounted a covert program of surveillance against them and their contacts in the press. Gradually, during the 1760s, the government pulled back from the espionage system it had used to infiltrate the press, although it kept on intercepting and reading mail. At this latter task, it was highly sophisticated. Not only was mail opened carefully, but seals were engraved and reaffixed on the envelopes so that the recipients would not know that their letters had been read. The government also, using bribes or threats, induced printers to provide the names of writers. Yet letters and broadsheets poured out in profusion, so that few opinions were ever really private and few secrets or identities remained restricted for long.

Grenville had little reason to trust the colonial agents; but he had no other source of more or less informed opinion, so he decided to lay his tax plan before them. After discussing it, he challenged the assembled group to come up with a more acceptable scheme. In reply to his question "If a different form of tax or levy were proposed, could the colonies agree upon how to apportion it among themselves?" the colonial agents admitted that they could not. Worse, as Grenville pointed out, any other apportionment they might agree upon would quickly come to seem unfair, since the colonies were growing at different rates. Consequently, he assumed that they would use their disagreement over fair shares to avoid the tax altogether. He had won his argument, but, cautiously, he delayed parliamentary action for a year, urging the colonial legislatures, through their London agents, to propose measures which they would find more acceptable and which would yield comparable revenue. Not a single proposal came from the colonial legislatures. Thus he introduced the Stamp Act on February 6, 1765, and it was passed in Parliament by a vote of 245 to fifty.

Grenville had been misled. The agents did not represent colonial opinion on the stamp tax. Even Benjamin Franklin thought that after an initial period of annoyance, the colonists would accept the new system. How, we must ask, had the conflict become so heated and so focused on this apparently innocuous issue?

Franklin himself partly illustrates the answer. He had spent years trying to guide the British government toward accommodations that would keep the American colonies under the hegemony of the king while also protecting colonial institutions. One way that he thought might work was to allot seats in Parliament to the American colonies, as had been done for Scotland and Ireland; one way he was sure would not work was to compel the colonies to pay money without their consent. That, he wrote, "would be rather like raising Contributions in an Enemy's Country." Whatever happened, Franklin argued, the colonists did not deserve to be treated "as a conquer'd People"; they should be treated "as true British Subjects." Neither the British nor the Americans were willing to listen.

On the British side, the problem was not only a lack of accurate intelligence but the perennial problem of analyzing intelligence, the mind-set by which information is judged. Such information as the British had did not fit into their concept of the relationship of the "daughter" to the "mother." The colonists' claims seemed to them specious: how could a group of mere outcasts, living in small, remote, dependent, mutually hostile colonies, claim more political rights than the inhabitants of England? Such a claim was worse than absurd; it was also dangerous. If it were admitted, it might further incite revolutionary tendencies, already evident under the leadership of John Wilkes, in England itself.

The colonists' view was, on the contrary, prideful. They were proud that they had built and were running political and civic institutions, proud that they had hacked farms and towns out of the wilderness, proud that they had grown prosperous and increasingly numerous. They felt that they were better judges of the situation in America and of how to play their role in the British empire than ministers in London or than the frequently inept, often corrupt, and financially dependent royal governors sent to oversee them. As they looked at the proposed tax, they found several things particularly obnoxious.

The first problem was that for Parliament to levy a tax was an innovation. The colonists had been used to raising, by acts of their own elected legislatures, the revenue needed for their rudimentary and inexpensive governments and their mostly volunteer militias. When they incurred extra expenses, as they did during the French and Indian War, they again raised the money themselves. True, the British government had later repaid them about 60 percent of those wartime expenditures. But they saw the repayment less as a cause for gratitude than as a confirmation of the proper direction of flow of revenue: from, rather than to, London. As Adam Smith observed, the colonists never even considered paying money *to* Great Britain.

The colonists' second cause for apprehension was that the workings of the Stamp Act would intrude on activities they regarded as private. A stamp would be required not only for documents relating to legal matters (deeds, mortgages, leases, bills of sale, etc.)—instances in which it might have been disliked but would ultimately have been accepted—but also for appointments to office, college diplomas, and, still worse, newspapers, pamphlets, books, calendars, games, and playing cards. People in the colonies felt that suddenly their every activity, even their leisure, was to become an issue before tax collectors.

The third reason for complaint was that stamps could bought only with "hard" currency: that is, sterling rather than colonial currencies. Hard currency was in desperately short supply in North America because Britain bled the colonies of specie. Money was so hard to get that some colonists thought the intent of the Stamp Act was to freeze America in a primitive condition by blocking its growth. Even Benjamin Franklin pointed out that the Stamp Act could never have worked because within a year, no one would have had the money to buy the stamped paper.

Finally, many Americans denied the justification for any tax. The war with the French was over, and most of the Indians had been driven away. No intervention by Britain was needed, they argued, because colonial militias and settlers would dispose of those Indians who remained. In any event, the redcoats were of questionable value against the Indians, as Franklin had observed. Moreover, the colonists had already developed antipathy to a standing army, and this attitude would color their politics for years to come. As he so often did, John Dickinson expressed their feelings:

> What then must be our chance, when the laws of life and death are to be spoken by judges totally dependent on *that crown,* and *that kingdom* . . . *backed by a* STANDING *army*—supported out of OUR OWN pockets, to "assert and maintain" OUR OWN "dependence and obedience."

Strident denunciations of the act were begun by young Patrick Henry's speech before the Virginia House of Burgesses: "If this be treason . . ." But a public opinion poll would have shown that most Americans were not so radical. When Henry's remarks were printed in the newspapers of Boston and New York, they shocked many readers. Even one of the leading militants in Boston, the lawyer James Otis, publicly denounced them as treasonable. Sober men wanted to calm the tumult.

Opposition did not die down, but no one knew what to do. So, on June 8, less than two weeks after having learned of the act, the Massachusetts House of Representatives decided to consult with its counterparts in the other colonies. To do this, it urged them to send representatives to New York for a "Stamp Act Congress." Early in October, the legislatures of nine of the colonies accepted the call. At the meeting, the twenty-eight delegates vied with one another to express their "warmest sentiments of affection and duty" to "the best of sovereigns, [and] to the mother country," and "all due subordination to that august body the Parliament of Great Britain." But, at the same time, they were adamant "that no taxes be imposed on them but with their own consent."

When Franklin realized how badly he had misjudged the mood in America—mobs in Philadelphia nearly burned down his house—he met with the new president of the Board of Trade, the earl of Dartmouth, and suggested a way to back out of the unwanted confrontation: "suspend" the stamp tax and then, when everything had quieted down, allow it to disappear silently. Thus the issue of the right of the British government to create such a tax in America would never have to be raised. His proposal was not accepted—Britain was not ready to back down, and the Americans had become aroused.

Their arousal took several forms. One reaction was to boycott British goods—a continuation of the "nonimportation" agreement with which the colonists had met the American Revenue ("Sugar") proposed by Grenville the previous year. The intent was to convince English merchants that their lucrative trade with the colonies would dry up unless they persuaded Parliament to rescind the Sugar Act. Campaigns were organized to get the citizens to avoid even using English cloth in clothing, but the boycott had proved impossible to enforce. Americans were ambivalent. Not only did they like British goods; they were loyal to the monarchy. George Washington alludes in his diary to spending a day helping to draw up the Non-Importation Association of the Virginia House of Burgesses and then spending the evening drinking toasts to George III, the royal family, and a lasting union. Not surprisingly, the colonists' threats were soon seen as rather paltry gestures.

Paltry gestures were not enough, so the radical political leaders in the colonies decided on more aggressive measures. Although no representatives from North Carolina attended the Stamp Act Congress in New York, since the royal governor had prorogued the colony's assembly before it could elect delegates, the people themselves quickly took extralegal measures. On November 16, 1765, some 300 men, calling themselves "sons of liberty," seized the local stamp master-designate and, with flags flying and drums beating, paraded him to the courthouse in Wilmington. There they forced him to sign his resignation. No stamps were ever sold in North Carolina.

New York did not initially oppose the Stamp Act, but it quickly took its cue from Boston, Newport, Annapolis, Albany, and Philadelphia, all of which were scenes of serious demonstrations. By the time the first cargo of stamped paper arrived in New York, it had to be spirited into Fort George because an angry crowd estimated at 2,000 had gathered to demonstrate. A few days later, another throng assembled, this time armed, and threatened to storm the British fort. The commander of the fort prudently forestalled what might have been a bloody confrontation by turning the stamp paper over to the city officials. On the same day, a group of merchants who had been "trading to Great Britain" met at Burns' City Arms Tavern and resolved "to import no more goods from the mother country while the Stamp Act remained in force." In December, a large crowd carried the consignment of stamped paper to the common and burned it.

The British reaction was equivocal. Misreading the reason for the colonists' anger, the government tried to placate them by assuring them that all the revenues raised by the Stamp Act would be used in the colonies. What the British meant to convey was that the tax would not further drain the already small supply of specie. What the colonists heard was quite a different message. They understood that their own money would make it possible for British troops to be stationed permanently among them. Horrified at that prospect, the assemblies of New York and Massachusetts joined South Carolina and Pennsylvania in declaring their opposition. So what had begun as a question of how the burden of debt and expenditure could be fairly divided between the mother country and its colonies turned into a challenge to Parliament's authority to legislate at all for the colonies. This was, in the clearest sense, a problem that has arisen throughout history in empires. I call it the fatal flaw of colonialism: whatever their beginnings, colonies will ultimately, as they grow, assert their right to autonomy.

It was in Boston that the Stamp Act was most violently opposed. A small group of men, calling themselves the "loyal nine," began to meet secretly to plan attacks on the appointed administrators. Boston—the birthplace of the leading agitator, Samuel Adams—was no stranger to violence. In 1689, Bostonians had hijacked a ship of the Royal Navy and attacked the British fort. That was the first of a long series of riots. When there was no immediate provocation or cause, Bostonians used annual celebrations as occasions for staging fights.

In the 1760s, two street gangs were active: one in the North End, led by a lawyer, Samuel Swift who was a graduate of Harvard College; and the second in the South End, with a twenty-eight-year-old shoemaker, Ebenezer MacIntosh (or Mackintosh), as captain. On designated occasions, these two bands clashed, amid great fanfare, each with a float to advertise itself, and, after flailing one another with staves and clubs, pulled down fences, broke windows and otherwise terrorized the neighborhood. Then, sometime after 1764, the gangs merged and—more significantly—they adopted quasi-military discipline (the governor called them a "trained mob") under MacIntosh. He was given a gaudy uniform, the contemporary equivalent of a bullhorn, and, to support himself, a sinecure as the town "sealer of leather." All this seems to have come from the Caucas Clubb, the group that acted as a steering committee for the town meeting. Although the clandestine groups naturally kept no records, there is an illuminating coincidence.

This coincidence involved the then still shadowy figure of Samuel Adams. He was a sort of godfather to the Caucas Clubb, and he also flitted among the various other political clubs in Massachusetts, including the "loyal nine." He and the other like-minded agitators were careful not to create or leave written records that might incriminate them, and possibly for this reason they decided to effect their resolutions on the street. The agent they chose was MacIntosh.

MacIntosh must have been unwilling, because Adams, who was Boston's tax collector, arranged to pressure him into cooperating. Adams slapped a warrant on him for back taxes and ordered the sheriff to arrest him and confiscate his property. What passed between Adams and MacIntosh is unknown, but suddenly, although MacIntosh had made no move to pay even a penny of the back taxes, Adams ordered the sheriff to drop the charges. MacIntosh then assumed the leadership of the unified gang and directed it against the officials administering the Stamp Act.

On the evening of August 14, 1765, MacIntosh led his men in an attack on the stamp collector-designate, Judge Peter Oliver. They first hanged Oliver in effigy; then, after burning a building that they said he had planned to use to sell stamps, they tore down his house. Oliver himself had fled. Finding his wine cellar, the mob drank all the contents and, then warmed by the spirits, ransacked neighboring houses. Arriving on this dismal scene, Thomas Hutchinson, who was the lieutenant governor and chief justice, bravely but imprudently ordered the crowd to disperse. As he started to speak, one of the mob yelled out, "To your arms, my boys!" Pelted by a volley of stones, Hutchinson fled "through backyards and alleys under cover of the night." The royal governor made no such brave sally. He retreated all the way out to Castle Island at the entrance of Boston Harbor to take refuge with the British garrison. He was the second royal governor to be chased out of Boston.

Terrified, Oliver immediately resigned his commission. Satisfied that they had defeated Oliver, the mob picked off one man after another who they thought was implicated in the Stamp Act and then turned back to Hutchinson, who was Oliver's brother-in-law and, who, ironically, had long opposed the Stamp Act. What damned him was that he had been given an appointment fervently sought by the radical leader Bostonian James Otis Jr. for his father. The driven, ambitious young Otis, then still a student, was quoted by one of the men in his circle as saying that "If his father was not appointed judge [in place of Hutchinson], he would set the whole province in a flame, though he perished in the attempt." John Adams wrote that the younger Otis was "raving Mad—raving vs. Father, Wife, Brother, Sister, Friend &c." Some Bostonians deplored these violent aspects of his personality, but Samuel Adams's cabal found them useful because Otis focused anger on the establishment's leading figure.

Otis's technique was to avoid discussing the issues as set forth by Hutchinson, instead to hammer away in every forum he could find—the courts, the press, and public meetings—on Hutchinson's presumed "corruption" and opposition to "natural equity." Obviously driven by what John Adams later recalled as "a flame of fire," Otis and his close associate Samuel Adams played a role comparable to that of Robespierre in the later French Revolution, inflaming the public and brushing aside the moderates.

A moderate and a rational man, Hutchinson was the perfect target for an Otis. Hutchinson believed that the colonies would eventually achieve a strong, separate national identity, but he also believed that, as it existed in the 1760s, America was too weak to stand alone in a dangerous world of predatory great powers like Spain and France. Events after the Revolutionary War would show the wisdom of his evaluation. Hutchinson was also a serious man. As one of the earliest American scholars, he was already at work on the massive history of Massachusetts that was to be his lifework. When the mob ravaged his house in 1765, many of the documents he had spent years collecting were lost. The leaders of the attack were disguised, but MacIntosh was recognized, and along with a few others was subsequently arrested. If there was any doubt as to the sponsors of the mob, it vanished as MacIntosh was bailed out by members of the "loyal nine."

The "nine" then turned back to Oliver. He was too visible a target to be neglected even after he had resigned; so they sent MacIntosh to his house to "escort" him to the "Liberty Tree," a huge elm near the Common, where before a crowd estimated at 2,000, or about one person in each three Boston adults, he was forced to swear before Justice of the Peace Richard Dana, a close collaborator of Samuel Adams, to have nothing further to do with stamps. In case Oliver was in any doubt about the gang's message, it was spelled out in a note fixed to one of the effigies:

> A goodlier sight who e'er did see?
> A Stamp-Man hanging on a tree!

Samuel Adams

Samuel Adams, who was to go on to distinguished service during the Revolution, was careful to keep well in the background, often pleading ignorance of what had been done or who had done it. Also, he described even spectacularly violent acts as schoolboy pranks—one was said to be merely a "Frolick of a few Boys [raiding a garden] to eat some Cherries." But there is much circumstantial evidence that he was the movement's organizer and commander.

Probably Samuel Adams was trying to provoke a British response, but the British government was reluctant to get involved. In the absence of a civil police force, only British troops might have restored order. Some troops did arrive in 1765, when a ship trying to get to Quebec was forced into the port by bad weather, but they were spirited out of sight to Castle Island. The first contingent of troops sent purposefully to Boston did not arrive until nearly three years later, in 1768, after a series of attacks had forced the royal customs officials to flee town.

Meanwhile, terrorist groups proliferated. In addition to the "sons of liberty," three groups known at least by name were the "Mohawk River Indians," the "sons of Neptune" and the "Philadelphia Patriotic Society." Throughout the colonies, even in then remote Georgia, groups formed under a variety of names to mount violent opposition to parliamentary taxation. If they were the shock troops, their way was prepared by the churches.

America's more than 3,000 churches offered the one institution in which their usually ethnically homogeneous and like-minded parishioners could feel completely secure. Thus nearly every church became an active center of political education, with the preacher leading the congregation in forming "public opinion." Each preacher vied with others for attention by having his sermons printed and circulated. So, while there were disagreements among them, they constituted a virtual industry of propagandists and agitators. What they had to say found a ready market because they were the providers of information—most of the other sources of information and entertainment that we think of today as normal were then lacking. Moreover, the public was ready for the churches' message because, since the 1730s, one of those waves of religious sentiment to which Americans are particularly susceptible had swept across the colonies. It was known as the "Great Awakening." A form of evangelical Calvinism, this movement reached its peak under the leadership of a fiery revivalist, George Whitefield. So vast was his appeal as he moved across America that no church was large enough to hold his audiences. His deeply emotional message was astonishingly successful not only among illiterate farmers but also among professors and students in the young American universities. Unquestionably, Americans hungered for religion—particularly when it was combined with politics. And everyone was ready to believe in British evil.

Ironically, far from being capable of an evil plot against American liberty, Britain then had a particularly weak government. George Grenville had died, and the king's henchman Lord Bute had been forced from office. The new prime minister was a Whig, Charles Watson Wentworth, second marquis of Rockingham, who did not have the support of the king and whose five colleagues had never before headed ministries. Rockingham bowed before the colonial challenge and repealed the Stamp Act on March 4, 1766. That was a sign of weakness; an even more obvious sign was that he could not do it cleanly. To win parliamentary support, he came up with a face-saving formula: while voiding the Stamp Act, he got Parliament to pass a Declaratory Act, modeled on an act passed for Ireland in 1720, "for the better securing the dependency of his Majesty's dominions in America upon the crown and parliament of Great Britain." Under the Declaratory Act, Parliament proclaimed its right to levy any taxes it wanted at any time in the future. Thus the issue raised by the Stamp Act was not settled but merely postponed. At the time, this move seemed both clever and successful. Americans were so relieved to be rid of the Stamp Act that the legislatures of both New York and Virginia gratefully voted to erect statues of their beloved sovereign, George III. In London, though, Rockingham got no thanks. Taking advantage of what appeared to be the dawn of a gratifying new era of loyalty and political placidity, the king dismissed him and turned to William Pitt (the Elder) to form a stronger government.

Pitt was no favorite of the king and was known to be ambivalent toward the colonies. So the king fatally undermined Pitt's authority in the one place it really counted, the House of Commons, by kicking him upstairs to the House of Lords. Pitt had been a major political force in the Commons, where he was known as the "Great Commoner" and where he was legendary for his fine oratory; but as the earl of Chatham he had to give up that base of power. Although the House of Lords was then more significant than it later became, the Commons, as David Hume observed at the time, was the "great scene of business." From his segregated position, Pitt, now Chatham, could no longer effectively build support for the administration, from which he virtually retired. He left to his successors the dilemma no one had solved: how to get the colonies to contribute money without actually taxing them. The government's answer would soon emerge, when Charles Townshend, the leader of the Commons, proposed a series of acts including an act to raise revenue that the Commons passed on June 29, 1767.

In this revenue act, the government sought to avoid the difficulties of the Stamp Act by levying duties only on "external" items—goods that were being shipped into the colonies from abroad. But embedded in the act was something far more dangerous to the colonists' aspirations than anything the Stamp Act had threatened: the revenue raised was not only to cover a part—about 10 percent, as it turned out—of the cost of the British military establishment in the colonies but also to pay the royal governors. What the British failed to see or chose to overlook was that just as Parliament had struggled for generations to subject the king to its financial control, so the colonial legislatures had used the power of the purse against royal governors and judges. To lose that power was to lose "an essential part of our constitution . . . without which we cannot continue a free state."

Disturbing as this provision was, some colonists believed that it disguised an even more sinister intent. John Dickinson pointed out, in one of the most widely read of all political pamphlets of the period, that since England had closed America to all sources of goods except those from England and had also prohibited manufacture of many necessities in the colonies, it could charge any amount it wished for imported goods. Thus the "Townhsend duties" should be seen as the first step toward a tyranny in which "the tragedy of *American* liberty is finished."

In his last foray into the dispute, the dying former prime minister, William Pitt, Lord Chatham, famously shouted out:

> It is my opinion, that this kingdom has no right to lay A TAX upon the colonies. . . . The *Americans* are the SONS, not the BASTARDS of England. . . . The COMMONS of *America*, represented in their several assemblies, have ever been in possession of this their constitutional right, of GIVING AND GRANTING THEIR OWN MONEY. They would have been SLAVES, if they had not enjoyed it.

That stormy session of Parliament was also the occasion when Isaac Barré, an Irish-born member, coined his ringing phrase "sons of liberty," which became the American radicals' favorite name for themselves.

Chatham and Barré failed to sway Parliament; but it was to be swayed, over a period of three years, by popular resistance and boycott of British goods in the colonies.

All the duties, except the one on tea, were repealed in April 1770. Tea was kept because selling it played a significant role in financing the conquest by the East India Company of Britain's other great colony, India; and because, as George III wrote to his prime minister, one tax must be maintained to protect the principle of government authority. "We must not retreat," said the king. "By coolness and an unremitted pursuit of the measures that have been adopted I trust they will come to submit. I am clear there must always be one tax to keep up the right, and as such I approve of the Tea Duty." But retreat the British had done, and the tax on tea would not so much "keep up the right" as provoke outrage and foster rebellion.

From these events, the colonists had learned that they could veto parliamentary action by resolute and violent resistance. This was a lesson they would not forget.

CONCLUSION

Stumbling Toward War

CHAPTER 16

Foreign Friends and Fellow Sufferers

By the 1760s, Americans were coming to believe that their cause was universal: the whole world known to them seemed to have sunk into despotism and corruption; and in the struggle against these ills, they found friends and allies wherever they looked. Far away, in Spain, Ireland, Corsica, and Poland, they thought they could see the faint glow of liberty. The strongest light came from England.

England was then in the early phase of the industrial revolution, and a new group of men was coming to the fore. They did not yet form a social class, but they had begun to find common interests and beliefs. Of these, the most important were that the government was inhibiting economic progress and that the distribution of political power, effected by a corrupt system, was unfair. Those beliefs resonated with Americans.

The men making up this new group were no more concerned with the well-being of the English lower class, who were as wretchedly poor as they had been in the previous century, than the Americans were with their own poor whites and black slaves. Both the English and the Americans had their eyes firmly fixed on the nobility grouped around the king; these nobles owned a large part of England's landed wealth and, by giving out government sinecures and buying election districts (boroughs), controlled Parliament.

In this "rotten borough" system, electoral districts were literally up for sale. Benjamin Franklin, who by 1768 had become thoroughly disillusioned with Britain, wrote to his son that "£4000 is now the market price for a borough. . . . This whole venal nation is now at market, [and] will be sold for about Two Millions; and might be bought out of the hands of the present bidders (if he would offer half a million more) by the devil himself."

It was not only the corruption of the system that galled the new kind of Englishman and Americans like Franklin; it was also that representation in Parliament was determined by an electoral process that had been handed down from ancient times and made no sense in the new England of the industrial revolution and the empire. The 513 members of Parliament were elected by very few voters; one borough actually had no voters at all. And since voting was restricted to men with an income of at least 2 pounds per annum and who owned farmland, 254 members represented only 11,500 voters, or an average of only forty-five each. Because balloting was open, the local political bosses, the larger landlords, could examine the votes of men who often were their tenants and either reward or punish them for their choices. Disillusioned with the system, voters customarily sold their votes to the highest bidders.

Benjamin Franklin registered his disgust, but it was not only Americans or even radicals who were disgusted. The duke of Richmond described one borough as a "new whore that is anybody's for their money." Many boroughs were not even that: by about 1750, voters for only about 150 seats even made a pretense of choosing. The Americans, who were excluded from voting for Parliament but were being told that they would be taxed by it, found the English system pure tyranny.

The system fed upon itself. Getting into Parliament cost a great deal of money; once there, members got no salary. Unless they were independently wealthy, they were virtually forced to accept the bounty of the king. He rewarded them well if they supported his choice of government ministers and programs; his bounty came in the form of a sinecure denoted by one of those quaint titles given to royal patronage: "lord of the bedchamber," "master of the king's tennis court," "taster of the king's wines," and so on. Such sinecures, which required no service other than voting for the government, might bring the holder, known as a "placeman," as much as 500 pounds a year. When King George's closest friend, Lord Bute, became first minister, he told Henry Fox to "fix" Parliament; Fox is said to have passed out about 25,000 pounds in cash bribes and to have appointed Bute's supporters to a number of high-paying sinecures. This was perhaps an exceptional episode, but the way was open for many forms of corruption. Men who could stand apart from the system were very rare. Few tried, since the "grandees had their jaws locked upon the profits and perquisites of the state."

This system had occasioned little criticism in the early eighteenth century; what changed was that the groups brought into being by the industrial revolution and the expansion of overseas trade, like the American colonists, had grown larger and more sure of themselves. They had begun to communicate with one another and to find a voice in public affairs. Men like Joshua Wedgwood (the potter), James Watt (inventor of a steam engine), and John Wilkinson (innovator of iron furnaces) were experimenting with new uses of power, new means of organizing labor, and new technologies. They were joined by larger-scale merchants who wanted scope for their products, regardless of whether it was in Europe or the American colonies. New means of capital formation came into being. Lloyds Bank, established in 1765, was a harbinger of new uses of money; the City of London already showed signs of becoming a financial powerhouse. The men behind these innovations were committed to a work ethic, to modernization, to efficiency, and to profit. They were not part of the court society, although some of them had become far richer than anyone at court; and they bridled at the stodginess and snobbishness of the nobility. Benjamin Franklin met some of these men at London's elite scientific "club," the Royal Society. They were not revolutionaries, as their French counterparts were becoming, but they did seek to modify the "system" to suit their interests. Individually, most of them were anxious to join the privileged elite; and many did so by buying estates from the poorer nobles and marrying their daughters to sons of titled families—a process that has been described as "the landed embracing the loaded." But on the way toward personal aggrandizement, they were prepared to use their new powers to hasten the process of change. They differed from the Americans in what they thought was the principal impediment to change. The colonists demonized Parliament, blaming it for all the actions they hated and for thwarting their growth and prosperity, but retained a sentimental loyalty to the monarchy. By contrast, the English reformers tended to blame the monarchy. They felt that they could handle the nobles, indeed could buy them, but the monarchy was beyond their reach. The monarchy was the "motor" that powered the system, dispensing the largess that kept everything running; it was, moreover, open to attack as foreign, being of recent German origin—but short of a French-style revolution it could not be directly attacked by these reformers. Others, however, could attack it.

The torchbearer of the new opposition was John Wilkes. An opportunist, he was quite willing to profit from corruption. But he also criticized the court and the nobility. It was this criticism that capitulated him to fame. In the spring of 1763, the *North Briton*—the newspaper he had founded with money from the brother of the sometime prime minister Lord Grenville—issued its forty-fifth and last edition, which suggested that the current prime minister, Lord Bute, was having an affair with the king's mother. The reading public, both in England and in the colonies, of course, was delighted; but the British government regarded the report as "seditious and treasonable" and decided to crack down on the *North Briton,* its owners, its writers, and even its printers. Since in those days journalists prudently hid their identities, the government sought to sweep up the whole lot by issuing "general warrants," on which the names of the accused were not given.

General warrants as a form of attack on the press had been used long before, in the "bad old days" of King Charles II; authorization for them should have expired when the Glorious Revolution brought William and Mary to the English throne, but they were retained. Although the form and means of execution were not the same, general warrants were regarded in England much as *lettres de cachet* were in pre-Revolutionary France, the ultimate symbols of a despotic government.

When Wilkes was arrested, he was treated very gently. He was carried to his interrogation in a sedan chair; and he was questioned not in jail but in the house of his neighbor, Secretary of State Lord Halifax. Then, after a good deal of confusion in which the government was made to appear inept as well as tyrannical, Wilkes was committed to the Tower of London. By this time, all London was following the affair almost minute by minute. An entourage that took on the aspects of a triumphal march followed when he was let out of the Tower to appear in court. Having found a man who was at the same time a member of Parliament and a victim of tyranny, the common people of London, the merchants, and even some major figures in the aristocracy flocked to his side. As he left the court he was greeted, as he was to be in the years ahead, with cries of "Liberty! Liberty!" He had become an overnight hero. His new status was affirmed a few days later when the court ruled that as a member of Parliament, he had been wrongfully arrested and was to be set free. He was escorted home by a crowd numbering in the thousands. And on a final note of triumph, he sued certain ministers and government agents and forced them to pay damages.

Greeted everywhere by large crowds shouting the new slogan, "Wilkes and Liberty," he made a triumphal tour of England. Elated by this sudden effusion of support, he reprinted his attacks on the government in a pamphlet. Also, probably only for the fun of it, he printed what was then regarded as an obscene and impious satire, a take off on Alexander Pope's *Essay on Man* (1734), entitled "Essay on Women." That was his downfall. Having failed to subdue him on political charges, the government went after him on moral grounds. In the course of the tumult, which mesmerized London, Wilkes had to fight two duels, in one of which he was seriously wounded. In fear of a vindictive court, he slipped away on Christmas Eve, 1763, for a delightful exile in France and Italy.

While he was abroad, the government of his former friend and current enemy George Grenville was replaced by one seemingly more favorable to him, under Lord Rockingham. Wilkes was happy to go home, and in 1768 he decided to run for Parliament on the issue of general warrants, the device that had put him in prison. Despite a rambunctious campaign, the like of which England had never seen, he lost. Undeterred, he ran again from a different borough, where he won by a large margin.

The government, already fearing public unrest over the worsening economy that followed the end of war in Europe, prepared the army to put down Wilkes's unruly followers. Viewing the tumult, Edmund Burke observed that since the "Great Commoner," William Pitt, fell from power, "there has been no hero of the Mob but Wilkes." He was right. Mobs ranging through London fastened on a new symbol: householders who supported Wilkes placed lighted candles in their windows, as did those opponents who did not wish to have their own windows smashed. London was alight as never before.

Again hauled before a court, Wilkes was convicted and committed to prison. To the "mob" that was the last straw: London went into a virtual siege, with riots every day for two weeks. The prison itself was attacked—in effect an English preview of the later French attack on the Bastille—and the huge, angry crowd dispersed only when Wilkes, realizing that his supporters had gone too far, personally asked them to go home. The next day an assembly of thousands—according to some reports, as many as 40,000—threatened to break into the prison and shouted, for the first time, slogans against King George III: "No Wilkes, no king! Damn the king! Damn the government! Damn the justices!" and frighteningly, "This is the most glorious opportunity for a revolution that ever offered!" Finally, the Guards regiment, composed mainly of soldiers from the king's ancestral land, Germany, opened fire, giving Wilkes's movement both its demons and its martyrs. London swayed on the brink of revolution. The "massacre of Saint George's Fields" on May 10, 1768, was a harbinger of the Boston "massacre" two years later on March 5, 1770. The parallels were later drawn by the American radicals: each incident had involved "alien" troops (German and Scottish in London, British in Boston) firing on civilians; and each resulted in the death of a young and presumably innocent boy who was given a huge, highly politicized funeral. In this event, as in others, the American radicals were hot on the heels of their English exemplars.

Wilkes was the exemplar par excellence. He was the pioneer in the great quest for "liberty." Americans showered him with presents as tokens of their respect. The "sons of liberty" in Boston sent him messages of support, and the South Carolina legislature sent a cash contribution of 1,500 pounds. Virginia voted Wilkes a contribution of tobacco; Massachusetts is said to have sent him several turtles. To Americans as well as Englishmen, Wilkes and liberty seemed inseparable. As one American wrote in 1769, "the fate of Wilkes and America must stand or fall together."

A number of Wilkes's followers among the merchants and industrialists in London created the Society of the Supporters of the Bill of Rights, which, among other things, ensured that Wilkes could not be silenced by being sent to a debtors' prison. These supporters were substantial, law-abiding men; but the mobs who prowled the streets were not. When some of the more violent were arrested, juries refused to convict them—just as a few years later, American magistrates would refuse to convict their American counterparts. In such ways, Wilkes's London in the 1760s was a trial run for Boston in early 1770s.

What the government could not get the courts to do, the cabinet persuaded Parliament to do: expel Wilkes. The king described his expulsion as "a measure whereon almost my Crown depends." On February 3, 1769, Parliament voted to remove Wilkes from his seat. On February 16, his constituents reëlected him; thereafter he was elected and expelled time after time until, wearying of the process, the government majority in Parliament declared his opponent the winner. Demonstrations broke out all over England, and about one-quarter of English adult males signed petitions of protest. One petition went on to urge the king "to banish from royal favour, trust and confidence for ever those evil and pernicious counsellors" whom it blamed for the attack on Wilkes.

By this time, petitions also included, at the behest of Benjamin Franklin's young colleague from Virginia, Arthur Lee, charges against the government for mismanaging the American colonies. Still other charges followed from the new manufacturing and commercial elite, which charged that government policy was "ruining our manufactories by invidiously imposing and establishing the most impolitic and unconstitutional taxations and regulations on your Majesty's colonies." Not surprisingly, when Wilkes was released from prison in April 1770, he was praised everywhere in the American colonies.

Wilkes was not to be stopped. Thwarted in his bid to return to Parliament, he captured a position on the London City Council and was elected lord mayor. Then, in 1774, the government gave up trying to find someone stronger to replace him in Middlesex, and he was again seated in Parliament. But by then, the steam had gone out of his campaign. While he occasionally supported the American cause, he was largely ignored. Those who had lionized him, particularly in the American colonies, read these events as proof that England was irredeemably corrupt and tyrannical.

Although they paid most attention to England, the colonists also looked farther afield for heroes. During the period of their fascination with Wilkes, they found or conjured up an even more romantic hero—the leader of the island of Corsica, "General" Pasquale Paoli.

To Americans, Paoli was a nearly perfect hero. He could be romanticized almost without limit, since he was in a place unknown to all but a very few Americans. In their mind's eye, the Corsicans became a proud, freedom-loving people, led by an intrepid hero in a fight against an ancient tyranny. And, if anyone had any doubt, this marvel had been enthusiastically set forth by the young James Boswell, Dr. Johnson's Boswell, who in 1766 had gone to Corsica at the urging of that other hero of the Americans, Jean-Jacques Rousseau. Suddenly, apparently from nowhere, Americans had found the "greatest man on Earth." They heard him speak to their hearts and loved every sentence of Boswell's account. On the fight for liberty: " 'Sir,' he said, 'if the event prove happy, we shall be called great defenders of liberty. If the event shall prove unhappy, we shall be called unfortunate rebels.' " On an army: "Every single man is as a regiment himself." On motivation: "My interest is to gain a name. I know well that he who does good to his country will gain that." Was he in touch with the struggles of others for freedom? Of course—while in exile in Naples he had learned English from Irish officers in exile. On independence: " 'We may,' said he, 'have foreign powers for our friends; but they must be *Amici fuori di casa* (friends at arm's length; literally, "outside the house"). We may make an alliance, but we will not submit ourselves to the dominion of the great nations in Europe. This people who have done so much for liberty, would be hewn in pieces man by man, rather than allow Corsica to be sunk into the territories of another country.' "

It was all there: simplicity, pride, heroism, nationalism, independence. What more could anyone want? In fact, everyone got a bit more, thanks to an American. The young American artist Henry Benbridge was commissioned by Boswell to paint Paoli's portrait, which he did, making Paoli look rather more like a contemporary British general than a Corsican guerrilla leader. The "sons of liberty" lapped it up. New England resounded with new political clubs known as the "Knights of Corsica." For Americans, Corsica became a symbol of the struggle between virtue and corruption and between liberty and tyranny. But disillusionment eventually set in: finally defeated at home, Paoli retreated to England, where he accepted a pension from the king and became a very tame former rebel. Once again, Americans found, corruption had overwhelmed virtue.

Virtue never had much of a chance in Ireland. As a colony of Britain, it showed what an imperial British future might hold for Americans. Its own parliament had been emasculated in 1720 by a Declaratory Act on which the Declaratory Act of 1766, dealing with the American colonies, was patterned. Powerless, the Irish parliament was denationalized. Of its 307 members in 1775, only forty-four had Irish names; the rest were either English or descendants of the Scots-Irish settlers. Only seventy-two of the members were elected; the rest were appointees of the English viceroy, styled the "lord lieutenant." Virtually all officials were non-Irish Protestants. Roman Catholics, who made up more than three-quarters of the population, were totally disenfranchised; and at least half of the Protestant Irish were also excluded by the Anglican sacramental "Test Act" of 1704, which imposed the Anglican creed as a prerequisite for the franchise.

Such wealth as Ireland had was drained away to England. As the English historian Roy Porter has written, "John Bull's other island, anglicized Ireland, was the most colonial part of Britain, being bullied and bled by a frequently absentee Protestant landlord class." As in England, lavish "pensions" or sinecures were the means by which the ruling establishment controlled the parliament. The contemporary American lawyer and journalist John Dickinson probably spoke for many Americans when he wrote that he was astonished to find "such a love of liberty still animating that LOYAL and GENEROUS nation . . . who have preserved the sacred fire of freedom from being extinguished."

Americans could detect no "sacred fire of freedom" in the ashes of what was by far the greatest of the British colonies, Bengal. No Wilkes or Paoli was found to speak for the people of India. There, as John Dickinson wrote, under the tyranny of East India Company, the same organization for whom the odious tea dumping in America had been arranged, showing

> how little they regard the Laws of Nations, the Rights, Liberties, or Lives of Man. . . . Fifteen hundred Thousand [Indians], it is said, perished by Famine in one Year, not because the Earth denied its Fruits, but this Company and its Servants engrossed all the Necessaries of Life, and set them at so high a Rate, that the poor could not purchase them.

The lessons of Ireland and India seemed clear to Dickinson: "In the same manner shall we unquestionably be treated . . . the parliament will levy upon us such sums of money as they choose to take, *without any other* LIMITATION, *than their* PLEASURE."

How did the American colonists react to these exotic episodes in the quest for liberty and in suffering under tyranny? The answer is partly conditioned by the passage of time: they were initially enormously stimulated, because the American press was filled with news from abroad and what they read accorded with their own experiences. Initially, also, they were encouraged. John Adams saw the hand of God in the upward course of mankind toward liberty—"the opening of a grand scene and design in providence for the illumination of the ignorant and the emancipation of the slavish part of mankind all over the earth." But then, as each of the heroic ventures seemed to collapse before their eyes, Americans came to believe that below the surface of those events they could see, there must be a deeper slough of corruption, stirred by some evil conspiracy whose ultimate aim was to enslave America. Unquestionably, the fear of a collapse of all moves toward liberty abroad played a significant part in convincing Americans that they had an almost divinely inspired task to defend what remained of freedom and decency in their world. As Franco Venturi has written, "The sense of an overhanging menace, of a corruption penetrating everything, almost as if it were a feared disease, carried many in the Britannic world toward reform, and in the colonies toward independence." So the Americans came to believe that in the final round of the struggle, they were alone. As John Adams put this feeling to Arthur Lee in 1771, "America herself under God must finally work out her own Salvation."

CHAPTER 17

"An Ungovernable People"

During the late 1760s misunderstandings, conflicting interpretations, and hostility built up between the British and the colonists. The colonists evaded, demonstrated against, or thwarted, one after another, the acts Parliament passed to regulate them. First came the Stamp Act, which London merchants and craftsmen saw as ruinous to their prosperity. Protests, riots, and boycotts in the colonies which cut into their revenues brought about the annulment of the Stamp Act in 1766. Victory against that act was seen throughout the colonies not only as proof of the validity of complaints against the "mother," but also as a guide to the means of enforcing demands. Massachusetts Chief Justice Hutchinson saw this clearly and warned that if Britain gave in to demonstrations against other taxes levied by Parliament, the colonies would be so encouraged that they might break away from the empire. But no single act seemed worth risking the tumult that a firm policy might entail. So, after years of agitation and despite Britain's need for revenue, the Townshend Revenue Act also fell, on March 5, 1770.

The most significant result of agitation in the colonies was the emergence of what might already be thought of as a national political elite. Men like James Otis of Boston and John Dickinson of Philadelphia learned to work together in the Stamp Act Congress; and through letters, articles, and quotations of their speeches in broadsheets and newspapers, they established the very category of national leadership. Soon they were joined in the public consciousness by a number of other elected representatives, including Patrick Henry of Virginia and Samuel Adams of Massachusetts. Less heralded, and leaving less of a mark in general histories, local leaders also came to the fore in each colonial legislature, in town meetings, and even in the unheralded gatherings of the militia troops and the innumerable political, religious, and social clubs where men living in small towns throughout America gathered to gossip, trade, and amuse themselves. Such informal discussion groups gave birth to a desire to take action both against the British and also against Britain's American allies, who were increasingly seen as turncoats or traitors. Focusing their anger on the "Loyalists," they began to set up "committees of safety," proclaim "resolves," and form themselves into paramilitary organizations. It was through this series of steps, both large and small, both national and local, that the revolutionary generation acquired the skills, the motivation, and the self-assurance that would make the Revolution both possible and, sooner or later, virtually inevitable.

Events in Boston moved earlier and more rapidly toward that end than events in New York and Philadelphia and set a "national" paradigm. This happened partly because Britain perceived Boston to be the most threatening collection of dissidents and partly—and perhaps even more importantly—because it was in Boston that the British tried most visibly and persistently to quell unrest. As events there showed, unrest thrives on repression, and thrives best on inept repression. Each confrontation there impacted upon the next and led to actions that pushed the two sides farther and farther apart. Collection of customs led the way.

Boston lived off the sea trade. A large proportion of its population was engaged as sailors, stevedores, shipwrights, chandlers, shipowners, and merchants. Many others who were not directly involved with the sea handled wares imported by the shipowners and merchants. No single issue commanded more attention from more people in Boston than maritime commerce. Britain had tried, sporadically and ineffectually, to regulate this commerce since the colony was established 150 years before; it failed for a number of reasons. First, it lacked the civil and military bureaucracy necessary to enforce parliamentary acts. It did not even command a police force. Its customs agents were few and had no means of ensuring compliance even when, as seldom happened, they managed to apprehend smugglers. There were other reasons, too. British officials could not resist profiteering; the colonial legal system that distant and distracted successive British governments had allowed to come into being actually protected those who violated the law; and the long and thinly settled coast lent itself to smuggling. Above all, the British failed because Bostonians believed that if the regulations were actually applied, their city would be ruined economically. Thus Britain was caught with the worst of all systems—one which those it attempted to govern regarded as tyranny and which it could not enforce.

Incident piled upon incident. Most episodes were settled before reaching a flash point, since smugglers usually managed to evade the occasional navy patrols or to bribe the nearly always compliant royal officials. Sometimes, however, an incident would explode in an ugly confrontation. On one occasion when the customs officers would not accept a bribe to let through a ship belonging to John Hancock that was smuggling wine, a mob attacked them with clubs, showered them with stones, and beat them nearly to death. Bostonians who were thought to have assisted the customs officers came to be regarded not only as informers but also as traitors—traitors because the colony's Puritan founders and their successors had always thought of Massachusetts as virtually a separate and religiously defined nation-state to which, rather than to Britain, residents owed loyalty. Bostonians who failed to put Massachusetts ahead of Britain became targets of mob violence. In using violence against such "traitors," Massachusetts set a pattern that would be vigorously followed throughout the colonies as events moved toward the Revolutionary War.

How the process functioned can be illustrated by the fate of a sailor, George Gailer, who was thought to have tipped off customs about Hancock's smuggling and was seized by a group calling themselves the "Liberty Boys." They dragged Gailer to a cart, stripped him naked, tarred and feathered him, and paraded him through the streets of Boston. Then they took him to the "Liberty Tree," where they made him swear never again to cooperate with customs. When they released him, Gailer tried to take action against them. He sought the only remedy that he thought was open to him: he went to court to sue them. He quickly found that he was wrong. The case, like most of those against the "patriots," never came to trial.

As much as the naked force they could apply in the street, Bostonian radicals had learned, the courts were a powerful and flexible weapon to thwart British laws, punish opponents, and protect their adherents. When smugglers failed to protect themselves with bribery, attacks by mobs, or threats of murder, they were sometimes, as Gailer tried to do with his tormentors, taken to court. There they had little to fear. Juries were drawn from the smugglers' fellow merchants or sympathizers and appointed by the town meeting. They usually dismissed charges or even issued writs against the arresting officials. Lacking any police power, knowing that access to fair trials was denied them, and often in fear of their lives, the collectors became objects of contempt. Long before the first shots were fired in the Revolutionary War, British authority had been defeated in the courthouse. Reasonably, most British officials concluded that they could not carry out their duties without military force.

The first demonstration of British military force came, appropriately, by sea when the frigate *Romney* of the Royal Navy sailed into Boston Harbor on May 17, 1768. Its arrival happened to coincide with the incident involving Hancock's ship. But far from suppressing unrest, it provoked more. Perhaps the captain was unaware of Hancock or the problem with customs, but in any case he desperately needed sailors to man his ship. Also, perhaps, he was unaware that his attempt to get them would violate a 1707 law that specifically forbade impressment by the Royal Navy in the American colonies. He was just doing what British naval commanders were accustomed to doing even in England. The Bostonians, however, were aware of the 1707 law and interpreted the event as proving the already widely held notion that Britain was not obeying even its own laws. Unaware or contemptuous of the mood of the town, the captain, astonishingly, compounded his mistake by allowing one sailor he had impressed to go ashore to collect his belongings and the pay owed him by his previous employer. That brought him to the attention of the town. The response was immediate. An angry mob gathered, showered the press gang with stones, and forcibly rescued the sailor. Infuriated, the *Romney*'s commander maneuvered his ship into a position that would have enabled it to bombard the town and held a drill in which he ran out the ship's cannon. He did not fire but, tipped off by the customs authorities who had taken refuge on the *Romney,* he sent a landing party to seize John Hancock's ship, *Liberty,* in reprisal.

This episode was a harbinger of worse to come. Reacting to it and other incidents, the British concluded that government in Boston had ceased to function. As the incoming British army commander, Major General Thomas Gage, despaired, "Government in Boston [is] in Truth very little at present." The governor, his council, and various royal officials were all powerless. They had no police or paramilitary force; and if they had attempted to rely upon the traditional "hue and cry" that normally kept order in British cities, they would have brought to the street the very people who were disturbing the peace. Summing up his observations, the British commander decided, as he wrote to the secretary at war in London on October 31, 1768, that only military force could "rescue the Government out of the hands of a trained mob, & to restore the Activity of the Civil Power, which is now entirely obstructed." He ordered two British regiments to Boston.

As bad as the situation was, moving in troops escalated the confrontation. As the British knew, mixing troops and civilians nearly always provoked incidents rather than pacifying the population. Moreover, Bostonians regarded the British move as another violation of law. The law to which they referred this time was the 1689 Bill of Rights, the charter on which William and Mary had become Britain's sovereigns. It forbade "raising and keeping a standing army within this kingdom in time of peace without consent of Parliament, and quartering soldiers contrary to law." To the British, the colonists' claim seemed far-fetched. To make the Bill of Rights—which had never been intended to apply to the colonies—relevant, the Bostonians had to redefine "this kingdom" to include the colonies and "Parliament" to mean "colonial legislature." Bostonians made this jump easily, since they regarded themselves as Englishmen and could point to Massachusetts's Second Charter (of October 7, 1691), which gave "the Great and Generall Court of Assembly" powers comparable to those of Parliament. The key issue was the General Court's authority "to impose and leavy proportionable and reasonable Assessments Rates and Taxes." When the Bill of Rights was read in this light, as of course the Bostonians did read it, importing troops without the consent of the legislature was unconstitutional.

What men of the generation of Benjamin Franklin, Samuel Adams, and Patrick Henry either did not understand or were unwilling to admit was that the relationship between the king and Parliament had evolved since the unfortunate reign of James II. The evolution was what the English philosopher of the Glorious Revolution, John Locke, had hoped would happen. That is, Parliament took advantage of royal weakness or inattention to acquire many of the prerogatives which had been exercised by the earlier Stuart rulers.

The Hanover kings who had replaced the Stuarts were by no means pleased by this transformation. By 1744, tension between King George II and Parliament had escalated to a flash point. When the king tried to act as the Stuarts had acted, Parliament went "on strike," and the king discovered that he was unable to rule without them. Seeking—gently and subtly but firmly—to make the king aware of and accept the new reality, the cabinet sent Lord Chancellor Hardwicke confront him. Hardwicke has left a dramatic account of their tense interview, during which the king said, "I have done all you ask'd of me. I have put all power into your hands and I suppose you will make the most of it. . . . Ministers are [now] the Kings in this country." Government had become "the king in Parliament." That was a very different structure from the one envisaged in 1689. Ironically but crucially, while this transfer of power from the monarchy to Parliament constituted a move toward representative government in England, it set in motion a move away from representative government in the colonies, since the colonial "parliaments" lost power to Parliament. The first legislation in this shift was the "Act of Union" (1707), which suppressed the Scottish parliament. It was followed in 1720 by a Declaratory Act that did essentially the same to Ireland. This trend was carried to the American colonies on March 18, 1766, when Parliament, while repealing the hated and unenforceable Stamp Act, counterattacked by declaring in the American Declaratory Act its right to legislate "in all cases whatsoever." Whereas Englishmen promoted Parliament to stem royal tyranny, colonists viewed the monarchy as a brake on parliamentary tyranny. This explains why nearly all the American leaders including George Washington remained loyal to the king even after the Revolution broke out. It was Parliament, not Britain, against which they were prepared to fight.

As Benjamin Franklin put it, "We have the same King, but not the same legislatures." The current challenge to that position had resulted from an unfortunate historical process: the "Colonies originally were constituted distinct States" but since the coming of William and Mary,

> the Parliament here [in England] has usurp'd an Authority of making Laws for them, which before it had not. We have for some time submitted to that Usurpation, partly thro' Ignorance and Inattention, and partly from our Weakness and Inability to contend . . . [but Parliament's claim] is founded only on Usurpation, the several States having equal Rights and Liberties, and being only connected, as England and Scotland were before the Union, by having one common Sovereign, the King. . . . Let us therefore hold fast our Loyalty to our King . . . as that steady Loyalty is the most probable Means of securing us from the arbitrary Power of a corrupt Parliament, that does not like us, and conceives itself to have an interest in keeping us down and fleecing us.

Many Americans, even the most radical, thought, like Franklin, that attempting to enforce the current parliamentary "usurpation" was tyrannical and dependent upon force. In America, Franklin had said, while British troops "will not find a rebellion; they may indeed make one." But, watching events in Boston, the British were convinced that rebellion was virtually there already. Even if the riots and demonstrations could not so far be legally defined as a rebellion, the town leaders certainly organized for one when they heard about the troop movements. Samuel Adams was quoted as saying that Boston could count on 30,000 armed men pouring into the town from outlying areas to "destroy every Soldier that dares put his foot on shore."

In fact, the 800 or so incoming British soldiers landed in the late fall without incident. They put on an impressive parade through the town, but at the end of the parade, their troubles began: there was no place for them to encamp. Under the terms of the 1765 Quartering Act, soldiers should be put up in "public houses"; if such buildings were not available, the governor was required to turn over government buildings or to rent space. In Boston, few government buildings existed and none had facilities for quartering troops. Even the shoddy, dilapidated "poorhouse" was fully inhabited, and, with winter approaching, the inmates were understandably reluctant to move (and were advised by opponents of the British that they need not); so when ordered to leave, they hunkered down. To try to force them out, the poorhouse was surrounded by troops. After a few days, though, the troops withdrew and the inmates kept their building. With soldiers making no progress, the fleet was called back into service. Attempting to overawe the town, the ships that had brought the soldiers took up a formation that would allow them to devastate the town by cannon fire. That also produced no movement. Irresistible force had met an unmovable object.

As fall progressed into winter, finding adequate places for the troops to sleep became urgent. If they slept in tents on Boston Common they would risk freezing to death. This prospect was so terrifying that almost a whole company deserted within two weeks of their arrival. Desertion probably began spontaneously, but soon the Bostonians began to encourage it. Soldiers could not venture out onto the streets without being cursed, pelted with lumps of snow or jagged pieces of ice, waylaid, or robbed. More subtly—and from the military commanders' point of view more dangerously—off-duty soldiers were approached individually or in small off-duty groups and told that if they deserted, they would be given shelter, food, civilian clothes, money, and even plots of land. The British regiments slowly melted away. Regular, if draconian, punishment by lashing and even a spectacular public execution of a deserter failed to stanch the flow.

Almost imperceptibly, but steadily, the confrontation between British soldiers and Bostonian civilians was producing hatred. Each side regarded the other as a cowardly but violent, loud but hypocritical, formal but illegal enemy. Consider the following incident, which involved racism.

Listening to the American litany of liberty, one British officer grew infuriated and, warmed by too much liquor, he accosted a group of blacks and yelled at them, "Go home and cut your Masters Throats; I'll treat your masters, & come to me to the parade; & I will make you free, & if any person opposeth you, I will run my Sword thro' their Hearts." However much they yearned for their freedom from the British, the Bostonians did not welcome the vision of liberty for blacks. They reacted swiftly, arresting the officer and cracking down on Boston's small black population. The "watch," the closest thing Boston had to a police force, was ordered to "take up all Negroes whom they shall find abroad at an unseasonable Hour." One city official, Justice of the Peace John Hill, was so infuriated by the sight of a black drummer leading a group of soldiers that he shouted, "You black rascal, what have you to do with white people's quarrels?"

The second set of confrontations was among whites. British soldiers were often taunted by crowds of Bostonians. Provocation was general and continual; insults occasionally drew blows, and blows drew crowds. When the crowds got large enough, they challenged the soldiers to fight, yelling obscenities and crude nicknames. Soldiers were caught immobile between painful choices: if on duty, they could not desert their posts, but without orders they dared not defend themselves. Spat upon, screamed at, occasionally physically assaulted, they stood and suffered. They had learned the power of the crowd, and the crowd also had learned its power. If soldiers used their weapons even to defend themselves they could be charged with assault, and in colonial courts they could be sure of being convicted. Astonishingly, to our eyes, the British allowed colonial civilian courts to try uniformed soldiers and officers for acts occurring even during the performance of their duties.

The reason they felt they had to do this was that they had to assert that the soldiers were in Boston not as an army of occupation but as a peacekeeping force to support the civil authorities in keeping order. The logic was impeccable, but the reality was far different. The civil authorities opposed the military authorities every step of the way. Records of hearings make the civil authorities' bias clear. And they had powerful statutes to apply. If a soldier was convicted, the law in Massachusetts allowed the court to assess triple damages and gave the magistrate authority to accept or reject payment. If the soldier could not raise the fine, or if the magistrate refused it, he could be imprisoned or even sold as an indentured servant. Often the man's regiment would raise money to pay the fine, but in at least one case, the amount demanded was too much and the soldier was sold "as a slave" for a three-year period. The British commander, General Gage, was furious, and, inevitably, his soldiers cracked under the strain. One result was the Boston "massacre" of March 1770: it made the confrontation between the British and the colonists irreversible and also dramatized the events in Boston for all the other colonies.

The dramatization was crucial. In the 1760s and 1770s, the colonial societies were quite diverse. Religion, language, modes of life, and local interests separated them. Laws; a system of courts, punishments, and jails; schools; and rudimentary public services had gradually matured, separately, leading to significant variety among the colonies. This separation was just beginning to be bridged by the burst of what we would call revolutionary "agitprop" emanating from the committees of correspondence that Samuel Adams was organizing. His effort was prodigious: "Soon nearly eighty of these furnaces of propaganda were ablaze." But at a time when most roads were hardly more than trails, it was often easier to travel to England than from one colony to the next. So, to a large extent, the eyes of the colonists were still on England. Few, of course, actually went back to England—although many influential men in certain categories, particularly lawyers, had lived and studied there. But colonists widely aped English styles, argued over English politics, and read English books.

Books, ironically, served less as a bridge between England and America than as a dam: the vision of the mother country that the colonists read in their books harked back to the century of Cromwell, the Glorious Revolution, and John Locke, whereas the contemporary British government had acquired—through the change of dynasty, the stressful wars of the eighteenth century, and the beginnings of the industrial revolution—quite a different cast. The English had come to see even their own 1689 Bill of Rights as suspect or subversive, not to be circulated in the colonies. And, like all revolutionaries, the Americans thought of themselves as conservatives, as trying to hold onto the mores of a previous age in which relations had been better, fairer, and more free. Their "England" was not contemporary England. They were harking back to a past that Britain had, by then, put aside.

Letters, like those of John Dickinson, circulated widely among dissident groups. But dissidents were not the only readers. Members of the New York Sons of Liberty warned their counterparts in Connecticut that a "Villin" had "Brouk open" a letter and urged them to be "Very Cautious." This incident is proof of a growing sense of alienation from British authorities: "*we* cannot trust *them.*" By treating the colonists as subversives, Britain succeeded only in convincing many that they were.

Subversion meant disobeying British laws and came to focus on tea. When the "Townshend duties" were suspended in 1770, the duty on tea remained in force, as King George had insisted, "to keep up the right" to tax. This created a serious problem for the British East India Company, which, over the next three years, built up a huge inventory of tea in its warehouses: some 17 million pounds. Because so many members of the English ruling class were investors in the company and because the British government relied upon it to control the emerging colony of India, Parliament passed the Tea Act of 1773, which made "British" tea more competitive but also prevented colonial merchants from engaging in the trade. As John Dickinson and others warned, when the British created a monopoly, even with what appeared to be beneficial effects such as reducing prices, they opened the way for future abuse. Tea thus came to be seen as a new, sinister plot against America. More pointedly, merchants who wished to profit feared that "dumping" cheap tea would ruin the trade.

The Boston Tea Party of December 16, 1773, *the* tea party in the history books, was only one of at least half a dozen such parties that took place along the Atlantic coast. Beginning in Philadelphia, groups sponsored and incited by merchants forced the agents of the British East India Company to resign—just as, previously, they and other groups had forced the designated sellers of revenue stamps to resign. Successful though these "parties" were in expressing anger, they alienated all but a few of America's friends in Parliament and among Englishmen generally. Daily, the press in London was filled with anti-American articles. Benjamin Franklin, Massachusetts's agent, was horrified; with regard to propaganda, the attacks had pulled the rug out from under his feet. Seizing the moment, Prime Minister Lord North persuaded Parliament to pass a series of laws, which the colonists called the "Coercive" or "Intolerable" Acts, aimed at punishing the colonial leader, Boston.

The first of the acts, in punishment for the destruction of British-owned tea, effectively closed the port of Boston—a virtual death knell to the city, which depended upon sea trade for its existence. The effect was compounded by the fact that Boston was then in the wake of a prolonged depression.

The other acts seemed to bear out the widely held notion that Britain was determined to destroy not only American liberty but American society. The Massachusetts Government Act enlarged the powers of the royal governor at the expense of the legislature. The Administration of Justice Act moved beyond the 1764 American Revenue (or "Sugar") Act, which had provided for trial without jury before the "vice admiralty" court for certain offenses. Colonists had already regarded that act as particularly obnoxious, since the judges were awarded part of any confiscated property and so had a "pecuniary temptation" to find against the defendant. The new act, of 1774, went much farther: it resurrected a centuries-old practice (dating from the time of King Henry VIII) of remanding to England for trial those charged with sedition. The new Quartering Act plugged the loopholes in the previous act and ordered American house owners to billet British troops when required. Finally, the Quebec Act stripped Massachusetts, New York, Connecticut, and Virginia of territories they claimed beyond the "interdiction" line established by the Proclamation Line of 1763, that is, beyond the western mountains, and awarded them to Quebec, whose Catholic status was affirmed. This act thus raised in the colonial context the very issues that had convulsed English society at the time of the Glorious Revolution, and it shattered the American dream of cheap land.

Reaction from the colonists came swiftly. Communities and "secret committees" of the several state legislatures began buying and storing arms and ammunition. Mutterings of anger could be heard throughout the colonies. America was seething. Throughout the South, where the personal debts of planters were already onerous, moratoriums were declared. On the frontier, the Quebec Act radicalized a whole new echelon of Americans who were determined to acquire land. The Virginia House of Burgesses, spurred on by the committees of correspondence, called for the assembling of the First Continental Congress.

Delegates to the Continental Congress came from all the colonies except Georgia, but public opinion there had nevertheless been radicalized. The *Georgia Gazette* told its readers they had "sucked the love of liberty at the same breast" as other Americans and Englishmen . . . [but] if we are no longer allowed the rights of Britons, WE MUST be Americans." The royal governor dissolved the legislature, so Georgia alone among the colonies sent no delegates.

The fifty-six delegates who arrived in Philadelphia were the most remarkable assembly that had ever taken place in America. The former British prime minister Lord Chatham (William Pitt) described it as "the most honourable Assembly of Statesmen since those of the ancient Greeks and Romans, in the most virtuous Times." His opinion was not shared by John Adams, a participant, who found the assemblage composed of "one third Tories, another Whigs, and the rest mongrels." They included George Washington, Patrick Henry, and Richard Henry Lee of Virginia; John Jay of New York; John Dickinson and Joseph Galloway of Pennsylvania; John Hancock, Samuel Adams, and John Adams of Massachusetts; and many of lesser known but impressive men.

John Dickinson emerges from his own writings as one of the most attractive participants. After studying law in London, he played a leading role in the assemblies of Delaware and Pennsylvania, so he came to the Congress as a highly experienced statesman. His *Letters from a Farmer in Pennsylvania* became one of the most widely read books of the eighteenth century; it was reprinted in almost every newspaper of the time, not only in America but also in England and Ireland. It even was discussed in the Italian newspaper *Gazzetta Universale* (where the very sophisticated Pennsylvania "farmer" was misunderstood to be an American peasant). Dickinson put Anglo-American relations in the context of English judicial and political history, with frequent allusions to the classical experience. For him, as for many of the colonists, the bedrock of liberty was the Glorious Revolution of 1688–1689 and its Bill of Rights. Dickinson's political program was to conserve those rights despite the turmoil of his age.

During the Second Continental Congress, Dickinson wrote the conciliatory "Olive Branch Petition" urging the British to compromise; his proposal was adopted by the congress against the strident opposition of the Massachusetts delegation. Finally, as events proceeded toward war, he wrote the "Declaration of the Causes of Taking Up Arms," but he opposed the Declaration of Independence, fearing, as he wrote, that:

> When the appeal is made to the sword, highly probable it is, that the punishment will exceed the offense; and the calamities attending on war out-

> weigh those preceding it. . . . Torn from the body, to which we are united by religion, liberty, laws, affections, relation, language and commerce, we must bleed at every vein.

Both in personality and in experience, Dickinson opposed the Boston radicals. He wrote:

> I hope, my dear countrymen, that you will, in every colony, be upon your guard against those who may at any time endeavor to stir you up, under pretenses of patriotism, to any measures disrespectful to our Sovereign, and our mother country. Hot, rash, disorderly proceedings, injure the reputation of the people as to wisdom, valor, and virtue, without procuring them the least benefit.

That statement, in all but name, was addressed to Samuel Adams.

Samuel Adams, as noted earlier, is a more shadowy figure than Dickinson. A superb political tactician, he has been described as the Lenin of the American Revolution. Hiller Zobel portrays him as a violent, scheming, dedicated manipulator who was willing to engage in any acts that would lead toward the single goal of American independence. The historian William Appleman Williams is more favorably inclined but admits that Adams's "penchant for destroying his correspondence and other records can be interpreted as the careful habit of a dedicated revolutionary. In any event, he clearly sought independence after 1769." Pauline Maier finds quite a different Adams. She argues that "he wanted reform within the context of empire in the early 1770s," always counseled patience, and "never justified force as a first response to oppression." Whatever his personality, motivations, and tactics, he was, as Thomas Jefferson called him, "truly the *Man of the Revolution.*"

While Samuel Adams was the leader of the "left," Joseph Galloway was the leader of the "right." Like Adams, Galloway was a highly experienced politician who, as speaker of the Pennsylvania Assembly, had learned the subtle art of leading men. Elected to the First Continental Congress, mainly with the support of Pennsylvania's wealthy Society of Friends (Quakers), he set out a conservative program. The day after he agreed, probably as a tactical measure, to the resolution not to import English goods, he astonished the assembly with a bold plan of accommodation that would have given Americans more or less the status of a dominion. It was narrowly defeated; and a month later, on October 22, it was deleted from the record of the proceedings. Angry and dispirited, Galloway went to New York, whose delegates had supported him, to try to turn the tide in favor of his plan. There he published a tirade against the congress, accusing it of being a Presbyterian plot to hustle America into war, and warning that if America "denies the authority of the mother-state . . . She must in all probability soon become the slave of . . . [French] Popish bigotry and superstition." Like the British government, Galloway endorsed the concept of the "king in Parliament" as the legitimate government for America. From London, Benjamin Franklin, who had become completely disillusioned with "the extream corruption prevalent among all orders of men in this old rotten State" wrote Galloway on February 25, 1775:

> I apprehend, therefore, that to unite us intimately will only be to corrupt and poison us also. . . . However, I would try any things, and bear any thing that can be borne with safety to our just liberties, rather than engage in a war with such near relations, unless compelled to it by dire necessity in our own defence.

Despite their growing differences, Franklin, who remained in close and friendly contact with him, urged Galloway to remain in political life but sought to convert him to support of the decision of the First Congress. After being elected to the Second Congress, Galloway found himself increasingly isolated, and he resigned on May 12, 1775. Unable to join or to stand aside, he made his choice. During Washington's retreat from New York, when it appeared that the revolutionary cause was lost, he took refuge with the British army and subsequently went to England.

In addition to the Americans, two Englishmen hovered in spirit over the meetings in Philadelphia. Although he had died seventy years before, John Locke was still a major personality in America in 1774. No account of that period can be complete without his presence. More than any other man, Locke embodied the heritage that even the most radical Americans sought to retain; his treatises on government defined political thought in America in the years before the Revolution. The particular idea in Locke's writing on government that most influenced Americans was that government grew out of a two-step process. In the first step, people contracted among themselves to create a community. For the people in Massachusetts, this had been an actual event. When the 101 pilgrims came to Plymouth Rock in 1620, they formed "a civil Body Politick" in the Mayflower Compact. That is, they actually did what Locke imagined to be the emergence of society from the "state of nature." Implicit was a second step in which a "contract" was formed between the sovereign and the society. The British Bill of Rights, which set out the terms on which William and Mary became the sovereigns, was essentially such a contract. As "fellow Englishmen," Americans regarded this document, augmented and affirmed by their charters, as also their agreement with the imperial government. Locke was the inspiration of virtually all Americans no matter how divided they were by other beliefs.

The second figure could hardly have been more different. Although Americans continued to proclaim their loyalty to him even after the fighting began, George III is usually cast as the villain of the American Revolution. True, he was the titular head of the British government and wanted to suppress what he regarded as treason, but a less likely candidate for demonization would be hard to find. George had been a retarded child: he could not read until he was nearly a teenager, and even as an adult he had trouble writing. Lonely and culturally confused, the first of his German family to be raised as an Englishman, he desperately sought emotional attachments to older men to replace the father he had lost as a boy of twelve. Almost overwhelmed by a feeling of inadequacy, he tried to make up for it by stubbornness. Finally, George early showed the signs of manic depression that would eventually—in 1811, long after the Revolution—incapacitate him.

Abstractly, George III "loved" his people, but he struck no chord of affection among them. Even when he was riding through London on the way to celebrate his marriage, he was booed and hissed by the crowds. He knew little of America and understood even less, but, ironically, he was almost as much annoyed by Parliament as the American colonists were.

Reacting to Parliament's infringement of what he regarded as his role in the British constitutional system, he twice wrote (but did not send) letters of abdication. For him, abdication would have been an action roughly analogous to the American Declaration of Independence. Up to the outbreak of war, the colonists believed or affected to believe that the king wished them well but was misled by his ministers, isolated, and ill informed. They were wrong. As his confidential writings (which of course they did not see) make clear, George was early on determined to suppress the rebellion. As he wrote to Lord North while the Continental Congress was in session, "The New England governments are in a state of rebellion, blows must decide whether they are to be subject to this country or independent."

Meanwhile, in Philadelphia, the delegates met to try to make that decision in what they thought were secret sessions, but the British were informed of every opinion they expressed. Their first task was to meet and sound one another out. John Adams recounts an endless round of breakfasts, lunches, dinners, drinks at houses and taverns, and rambles through the city, even to the "Cells of the Lunaticks." He also has left us vignettes of the delegates: John Dickinson was "a Shadow—tall, but slender as a Reed—pale as ashes"; Richard Henry Lee was "a tall, spare, Man . . . a masterly man"; Edward Rutledge of South Carolina was "young—sprightly but not deep"; Cæsar Rodney of Delaware was "the oddest looking Man in the World . . . his Face is not larger than a large Apple"; and Tom Paine raised the toast "May the Collision of British Flint and American Steel, produce that Spark of Liberty which illumine the latest Posterity."

The formal meetings began on September 5, 1774, at the City Tavern, from which the delegates walked to Carpenters Hall. There they elected a chairman, examined one another's commissions, and discussed procedure. Since they had no precedents to follow, they cautiously felt their way through "such a Field of Controversy as will greatly perplex us." Finally, they agreed that each colony would have only one vote, that the meetings would be secret, and that, after debate, resolutions would be unanimous. Virginia had insisted that the only matters to be discussed would be acts by Parliament after the end of the French and Indian War in 1763. This focused attention on taxes, customs duties, and the "Intolerable Acts." The delegates also affirmed their devotion to King George, the "Most Gracious Sovereign," reserving their opprobrium for his ministers as "designing and dangerous men . . . prosecuting the most desperate and irritating projects of oppression."

Hardly had they begun their serious discussions when, on September 6, a rumor as powerful as it was untrue was rushed into town by an "express," seventy hours out of Boston: the messenger reported that the British fleet had bombarded Boston and General Gage had begun a "horrid butchery." Rising to his feet, Patrick Henry proclaimed, "Government is dissolved. . . . We are in a State of Nature. . . . I am not a Virginian, but an American." For the next several days, all was panic; a general call to arms brought streams of volunteers off farms and out of villages, ready to do battle with the "cursed British." The "Powder Alarm," as this incident came to be known, set the emotional context of the congress.

The first substantive business of the congress was agreeing on means to put pressure on the British. The delegates remembered that their most successful tactic had been to curtail purchases from English merchants, manufacturers, and artisans, who in turn had petitioned their government for a change of policy. It was at this point that a copy of the "Suffolk Resolves"—the acts of the assembly that took the place of Massachusetts's banned town meetings—arrived, strongly urging an end to all trade with England until the "Intolerable Acts" were repealed. These resolves sanctioned taking hostages if any colonials were arrested by the British; the resolves also urged military preparedness. The congress immediately endorsed the resolves. The "Nonimportation Agreement" was the easy part.

The hard part was deciding whether the colonies should export their produce to England. Representatives of the northern colonies, having little to lose, argued for a complete ban; but the southern agricultural colonies, which depended absolutely on exports, were greatly disturbed. South Carolina's delegates stormed out of the meeting; they returned only when rice was excluded from the ban. The differences between North and South thus illustrated would become a central theme of American history throughout the following century.

It was at this point that Joseph Galloway's plan to keep America in the empire was announced, debated, and narrowly defeated. The moderates at Philadelphia fell silent as the radicals pressed ahead. Even the radicals, however, felt the pressure from the hinterland as little communities all over America pressed "resolves" urging sterner resistance to Britain or a complete break.

CHAPTER 18

Casting the Die

As we observe Great Britain and the American colonies on the brink of war in 1774, we confront two questions. First, why did the British and the colonies not reach some sort of accommodation that would have headed off the Revolution and kept the colonies in the British Empire? Second, what would a knowledgeable, neutral observer have thought would happen if they did not? To approach these questions in the context in which contemporaries saw them is difficult for us, because knowing what happened, we tend to think it was inevitable. Contemporaries, of course, did not; to see what they saw takes a leap of imagination. To bring some order into what was at the time great confusion, I will divide my account into parts: on the one hand, what the participants saw as the issues and how they viewed themselves; and on the other hand, what their relative capacities and weaknesses were.

First, how did the British see the issues? The British government had many critics both at home and in the colonies. The electoral system produced a Parliament that imperfectly represented the population; whole cities were not represented at all. A sanctioned system of buying favors further distorted the political structure. Yet despite growing criticism, no one in England or abroad denied that as it existed the government had the right to speak for Britain.

At the apex of the government was the monarch. While George III was not so absolute as most of his predecessors had been, he was able to appoint or dismiss the chief ministers. A significant portion of the members of Parliament were beholden to him for their "places," the sinecures that he controlled and that provided the income upon which they relied. Consequently, aspiring politicians tended to follow his lead. Moreover, among his powers was appointment of officers in the armed services, so that, in a fundamental sense, military power was also his.

Because the reigning family was of recent German extraction and both England itself and the empire were growing increasingly complex, the monarch relied upon a privy council to advise him on major issues. Its members were tied to him not only legally but also by religion, social mores, economic interests, and other attributes. Disagreements occurred, but they were within a fairly narrow spectrum and were usually easily adjusted.

Over the previous century, Parliament had filled some of the spaces left unattended by distracted, alien, or weak monarchs. While it contained no representatives of the poor, and represented only the owners of land, it did on occasion offer a platform for men like Edmund Burke to offer alternative policies and a sanctuary for radicals like John Wilkes to criticize the government. The king had the option, which he used against William Pitt, to emasculate such men by "promoting" them to the less powerful House of Lords or, as he tried (and failed) to do with Wilkes, to harass them through the judiciary. But Parliament was growing in power and privilege at the expense of the monarchy. Ironically, George and the American colonists agreed in resenting its new assertions of power.

But George was firm in upholding the right of Parliament to enact legislation for the colonies. As he wrote to his prime minister, Lord North, on September 11, 1774,

> the dye is now cast, the Colonies must either submit or triumph; I do not wish to come to severer measures but we must not retreat; by coolness and an unmerited pursuit of the measures that have been adopted I trust they will come to submit; I have no objection afterwards to their seeing that there is no inclination for the present to lay fresh taxes on them, but I am clear there must always be one tax to keep up the right, and as such I approve of the Tea Duty.

Beneath the monarch and his ministers was a bureaucracy that had, like Parliament, also grown in complexity, reach, and power over the previous century. Much of what we think of as government was still "farmed out" to private contractors, as it had been when the American colonies were established. Britain's most important colony, Bengal, was the preserve of the British East India Company, and many domestic functions were carried out by private contractors. But the number of civil servants had multiplied; and bureaucracy, which in Tudor times had been merely the households of ministers, had become institutionalized. By the 1770s, Britain commanded the services of thousands of men in the military, diplomacy, intelligence, espionage, customs, excise, and other activities. In short, Britain had followed Spain and France in acquiring a large, functioning, and centrally directed government.

Those attributes were precisely what the American colonies did not have. "America" had no counterpart to George III or even to Parliament; it had no bureaucracy or army. In 1774, it was not America, a single entity, but a loose collection, only in the process of becoming a coalition, of scattered societies.

In recent years the colonies had been divided, often bitterly, from one another: Georgia and South Carolina disputed their border; North Carolina had suffered what amounted to a civil war with the "Regulators"—suppressed but nursing economic, geographical, and political grudges against the affluent coastal part of the colony. Virginia and Pennsylvania actually fought over their claims to Indian lands in the interior. Pennsylvania and New York were at loggerheads over their interior lands. Massachusetts was hostile toward "breakaway" Rhode Island. These divisions surfaced when, near the end of the seventeenth century, King James II had tried to create a degree of unity in the "Dominion of New England"; so little interested were the colonists in unity that they quickly used James's overthrow to abolish even this anemic form of it. During the French and Indian War, the colonies failed to cooperate even when threatened by invasion; indeed, pushing local interests to the fore, some aided the French by covert sales of supplies and food. Except for the short-lived and limited Stamp Act Congress, there was virtually no attempt at coordinating attitudes and policies. "Were these colonies left to themselves tomorrow," as the Massachusetts radical leader James Otis opined in 1765, "America would be a mere shambles of blood and confusion."

Into the 1770s, there was no central organization of these virtually warring states. It was in an attempt to meet this challenge that representatives of the separate legislatures met on October 14, 1774, as the First Continental Congress, and formed the "Association." The Association was very far from being a government. Its sponsors, the colonial legislatures, had given it no executive authority. It had no treasury and no independent means of acquiring one; and it had no sinews of power—no military, diplomatic, or intelligence services, indeed no staff of any kind. Worse, it had no generally recognized, experienced leadership.

So, who could speak for America? Since America had no "government" or coherent national movement, Britain could only hope to gauge public opinion from the babble of voices reaching London. Was consensus to be found among "agitators" or "loyalists," merchants or frontiersmen, Virginia Anglicans or New England Nonconformists? Each listener had his own sources. Consequently, like every colonial "mother" known in history, Britain listened most closely to the voices of those "daughters" who sounded most "sensible" and disregarded those whose tone and content seemed subversive or even impolite.

We now know a great deal more than contemporaries did about the makeup of the American population in 1774. Of the approximately 2.5 million people in "British America," one person in six was a black slave. Although black slaves had no means to make their voices heard, the British concluded on the eve of the Revolution that slaves hated their masters, would identify their masters with the American cause, and so at least would threaten to rebel. (This was the "intelligence appreciation" on which Lord Dunmore based his emancipation proclamation.) An equal number of the colonists were newly arrived Germans. They had not yet found their voice in American affairs, but they stood, or appeared to stand, apart from American politics and its dispute with England. They might even, a contemporary would have reasonably concluded, be inclined to be pro-British, since the British king was himself of German background and a large part of the British army sent to America was German. At least there was no reason to think that these Germans would oppose Britain. What of the British Americans? Demonstrations in the American towns had evinced deep and bitter splits. The loudest, most active, and most violent

Americans were the "radicals," but they did not necessarily have the greatest staying power. And in numbers, as we now know, roughly half of the English-speaking colonists were either pacifist, neutral, or actively pro-British. Many of the waverers, particularly in the South, were alienated only when the British tried to incite the blacks against their masters. Ironically, Lord Dunsmore's emancipation proclamation did more to unify opposition than the Stamp Act.

One group does appear, but perhaps only in hindsight, to have been solidly anti-British: the colonists who lived on the frontier. Many were Scotch-Irish Presbyterians of whom 250,000 had arrived in the half century before the Revolution. But did they really count? They were widely scattered, poor, and mostly illiterate. I imagine that until about 1770, an outside observer would have thought that their only interests were grubstaking newly acquired land and driving away the neighboring Indians. The War of the Regulation had shown the limits of their cohesion and power: the victors had supported the royal governor, and the defeated had promised undying allegiance to the crown. In 1770, a reasonable observer would have either discounted them or thought they would be loyal to the crown. Far from the agitation of towns like Boston, they became anti-British only on the eve of the war, when Britain tried to rescind titles to disputed lands. We now know what was then probably not known in London, Boston, or Philadelphia: that in 1774 and 1775 nearly a hundred little groups of them would push vigorously for American independence.

In short, a knowledgeable British observer could reasonably conclude, the unity for which the rebel leaders had striven with their committees of correspondence, their "sons of liberty," their "tea parties," and their boycotts was a very long way from being achieved. There was no "America." Singular. America was, in fact, still a myth.

Also a myth was the Americans' quest for independence, as we think of "the spirit of '76." Even after the outbreak of fighting at Lexington and Concord and Bunker's (or Breed's) Hill, the leading "patriots" were determined to remain in the empire as loyal subjects of George III. With muskets already in hand at Lexington and Concord, they still thought of themselves not as quitting the empire or opposing its sovereign but only as remedying the wrongdoing of Parliament. John Adams and other members of the Continental Congress damped down the voice of those urging independence and sent what became known as the "Olive Branch Petition" to King George, imploring him "to point out some mode of accommodation." As George Washington later wrote, "When I took command of the army, I abhorred the idea of independence." Even seven months after he had been appointed commander, Washington led his staff at each meal in drinking to the health of the king; from other leaders, petition after humble petition was addressed to "our lawful and most gracious Sovereign King George the Third." It was not until April 12, 1776, that the first colonial assembly, North Carolina's, instructed its delegates to the Continental Congress to vote for independence. America was at war long before admitting it.

Just as the parties to the conflict were asymmetrical, so their definitions of the issues differed. George III and his ministers heard the colonists say that they rejected the right of Parliament to legislate for them. What the king believed the colonists meant was that they wanted independence. On independence, the position of the king was both consistent and clear: he would not contemplate destruction of the empire. As he put it, "if any one branch of the Empire is allowed to cast off its dependency . . . the others will infalably follow the example [and] the state will be ruined." That is, he believed in what has been called in our time the "domino effect." Ireland and the highly profitable West Indies, he feared, would follow the American lead. American independence would bring "total ruin, [since] a small State may certainly subsist, but a great one mouldering cannot get into an inferior situation but must be annihilated." As the king saw it, surrendering to the demand of the radical minority of Americans would be tantamount to state suicide.

From the time he became king until the last shot was fired, George clung tenaciously to the widely believed mercantilist interpretation of economics: to prosper, a state must create for itself a monopoly area from which it excludes all foreigners. That was what Britain was doing in India and was trying to do in America. Britain needed raw materials and a market for its industry. Therefore, allowing the American rebels to escape Parliament's control of trade (the Navigation Acts) would be commercially ruinous. "Then this island would be reduced to itself," the king said, "and soon would be a poor island indeed." Americans at the time, and most later historians, saw this rigidity as a sign of George's stupidity and stubbornness. With hindsight, historians have pointed out that the king's fears were not realized, since after becoming independent in 1783, America made an even larger contribution to English prosperity than before. But, of course, neither George nor anyone else could have prophesied this in 1774. While some Englishmen then disagreed with government policy, most articulate Englishmen did not. Even though the government was criticized for its corruption, ineptness, and tyranny, it was generally warmly supported in its hard-line policy toward the Americans by the early 1770s. Many who had favored the Americans before the Boston Tea Party of December 16, 1773, subsequently became part of what one perceptive English observer, Edward Gibbon, called a nationwide "clamour" for a hawkish policy.

As the hawkish policy took hold, the king opened Parliament in December 1774 with a speech that John Wilkes considered an "American death warrant." King George then refused to accept further petitions from the colonists, even the "Olive Branch" of the Continental Congress in the summer of 1775, and declared his "firm and steadfast resolution to withstand every attempt to weaken or impair the supreme authority of the Legislature over all the dominions of the crown." In short, there was to be no negotiation. That cut the ground out from under the "moderate" nationalists like John Dickinson. Only then would they begin, slowly and reluctantly, to turn against the king.

The clearest American definition of the fundamental issue was not set forth until July 6, 1775, by representatives of "the United Colonies of North America, now met in Congress at Philadelphia." It was a rejection of Parliament's authority to "make laws to bind us in all cases whatsoever." If that right was allowed to stand, the delegates asked,

> What is to defend us against so enormous, so unlimited a power? Not a single man of those who assume it is chosen by us; or is subject to our controul or influence; but on the contrary, they are all of them exempt from the operation of such laws, and an American revenue, if not diverted from the ostensible purposes for which it is raised, would actually lighten their own burdens in proportion as they increased ours.

That is, Parliament, which was already unrepresentative even of the English, would be led further astray: self-interest (English as opposed to imperial interest) would lead it to destroy Americans' prosperity and freedom. Inevitably, as noted earlier, colonialism contains this fatal flaw: sooner or later, the interests of the colonizer and the colonials will diverge beyond reconciliation.

What would happen then? That is, what would a knowledgeable, neutral observer have predicted if words gave way to bullets? Had there then been something like the U.S. National Intelligence Council or the Policy Planning Council, its estimate might have included the following assets and deficits.

First, consider Great Britain. It had a population of approximately 6.5 million and a yearly government revenue of about 11 million pounds. It had the most powerful military combination of any European power. Its fleet, comprising 30,000 men on 230 ships, could land its 18,000-man, highly professional army almost anywhere to achieve "theater superiority." It had augmented its own land army with troops rented from three German states; and, with its already formidable and growing economic power, recently augmented by its conquest and rape of the rich Indian province of Bengal, it could afford to hire many more.

The contrast with America could hardly be more striking: scattered and divided, the colonies had a population that aggregated only about one-third as many as Britain's. True, several colonial towns were home ports for armed merchant ships, but none had a single ship that could contest even a Royal Navy corvette, much less a frigate or ship of the line. No ship captain had ever commanded in a naval engagement. On land, no officer in any colonial militia had ever commanded a sizable number of troops in a serious engagement, except George Washington, whose engagement had ended in a humiliating surrender. There was neither a system of cooperation between the separate militias nor anything resembling a standing army. When Washington arrived in Cambridge the first week of July 1775 to take command, he found that his "army"—actually just a collection of the town militias of Massachusetts—had assembled on the understanding that it could disband, regardless of what was happening around it, a few weeks later; as commander, Washington had no funds, no commissary, no staff, no chain of command, indeed no command at all.

Regarding equipment, the situation was even worse. Depots of British arms were being illegally seized, but Britain had retaliated by embargoing the supply of military equipment to the colonies. Even at the time of the battle of Lexington and Concord, there were said to be only 21,549 firearms in the whole Massachusetts Bay colony. As late as July 1776, about one soldier in four had no arms. In any event, few of Washington's "soldiers" knew how to shoot. Realizing this, the Continental Congress had urged the several colonies to begin to train military forces; but, as the British knew, the process of turning civilians into soldiers was long and costly.

To make matters worse, the colonists faced the probability of a conflict on two fronts. In the interior, the Indians, while not included in colonial politics per se, probably could be incited by the experienced and skilled cadre of agents Britain had deployed among them to take revenge on the colonists. While the Indians had little reason to be grateful to Britain, they were swayed by gifts and perhaps by Britain's previous attempts to reverse (in the Easton treaty of 1758) or halt (in the royal proclamation of 1763) the invasion of their lands by the American colonists. Certainly the Indians—whom George Washington called "a cruel and blood thirsty Enemy upon our Backs"—had no reason not to attack the Americans, who had always been their most dangerous enemy. Indians' military importance, while declining, was still formidable enough to distract, terrorize, and weaken the colonists along the frontier.

In foreign affairs, the British and Americans were almost incomparably different. The colonists had no foreign relations and no cadres of experienced diplomats or analysts. Worse, they did have attitudes that were diametrically opposed to their quest for independence. If a poll had been taken in 1774, France would surely have ranked as more disliked than any other power except Spain. The idea that the two Catholic powers might be the salvation of the Revolution would have struck contemporaries as laughable. Yet, of course, that is what happened.

Britain, on the contrary, was a major world power with interests and objectives spanning Europe, Africa, and much of Asia. American historians have naturally focused on the American aspects of the crisis of 1774–1775, but the British were accustomed to dealing with complex issues of foreign affairs. Britain employed an experienced bureaucracy, which had become skilled in intelligence and diplomacy.

Of course, America's potential revolt was only one issue with which Britain had to contend. Others that could not be neglected included the following. Close to home, both Ireland (where riots had prevented the dispatch of British troops to the New World in 1766) and Scotland (which was still suffering from the suppression of the Highlanders after the battle of Culloden) remained insecure. The British-occupied Caribbean "sugar islands," then a major source of revenue, were convulsed by a Carib Indian revolt (which did not end until 1796) and always seemed on the point of an uprising by black slaves. From its "bridgehead" in Bengal, Britain was expanding into the still formidable Mughal Empire and was beginning to engage the even more powerful Marathas, who ruled most of the center of the Indian subcontinent. In Europe, relations with Spain had long been strained (over Ireland, the Caribbean, Holland, and Spain's espousal of the English Catholic cause) and had steadily deteriorated following the end of the Seven Years' War. Finally, rich, powerful, and populous France was thirsting for revenge on Britain for the humiliating loss of much of its empire in Asia and North America and was potentially allied (under the so-called family compact) with its fellow monarchy, Spain. That was a particularly worrying problem, since such an alliance would give the two powers a naval force which would be larger than Britain's and which might enable the large French army to invade England. Even if invasion could be fended off, the 80,000 French-speaking Catholic Canadians were a potential threat.

To cope with all these issues simultaneously would have challenged a far more sophisticated government than Britain then had. And it had to deal with these matters virtually alone. In the 1770s, Britain had no European friends or allies. The central doctrine of later British foreign policy, the balance of power, was not even a dream. Thus, ironically but perhaps not so surprisingly as it might then have seemed, the isolated, inexperienced, ill-equipped Americans would be, relatively, stronger than Britain in international affairs.

Finally, consider the economy. Britain was already beginning its industrial revolution and was the most dynamic manufacturing and commercial nation in the world. In contrast, the economy of the colonies was rudimentary, fragile, and largely restricted to primary products. Industry hardly existed, particularly in those areas essential to warfare. Britain had retarded the growth of colonial mills capable of turning iron into steel, and only three sizable shops were capable of making firearms. Lead was in such short supply that when fighting began, the colonists had to melt down statues, window weights, and even gutters from houses. Flint, essential for firing a musket, was desperately short. Only in 1776 was a good vein of stone located, and even then few residents in the colonies knew how to shape it to strike a spark. No one in the colonies knew how to turn saltpeter (sodium nitrate) into gunpowder. Such powder as was needed for hunting was customarily imported from Britain. Washington's army in the summer of 1775 had only about half a pound per man. Getting gunpowder would remain a serious problem. Even the paper required to make a musket "cartridge" was imported from Britain. As the British knew, the hunting rifles used on the frontier could not be fitted with the weapon the British regarded as basic to infantry tactics, the bayonet; so "Kentucky long rifles" could be largely discounted from the already scanty inventory of colonial firearms. Even shoes were not being produced in the quantity an army would require. And the transport and communication system, which relied heavily on coastal shipping, was vulnerable to the strongest of Britain's weapons, its navy.

Thus a knowledgeable observer would not have been wrong to conclude that such forces as the colonies might assemble would be ill equipped, scattered, without trained leadership, and almost immobile. To experienced British officers, this conclusion was self-evident. Well might Major General James Grant tell his fellow members of Parliament on February 2, 1775, that the Americans would not, indeed could not, fight and that with just 5,000 British soldiers he would undertake to march from one end of the continent to the other. One of his colleagues went even further. In the hearing of Benjamin Franklin at a private house in London, General Clarke said that "with a Thousand British grenadiers, he would undertake to go from one end of America to the other, and geld all the Males, partly by force and partly by a little Coaxing."

In summary, no prudent Englishman believed that there was any significant possibility that "America," a concept yet to be born, could survive apart from Britain. That it would try to fight Britain was, as one of Britain's most distinguished officers, Major Patrick Ferguson, would later write, "a mystery indeed." Mysterious it was, but King George seemed almost relieved when the uncertainty ended and war seemed inevitable. As he wrote to his prime minister, Lord North, "I am not sorry that the line of conduct seems now chalked out. . . . blows must decide whether they are to be subject to this country or independent."

Some colonists continued to hope that war could be avoided despite the fact that by the spring of 1775, moves preliminary to the war had already gained a momentum over which even sophisticated governments often lose control. The British government had already stationed troops in New York and Boston, and the colonial legislatures had begun seizing British arms depots. On June 15, 1775, the Continental Congress and the "Association" voted to establish a unified military authority under the command of George Washington. These moves did not have the sanction of the sovereign and so constituted acts of rebellion.

Finally, on July 6, 1775, representatives of the "United Colonies of North America, now met in Congress at Philadelphia" proclaimed the stark choice they faced and the decision they had made: "We are reduced to the alternative of choosing an unconditional submission to the tyranny of irritated ministers or resistance by force. The latter is our choice. We have counted the cost of this contest, and find nothing so dreadful as voluntary slavery."

Even then, the decision seemed negotiable. John Hancock as president committed the Congress to "reconciliation on reasonable terms, and thereby to relieve the Empire from the calamities of civil war." But this was a faint hope and one that came too late. As in crises in our times, we can see that events had acquired a sort of rhythm; each move made the next not only logical but virtually beyond the control of the participants. Thus, although hawks often continued to talk like doves, they were caught in a process that seemed almost inevitable. In the two centuries since the foundation of Jamestown, the colonies had grown up and apart. In an almost biological sense, they were no longer "daughters" who could be treated—or could act—as children. Events were beyond control, beyond even desire, and certainly beyond plan, and there could be no return. At Bunker Hill, the "labor pains of revolution" had begun. "America" was being born. Delivery would be long and painful.

Notes

INTRODUCTION: REVISING THE AMERICAN PAST

xi **notched sticks:** Although we cannot read them, these were the first American histories. On one described in the late nineteenth century by Colonel Luther P. Bradley, see Thomas Powers, "The Indians' Own Story," *New York Review of Books* (April 7, 2005). Also Peter Nabokov, *A Forest of Time: American Indian Ways of History,* (Cambridge: Cambridge University Press, 2002). On "winter counts" or tribal histories, see *http://www.wintercounts.si.edu.* Most surviving winter counts date from after the period I discuss in this book.

xii **from an "external" perspective:** In contrast, an excellent book by Jon Butler, *Becoming America: The Revolution before 1776* (Cambridge, Mass.: Harvard University Press, 2000), focuses on the colonies as they coalesced toward the end of the seventeenth century.

xiii **and then recorded them:** David Freeman Hawke (ed.), *Robert Beverley, The History and Present State of Virginia* (New York: Bobbs-Merrill, 1971); and Hugh Talmage Lefler (ed.), *A New Voyage to Carolina by John Lawson* (Chapel Hill: University of North Carolina Press, 1967).

xv **a new school of American historians:** *The Spanish Borderlands: A Chronicle of Old Florida and the Southwest* (New Haven: Yale University Press, 1921). Bolton's work was carried on by Edward H. Spicer, *Cycles of Conquest: The Impact of Spain, Mexico, and the United States on the Indians of the Southwest, 1533–1960* (Tucson: University of Arizona Press, 1962); and by, among others, David J. Weber, *The Spanish Frontier in North America* (New Haven: Yale University Press, 1992).

xvi **the attention of scholars:** John T. McGrath, *The French in Early Florida: In the Eye of the Hurricane* (Gainesville: University Press of Florida, 2000). See the review article by Amy Turner Bushnell of this and other works in *William and Mary Quarterly* (60:3, July 2003).

xvii **1585 by John White:** Paul Hulton, *America 1585* (Chapel Hill: University of North Carolina Press and British Museum Publications, 1984).

xvii **worked their lands:** Samuel Eliot Morison (ed.), *Of Plymouth Plantation, 1620–1647 by William Bradford* (New York: Knopf, 1959).

xvii **to original documents:** *The Genesis of the United States* (Boston: Houghton

Mifflin, 1890) and *The First Republic in America* (Boston: Houghton Mifflin, 1898).

xvii **ventures into social history:** A pioneer in the field was Alice Morse Earle, whose *Home Life in Colonial Days* was published in 1898. Richard Hofstadter, *America at 1750: A Social Portrait* (New York: Random House, 1971); David Freeman Hawke, *Everyday Life in Early America* (New York: Harper and Row, 1988); and James Deetz, *In Small Things Forgotten* (New York: Anchor, 1977) and *The Time of Their Lives: Life, Love, and Death in Plymouth Colony* (New York: Freeman, 2000) have been followed by a number of scholars. On houses of the early settlers, see Harold R. Shurtleff, *The Log Cabin Myth* (Cambridge, Mass.: Harvard University Press, 1939); and Abbott Lowell Cummings, *The Framed Houses of Massachusetts Bay, 1625–1727,* (Cambridge, Mass.: Harvard University Press, 1979).

xviii **eighteenth-century Virginia:** William Byrd's diaries have been published in several volumes since the first (Richmond: Dietz, 1941); and Rhys Isaac used Landon Carter's diary as the basis for his *Landon Carter's Uneasy Kingdom* (Oxford: Oxford University Press, 2004).

xviii **Benjamin Franklin provided us:** Leonard W. Labaree (ed.), *The Papers of Benjamin Franklin* (New Haven: Yale University Press) and *The Papers of George Washington* (Charlottesville: University Press of Virginia, several series).

xix **the frontier in American history:** Frederick Jackson Turner's most influential essays are collected in *The Frontier in American History* (reprint of the 1920 edition, New York: Holt, Rinehart, and Winston, 1962).

xix **a new generation of historians:** James Axtell, *The Invasion Within: The Contest of Cultures in Colonial North America* (New York: Oxford University Press, 1985); Richard White, *The Middle Ground: Indians, Empires, and Republics in the Great Lakes Region, 1650–1815* (Cambridge: Cambridge University Press, 1991); Eric Hinderaker, *Elusive Empires: Constructing Colonialism in the Ohio Valley, 1673–1800* (Cambridge: Cambridge University Press, 1997); Daniel K. Richter, *The Ordeal of the Longhouse: The Peoples of the Iroquois League in the Era of European Colonization* (Chapel Hill: University of North Carolina Press, 1992) and *Facing East from Indian Country: A Native History of Early America* (Cambridge, Mass.: Harvard University Press, 2001).

xx **their complex roots:** The best account I have found of the medieval period is Roland Oliver and Anthony Atmore, *Medieval Africa, 1250–1800* (Cambridge: Cambridge University Press, 2001). John K. Thornton has provided perhaps the best scholarship on "Atlantic Africa" in the period of the western slave trade in several works: *Africa and Africans in the Making of the Atlantic World, 1400–1680,* 2nd ed. (Cambridge: Cambridge University Press, 1998); "The African Background to American Colonization," in Stanley L. Engerman and Robert E. Gallman (eds.), *The Cambridge Economic History of the United States,* vol. 1, *The Colonial Era* (Cambridge: Cambridge University Press,

1996); and "Cannibals, Witches, and Slave Traders in the Atlantic World," *William and Mary Quarterly,* (3rd Series, 60:2, April 2003). Richard W. Harms provided a view of the central Zaire (Congo) region in his *River of Wealth, River of Sorrow* (New Haven: Yale University Press, 1981). C. W. Newbury, in *The Western Slave Coast and Its Rulers* (Oxford: Clarendon, 1961), gave an early and then rare look at African society.

xx **to the New World:** James A. Rawley provided a sober overview in *The Trans-Atlantic Slave Trade* (New York: Norton, 1981). Basil Davidson's *Black Mother* (Boston: Little, Brown, 1961) and Daniel P. Mannix's *Black Cargoes* (New York: Viking, 1962) are rather more polemic. Philip D. Curtin's *The Atlantic Slave Trade: A Census* (Madison: University of Wisconsin Press, 1969) was the standard for many years on the statistics of the slave trade. It has now been overtaken by a huge statistical study, *The Trans-Atlantic Slave Trade, 1527–1866: A Database on CD-ROM* (Cambridge: Cambridge University Press, 1999), which brings together information on over 27,000 voyages. It has been summarized and illuminated by a series of articles in *William and Mary Quarterly* (58:1, January 2001), of which some will be cited below; and by Stephen D. Behrendt, *Teacher's Manual* (Cambridge: Cambridge University Press, 1999), available online. It contains almost no information on individuals and little on groups. Two recent studies flesh out parts of the story: Randy Sparks, *The Two Princes of Calabar* (Cambridge, Mass.: Harvard University Press, 2004); and Robert Harms, *The Diligent: A Voyage through the Worlds of the Slave Trade* (New York: Basic Books, 2002).

xxi **accounts by blacks:** Most of the materials on American slavery are based on accounts in the nineteenth century; in the earlier period, the only materials are by whites. A very good account is by David Eltis, who played a key role in the huge study of slaving voyages, in *The Rise of African Slavery in the Americas* (Cambridge: Cambridge University Press, 2000).

xxi **in the 1930s:** *Colonists in Bondage: White Servitude and Convict Labor in America* (finally printed in 1947 by the University of North Carolina Press).

xxi **broadening this view:** Franco Venturi, *The End of the Old Regime in Europe* (Princeton: Princeton University Press, 1991); Pauline Maier, *From Resistance to Revolution* (New York: Random House, 1972); Bernard Bailyn, *The Ideological Origins of the American Revolution* (Cambridge, Mass.: Harvard University Press, 1967) and *Voyagers to the West* (New York: Knopf, 1986).

xxi **Economic affairs:** Particularly useful are John J. McCusker and Russell R. Menard, *The Economy of British America, 1607–1789* (Chapel Hill: University of North Carolina Press, 1991); and Stanley L. Engerman and Robert E. Gallman (eds.), *The Cambridge Economic History of the United States,* Vol. 1, *The Colonial Era* (Cambridge: Cambridge University Press, 1996). John J. McCusker, *Money and Exchange in Europe and America, 1600–1775* (Chapel Hill: University of North Carolina Press, 1978), deals with the complex monetary systems of the era. Also stimulating and informative is Alice Hanson Jones, *Wealth of a Nation to Be: The American Colonies on the Eve of the Revolution* (New York: Columbia University Press, 1980). The growth of colonial commerce is the subject of one of Bernard Bailyn's excellent histories, *The New England Merchants in the Seventeenth Century* (Cambridge, Mass.: Harvard University Press, 1955).

xxii **led up to the Revolution:** Bernard Bailyn, *The Ordeal of Thomas Hutchinson* (Cambridge, Mass.: Harvard University Press, 1974). Arthur M. Schlesinger, *Prelude to Independence* (New York: Random House, reprinted 1965 from the 1958 edition). Hiller B. Zobel, *The Boston Massacre* (New York: Norton, 1970).

xxiii **what they ate and how hard they worked:** James Deetz, *In Small Things Forgotten: An Archaeology of Early American Life* (New York: Anchor, 1996); Clark Spencer Larsen, "Reading the Bones of La Florida," *Scientific American* (June 2000); Ivor Noël Hume, *Martin's Hundred* (New York: Knopf, 1982).

CHAPTER 1: THE NATIVE AMERICANS

3 **William Strachey spoke for them:** Louis B. Wright and Virginia Freund (eds.), *The Historie of Travell into Virginia Britania (1612)* (London: Hakluyt Society, 1953), 53–55.

3 **they call Amerind:** Joseph H. Greenberg and Merritt Ruhlen, "Linguistic Origins of Native Americans," *Scientific American,* November, 1992. Criticism of his analysis is given in *Science* 242(23 December 1988), 1632–1633.

4 **by genetic studies:** Luigi Luca Cavalli-Sforza, *Genes, Peoples, and Languages* (New York: Farrar, Straus, and Giroux, 2000), 134, 136–137; and Luigi Luca Cavalli-Sforza and Francesco Cavalli-Sforza, *The Great Human Diasporas* (London: Perseus, 1995), 109–112, 172–175.

4 **Climatologists:** William J. Burroughs, *Climate Change in Prehistory: The End of the Reign of Chaos* (Cambridge: Cambridge University Press, 2005), 207–14 and passim.

5 **migrated southward:** Excellent studies of them are Francis Jennings, *The Ambiguous Iroquois Empire* (New York: Norton, 1984); and Daniel K. Richter, *The Ordeal of the Longhouse: The Peoples of the Iroquois League in the Era of European Colonization* (Chapel Hill: University of North Carolina Press, 1992).

5 **now mainly forgotten groups:** An early study of them is James Mooney, *The Siouan Tribes of the East* (Washington, D.C.: Smithsonian Institution, 1894). On the Catawba, see James H. Merrell, "The Indians' New World: The Catawba Experience," *William and Mary Quarterly* (3rd Series, 41:4, October 1984).

6 **learn some dialect:** Alden T. Vaughan, "Sir Walter Ralegh's Indian Interpreters, 1584–1618," *William and Mary Quarterly* (59:2) 2002, tells what is known about the first Indians trained by the English and about Thomas Hariot's study

of "the Virginian speche." On Englishmen who served as interpreters, see J. Frederick Fausz, "Middlemen in Peace and War: Virginia's Earliest Indian [Language] Interpreters, 1608–1632," *Virginia Magazine of History and Biography* (5, 1987), 41ff.

6 **Juan Nentuig:** Quoted in Edward H. Spicer, *Cycles of Conquest: The Impact of Spain, Mexico, and the United States on the Indians of the Southwest, 1533–1960* (Tucson: University of Arizona Press, 1962), 320–321.

8 **Obviously, Bradford did not care:** Samuel E. Morison (ed), *Bradford, Of Plymouth Plantation. 1620–1647* (New York: Knopf, 1959), 82–84.

8 **Robert Beverley:** *The History and Present State of Virginia,* originally published in London by Richard Parker in 1705; new edition edited by David Freeman Hawke (Indianapolis: Bobbs-Merrill, 1971), 102–105.

9 **could be eaten:** Information on the plants they harvested can be found in Harold E. Driver, *Indians of North America* (Chicago: University of Chicago Press, 1961), 21–22; and Jack Weatherford, *Indian Givers* (New York: Fawcett Columbine, 1988), 59ff.

10 **slowly made its way:** Bruce D. Smith, "Origins of Agriculture in Eastern North America," *Science* (246, 1566ff). Smith says that maize (*Zea mays*) or corn was introduced into eastern North America about A.D. 200 "perhaps as a high status or ceremonial crop, until after A.D. 800."

11 **comment:** *Eastern Shore Indians of Virginia and Maryland* (Charlottesville: University Press of Virginia, 1997), 39.

11 **William Strachey described:** Wright and Freund, (eds.), *Strachey,* 47–48, using almost the same words as Captain John Smith, *The General History of Virginia (1624),* Book 2, Chapter 1, reprinted in Louis B. Wright (ed.), *The Elizabethans' America* (Cambridge, Mass.: Harvard University Press, 1965), 171–173.

11 **Champlain similarly found:** *Algonquians, Hurons, and Iroquois . . . 1603–1616,* translation of *Les voyages de la Nouvelle France* (Dartmouth, Nova Scotia: Brook House, 2000), 186.

11 **George Percy described:** "Observations Gathered Out of a Discourse of the Plantation of the Southern Colony in Virginia by the English, 1606," from Purchas, *Pilgrims* (1625), Book 9, Chapter 2, reprinted in Wright (ed.), *The Elizabethans' America,* 168.

12 **Robert Beverley waxed lyrical:** Hawke, *Beverley,* 89–91.

12 **John Lawson found:** Hugh Talmage Lefler (ed.), *A New Voyage to Carolina* (Chapel Hill: University of North Carolina Press, 1967), 195.

12 **than those of Europeans:** William Peden (ed.), *Thomas Jefferson, Notes on the State of Virginia* (Chapel Hill: University of North Carolina Press, 1982), 199–200.

12 **their mode of living:** Mary Jemison described the simplicity and ease of Indian life as she experienced it in the eighteenth century in James E. Seaver (ed.), *A*

Narrative of the Life of Mrs. Mary Jemison, partly reproduced in Colin G. Calloway (ed.), *The World Turned Upside Down: Indian Voices from Early America* (Boston: St. Martin's, 1994), 74ff.

13 **Robert Beverley wrote:** Hawke, *Beverley,* 112. See also Paul A. W. Wallace, *Indians in Pennsylvania* (Harrisburg: Pennsylvania Historical and Museum Commission, 1993), 30; and William C. Sturtevent, "Notes on the Creek Hothouse," *Southern Indian Studies* (20, 1968).

13 **Champlain saw:** Champlain, *Algonquians,* 62. *A Relation of Maryland, 1635,* reprinted in Clayton Colman Hall, *Narratives of Early Maryland, 1633–1684* (New York: Scribner, 1909), 87.

13 **rather closely observed:** *Travels through North and South Carolina, Georgia, East and West Florida, the Cherokee Country, the Extensive Territories of the Muscogulges or Creek Confederacy, and the Country of the Choctaws* (London, 1792), quoted in Grace Steele Woodward, *The Cherokees* (Norman: University of Oklahoma Press, 1963), 37–38.

13 **William Fyffe:** Manuscript in Gilcrease Institute, quoted in Woodward, *The Cherokees,* 36. See also John Hulton, *America 1585: The Complete Drawings of John White* (Chapel Hill: University of North Carolina Press and the British Museum, 1984), plates 46, 47, 48.

13 **John Lawson remarked:** Hugh Talmage Lefler (ed.), *A New Voyage,* 176.

14 **the author wrote:** Reprinted in Hall, *Narratives,* 86.

14 **an early Spanish visitor:** James Alexander Robertson (ed. and transl.), *True Relation of the Hardships Suffered by Governor Hernando de Soto and Certain Portuguese Gentlemen during the Discovery of the Province of Florida . . .* (De Land: Florida State Historical Society, 1933), 75–76, quoted in John R. Swanton, *The Indians of the Southeastern United States* (Washington, D.C.: Smithsonian Institution, 1946, reprinted 1979), 386–388.

14 **as hot as Stoves:** Hugh Talmage Lefler (ed.), *A New Voyage,* 180.

14 **Dutch stoves:** *The History of the American Indians* (London, 1775, new edition edited by Samuel Cole Williams; Johnson City, Tenn.: National Society of Colonial Dames of America, 1930), 419–420, quoted in John R. Swanton, *The Indians,* 386–388.

14 **James Adair commented:** *The History of the American Indians* (London, 1775, new edition edited by Samuel Cole Williams; Johnson City, Tenn.: National Society of Colonial Dames of America, 1930), 419–420, quoted in Swanton, *The Indians,* 417–419.

15 **the East Coast:** One in New York was described by Jasper Dankers and Peter Sluyter (ed. and transl. Henry C. Murphy), "Journal of a Voyage to New York in 1679–1680." It was about 60 feet by 14 to 15 feet with the sides and roof made of reed and bark of chestnut trees attached to polls stuck in the ground.

15 **English visitors remarked:** In 1686, one visitor commented: "There are above 100,000 houses in Ireland which are now worth ten shillings each." An Irish

house, like an Indian house, "could be built in three or four days." J. H. Andrews, "Land and People, c. 1685," in T. W. Moody, F. X. Martin, and F. J. Byrne (eds.), *A New History of Ireland* (Oxford: Clarendon, 1976), 465–466.

15 **to the land:** "The Indians were essentially a nomadic race," "wayward and treacherous children of the forest," "worthless vagabonds." Franklin Ellis and Samuel Evans, *History of Lancaster County, Pennsylvania* (Philadelphia: Everts and Peck, 1883), 6, 7, 13.

15 **in 1635 admitted:** Reprinted in Hall, *Narratives,* 86.

15 **was described as:** "A Narrative of the Expedition of Hernando de Soto into Florida by a Gentleman of Elvas," *Historical Collections of Louisiana* (2:139), quoted in Lewis H. Morgan, *Houses and House-Life of the American Aborigines* (Chicago: University of Chicago Press, 1965, from the 1881 edition), 48–49.

15 **a population of:** Melvin L. Fowler, "A Pre-Columbian Urban Center on the Mississippi," *Scientific American* (August 1975); and Sally Chappell, *Cahokia: Mirror of the Cosmos* (Chicago: University of Chicago Press, 2002).

15 **by John White:** Hulton, *America 1585:* plate 32.

15 **described the government:** Reprinted in Hall, *Narratives,* 86.

16 **one modern historian:** Richard White, *The Middle Ground: Indians, Empires, and Republics in the Great Lakes Region, 1650–1815* (Cambridge: Cambridge University Press, 1991), 38.

16 **the whole community:** William Penn in 1683 on the Delawares, quoted in Paul A. W. Wallace, *Indians in Pennsylvania* (Harrisburg: Pennsylvania Historical and Museum Commission, 1993), 53.

16 **the Shawnee chiefs:** "Character, Manners, and Customs of the Indians" (c. 1701), in Henry Harvey (ed.), *History of the Shawnee Indians, from the Year 1681 to 1854* (Cincinnati: Ephraim Morgan, 1855), quoted in Frederic W. Gleach, *Powhatan's World and Colonial Virginia: A Conflict of Cultures* (Lincoln: University of Nebraska Press, 1997), 30.

16 **He reported:** William M. Darlington (ed.), *An Account of the Remarkable Occurrences in the Life and Travels of Col. James Smith* (Cincinnati, 1870), 147, quoted in Wallace, *Indians,* 53.

16 **the best observation:** Archer Butler Hulbert and William Nathaniel Schwarze, "David Zeisberger's History of the North American Indians," *Ohio Archaeological and Historical Quarterly* (19, 1910), 92–93, quoted in Gleach, *Powhatan's World,* 30.

17 **As Richard White:** White, *The Middle Ground,* 312–313.

17 **limits of chieftainship:** White, *The Middle Ground,* 299–300. The different interpretations of Pontiac are given by Eric Hinderaker in a review of Gregory Evans Dowd, *War under Heaven* (Baltimore: Johns Hopkins University Press, 2002), in *William and Mary Quarterly* (60:4, October 2003), 865ff. Dowd points out that Pontiac was a civil leader, an *ogema,* who was more or less thrust into the position of war leader by British policy.

17 **political organization:** Harold E. Driver, *Indians of North America* (Chicago: University of Chicago Press, 1961), 79.
18 **As Lewis H. Morgan wrote:** Morgan, *Houses,* 95.
18 **to incoming whites:** Calloway (ed.), *The World,* 95.
18 **the Moravian missionary:** *History, Manners, and Customs of the Indian Nations,* quoted in Morgan, *Houses,* 49.
19 **Lewis Morgan wrote:** Morgan, *Houses,* 45.
19 **lovingly detail:** Richard Hakluyt, *Principall Navigations, Voyages, Traffiques and Discoveries of the English Nation,* published as *The Voyages* (London: Dent, 1907), 6:121ff.
19 **would destroy us:** *A True Relation* (London, 1608), reprinted in Lyon Gardiner Tyler, (ed.) *Narratives of Early Virginia, 1606–1625,* New York: Scribner, 1907), 36.
19 **to the visitors:** "A Narrative of the Expedition of Hernando de Soto into Florida by a Gentleman of Elvas," quoted in Morgan, *Houses,* 48–49.
19 **the best that we have:** John Demos, *The Unredeemed Captive: A Family Story from Early America* (New York: Knopf, 1994), 153, quoting from *An Account of the Remarkable Occurrences in the Life and Travels of Col. James Smith* in Archibald Loudon (ed.), *Outrages Committed by the Indians* (London, 1808).
19 **French Jesuits saw:** Reuben Goldthwaites (ed.), *The Jesuit Relations and Allied Documents* (Cleveland, Ohio, 1896–1901), Vol. 64, 131. See also Claude Chauchetière, *Narrative of the Mission of Sault St. Louis, 1667–1685* (Kahnawake, Canada, 1981), 51, cited in Demos, *The Unredeemed Captive,* 150.
20 **Jonathan Carver visited:** *Carver's Travels* (Philadelphia, 1796), 171, cited in Lewis H. Morgan, *Houses,* 152.
20 **James Adair remarked:** Cited in Morgan, *Houses,* 46. See also J. Frederick Fausz, "Merging and Emerging Worlds," in Lois Green Carr, Philip D. Morgan, and Jean B. Russo, *Colonial Chesapeake Society* (Chapel Hill: University of North Carolina Press, 1988), 50.
20 **Wallace has written:** Paul A. W. Wallace, "Pathfinding in Pennsylvania," *Papers of the Kittochtinny Historical Society* (13, 1957), 269–270. A map of Pennsylvania trails is given in *Indian Paths of Pennsylvania* (Harrisburg: Pennsylvania Historical and Museum Commission, 1993).
20 **form of seashells:** Called cowries in Africa.
20 **more common *roanoke:*** The Narragansetts and Pequots "grew rich and potent by it," Morison (ed.), Bradford, 203.
21 **silver, beaver skins, and "wampum":** John J. McCusker, *Money and Exchange in Europe and America, 1600–1775* (Chapel Hill: University of North Carolina Press, 1978), 157.
21 **Lewis Morgan counts:** Morgan, *Houses,* 23.
21 **John Lawson observed:** J. Leitch Wright, Jr., *The Only Land They Knew: The*

Tragic Story of the American Indians in the Old South (New York: Free Press, 1981), 126–127.

21 **in the New World:** James Axtell and William C. Sturtevant, "The Unkindest Cut, or Who Invented Scalping?" *William and Mary Quarterly* (3rd Series, 37:3, July 1980), 455.

22 **sticks and firebrands:** Daniel K. Richter, "War and Culture: The Iroquois Experience," *William and Mary Quarterly* (3rd Series, 40:4, October 1993), 532–533.

22 **my family did:** Delilah Polk, who was taken prisoner in an Indian raid in what became Kentucky. Recounted in *Polk's Folly* (New York: Doubleday, 2000), 93–95.

22 **wrote in 1788:** Quoted in Daniel K. Richter, "War and Culture: The Iroquois Experience," *William and Mary Quarterly* (3rd series, 40:4, October 1983), 531.

22 **Benjamin Franklin:** Letter to Peter Collison, dated Philadelphia, May 9, 1753, Leonard W. Labaree (ed.), *The Papers of Benjamin Franklin,* vol. 4 (New Haven: Yale University Press), 432.

23 **dragged back:** James Axtell, "The White Indians of Colonial America," in James Axtell, *The European and the Indian* (New York: Oxford University Press, 1981).

23 **we can only imagine:** As in his remarkable attempt to reconstruct an Indian view of history, Daniel K. Richter does this in *Facing East from Indian Country: A Native History of Early America* (Cambridge, Mass.: Harvard University Press, 2001).

CHAPTER 2: THE FEARSOME ATLANTIC

25 **North African ships:** The word "carrack" is drawn from the Turco-Arabic *qaráqir,* which is the plural of *qarqur* and referred to long ships. About 1440 they began to replace the *barca,* a boat only about half the size, which was said to be "cumbersome." Edward William Bovill, *The Golden Trade of the Moors: West African Kingdoms in the Fourteenth Century* (Oxford: Oxford University Press, 1958), 115.

26 ***caravela redonda:*** Illustrated in numbers 275 and 277 in Björn Landström, *The Ship* (Garden City, N.Y.: Doubleday, 1961), 106–107. Ironically, when Columbus's *Santa María* struck a reef on Christmas day 1492, it could have been saved had it been a Mediterranean ship: curved on the bottom, such ships could be rocked back and forth, by moving men and cargo, to float free. But as an Atlantic hulled ship, the *Santa María* quickly broke up.

26 **pilots' instruction—*routiers* or *portolani:*** For two examples of a *routier* that later was worked out for the Caribbean, see Richard Hakluyt, *Principall Navigations, Voyages, Traffiques, and Discoveries of the English Nation, Published as The Voyages* (London: Dent, 1907), 7: Vol. 7, 224–266.

27 **the real world:** J. H. Parry, *The Discovery of the Sea* (Berkeley: University of California Press, 1981), 189–190.

27 **has survived:** G. V. Scammell, *The World Emcompassed: The First European Maritime Empires c. 800–1650* (Berkeley: University of California Press, 1981), 217.

28 **die miserably:** Carl T. Eben (trans.), *Gottlieb Mittelberger's Journey to Pennsylvania in the Year 1750* (Philadelphia, 1898), 20. Quoted in John Duffy, "The Passage to the Colonies," *Mississippi Valley Historical Review* (38, June 1951–March 1952), 22.

28 **freighted with fools:** Ebenezer Cook in 1708, quoted in Arthur Pierce Middleton, *Tobacco Coast: A Maritime History of Chesapeake Bay in the Colonial Era* (Baltimore: Johns Hopkins University Press, 1994), 6.

28 **survivors later reported:** "A particular note of the Indian fleet, expected to have come into Spaine this present yeere of 1591 with the number of shippes that are perished of the same: according to the examination of certaine Spaniards lately taken and brought into England by the ships of London." Hakluyt, *Voyages,* Vol. 5, 14–15.

30 **an ideal inventory:** In the promotional pamphlet *A Relation of Maryland* (1635), probably ordered by Lord Baltimore and reproduced in Clayton Colman Hall (ed.), *Narratives of Early Maryland, 1633–1684* (New York: Scribner, 1909), 97. The use of citrus fruit advocated by the Scottish naval surgeon James Lind gave rise to the nickname for British sailors, limeys.

30 **passengers were starving:** Arthur Pierce Middleton, *Tobacco Coast,* 20.

31 **about 75 feet long:** John R. Hale, *Age of Exploration* (New York: Time-Life Books, 1974), 89.

31 **70 feet long:** G. V. Scammell, *The World Encompassed,* 442–443.

31 **less than 80 tons:** David Beers Quinn, *England and the Discovery of America, 1481–1620* (New York: Knopf, 1974), 199–200.

32 **cold nott endure the sentt:** "A trewe Relacyon of the Procedinges and Ocurrentes of Momente Which Have Hapned in Virginia from the Time Sir Thomas Gates was Shipwrackte upon the Bermudes an° 1609 untill my Departure Outt of the Country which was in An° Dni 1612," *Tyler's Quarterly Historical and Genealogical Magazine* (3:4, April 1922), 281.

32 **1,500 gallons of it:** Hale, *Age of Exploration,* 82.

32 **had run out of beer:** Bert L. Vallee, "Alcohol in the Western World," *Scientific American* (June 1998), 61 ff.

32 **to reach Maryland:** Father Andrew White, S.J. "A Briefe Relation of the Voyage unto Maryland, 1634," in Hall, *Narratives,* 31.

32 **another speake:** Louis B. Wright and Virginia Freund (eds.) *William Strachey: The Historie of Travell into Virginia Britania, 1612* (London: Hakluyt Society, 1953), xix. Strachey's letter described the storm that wrecked his ship on the shoals of Bermuda.

33 **described in 1526:** Hakluyt, *Voyages,* Vol. 4, 22.

34 **optimistically put:** Sir George Peckham, "A True Report of the late discoveries . . ." in Hakluyt, *Voyages,* Vol. 6, 70–71.

34 **focus on the Atlantic:** Hakluyt gives one for the "West Indies, and for Tierra firma, and Nueva Espanna," Hakluyt, *Voyages,* Vol. 7, 224ff.

35 **a French traveler remembered:** Quoted in John Duffy, "The Passage to the Colonies," *Mississippi Valley Historical Review* (38, June 1951–March 1952), 22.

36 **swolne euery ioint withal:** Louis B. Wright memorably describes Hakluyt as "a parson who made the doctrine of colonial expansion a religion and preached the belief that it was part of the divine plan for England to seize an empire in North America." *The Atlantic Frontier: Colonial American Civilization 1607–1763* (New York: Knopf, 1951), 51.

36 **who had long understood scurvy:** William R. Polk, *Neighbors and Strangers* (Chicago: University of Chicago Press, 1997), 150.

36 **thirty died of smallpox:** David Hackett Fischer, *Albion's Seed: Four British Folkways in America* (New York: Oxford University Press, 1989), 421.

37 **Virginia in 1621:** Donald G. Shomette, *Pirates on the Chesapeake* (Centerville, Md., Tidewater, 1985), 7–8.

37 **city of Havana:** John T. McGrath, *The French in Early Florida: In the Eye of the Hurricane* (Gainesville: University Press of Florida, 2000), 27. De Sores was operating under a royal letter of mark and so was quasi official. That he was a Protestant rubbed salt into the Spaniards' wounds.

37 **in the Atlantic:** Marcus Rediker, *Between the Devil and the Deep Blue Sea* (Cambridge: Cambridge University Press, Canto Edition, 1993), 256–258.

37 **called *renegadoes:*** Christopher Lloyd, *English Corsairs on the Barbary Coast* (London: Collins, 1981).

37 **piracy on the side:** David Beers Quinn, *England and the Discovery of America, 1481–1620* (New York: Knopf, 1974), 206–207.

38 **English pirate recounted:** Hakluyt, *Voyages,* Vol. 6, 296ff.

38 **Anglo-American pirates:** Rediker, *Deep Blue Sea,* 274, 283.

38 **mouth of the Chesapeake:** Shomette, *Pirates* op. cit., 103–107.

38 **was kidnapped:** Seth Sothell in 1712 was taken by Algerian corsairs. His troubles did not end with this incident; he was overthrown by colonial insurgents in 1718.

38 **the Irish revolt of 1651:** Kerby A. Miller, *Emigrants and Exiles: Ireland and the Irish Exodus to North America* (New York: Oxford University Press, 1985), 20–21.

CHAPTER 3: SUGAR, SLAVES, AND SOULS

40 **they had reached Portugal:** J. H. Parry, *The Discovery of the Sea* (Berkeley: University of California Press, 1981), 96.

40 **gift for the royal children:** Felipe Fernández-Armesto, *Before Columbus: Exploration and Colonization from the Mediterranean to the Atlantic, 1229–1492* (Philadelphia: University of Pennsylvania Press, 1987), 198.

41 **much of Africa's gold:** Roland Oliver and Anthony Atmore, *Medieval Africa, 1250–1800* (Cambridge: Cambridge University Press, 2001), 65.

41 **10,000 black slaves:** Stephen D. Behrendt, Teacher's Manual for The Trans-Atlantic Slave Trade: A Database on CD-Rom, Edited by David Eltis, David Richardson, Stephen D. Behrendt and Herbert S. Klein (Cambridge: Cambridge University Press, 1999), 14.

41 **memorably noted:** *Algonquians, Hurons, and Iroquois . . . 1603–1616,* translation of *Les voyages de la Nouvelle France* (Dartmouth, Nova Scotia: Brook House, 2000), 2.

42 **to western Europe:** G. V. Scammell, *The World Encompassed: The First European Maritime Empires c. 800–1650* (Berkeley: University of California Press, 1981), 172.

42 **whose daughter he married:** Parry, *Discovery,* 96–97.

42 **and the Gold Coast:** P. E. H. Hair, "Columbus from Guinea to America," *History in Africa* 17 (1990), 115. Hair believes he made more than one voyage south along the African coast and must have seen Madeira, Porto Santo, the Canaries, and the Cape Verdes.

42 **next two centuries:** John Ramsey, *Spain: The Rise of the First World Power* (University: University of Alabama Press, 1973), 243. In return for allowing others to use the Azores, Portugal got a free hand in the areas it was most concerned with: West Africa, its route to India, and Brazil.

43 **sticks and stones:** Fernández-Armesto, *Before Columbus,* 207.

43 **he remarked:** In a letter written on board the *Niña,* February 15, 1493; quoted in Samuel Eliot Morison, *Christopher Columbus Mariner* (New York: Meridian, 1983), 151.

43 **the Canary Islanders:** Quoted in I. Bernard Cohen, "What Columbus 'Saw' in 1492," *Scientific American* (December 1992), 104.

44 **a whole society in miniature:** Parry, *Discovery,* 205.

44 **to the Caribbean:** Ramsey, *Spain,* 262.

45 **the soldiers':** Quoted in Charles Gibson, *Spain in America* (New York: Colophon, 1967), 39.

45 **as 8 million:** Bartolomé de las Casas, *The Devastation of the Indies: A Brief Account,* trans. Herma Briffault from the 1552 edition (Baltimore: Johns Hopkins University Press, 1992), 27. See also Kathleen A. Deagan, "Spanish-Indian Interaction in Sixteenth-Century Florida and Hispaniola," in William W. Fitzhugh (ed.), *Cultures in Contact* (Washington, D.C.: Smithsonian Institution, 1985), 283. Deagan puts the figure at 8 million. Also see Kenneth R. Andrews, *Trade, Plunder, and Settlement: Maritime Enterprise and the Genesis of the British Empire, 1480–1630* (Cambridge: Cambridge University Press, 1984), 117.

45 ***Brevissima Relación:*** Briffault: *Las Casas* 29 et passim.
46 **still alive:** Deagan, "Spanish-Indian," 288.
47 **senior Spanish official:** Quoted by Scammell, *The World Encompassed,* 356.
47 **brainwashing:** J. H. Parry, *The Spanish Seaborne Empire* (Harmondsworth: Penguin, 1973), 152–153.
47 **Columbus had described:** Quoted in Morison, *Columbus,* 151.
47 **weapon of war, horses:** Paul Quattlebaum, *The Land Called Chicora* (Gainesville: University of Florida Press, 1956), 118–119.
48 **South Carolina:** Most of what is known of these events is drawn from Gonzalo Fernández de Oviedo y Valdés, *Historia General y Natural de las Indias, Isles, y Tierra-Firme del Mar Océano* (Madrid, 1851–1855).
48 **succumbed to fevers:** David J. Weber, *The Spanish Frontier in North America* (New Haven: Yale University Press, 1992), 42–43.
48 **he later wrote:** Enrique Pupo-Walker (ed.) and Frances M. López-Morillas, *Álvar Núñez Cabeza de Vaca, Castaways* (Berkeley: University of California Press, 1993), 114; from the Spanish edition *Relación* (Madrid: Editorial Castalia, 1992).
49 **hunting Indians on horseback:** Letter by the archbishop of Mexico, Pedro Moya de Contreras, quoted in Weber, *The Spanish Frontier,* 51.
49 **iron chains and collars:** Weber, *The Spanish Frontier,* 51.
49 **grouped around churches:** Edward H. Spicer, *Cycles of Conquest: The Impact of Spain, Mexico, and the United States on the Indians of the Southwest, 1533–1960* (Tucson: University of Arizona Press, 1962), 288–306.
50 **actual or alleged Christian Indians:** Deagan, "Spanish-Indian," 298–299. With uncanny accuracy, the Spanish mission towns presaged the "praying towns" that New England Protestants would establish over a century later.
50 **archaeological studies:** Clark Spencer Larsen, "Reading the Bones of *La Florida,*" *Scientific American* (June 2000), 61ff.
50 **one of the Spanish viceroys:** Quoted by Weber, *The Spanish Frontier,* 123.
51 **colonists at Jamestown:** "O-pe-chan-can-ough: A Native American Patriot" in *Early Americans* (New York: Oxford University Press, 1981).
51 **de Soto reported:** David B. Quinn (ed.), *New American World: A Documentary History of North America to 1612* (New York: Arno, 1979), Vol. 2, 199, quoted in John T. McGrath, *The French in Early Florida: In the Eye of the Hurricane* (Gainesville: University Press of Florida, 2000), 22.
51 **practical settlement:** Matthew J. Connolly, "Four Contemporary Narratives of the Founding of St. Augustine," *Catholic Historical Review* (60:3, October 1965), 306.

CHAPTER 4: FISH, FUR, AND PIRACY

53 **lived in Greenland:** J. R. Hale, *Renaissance Exploration* (New York: Norton, 1968), 47.

53 **blocked by ice floes:** Brian Fagan, *The Little Ice Age: How Climate Made History, 1300–1850* (New York: Basic Books, 2000), 62.

53 **had been abandoned:** J. H. Parry, *The Discovery of the Sea* (Berkeley: University of California Press, 1981), 44.

53 **the cod:** Mark Kurlansky, *Cod: A Biography of the Fish That Changed the World* (London: Vintage, 1999).

54 **by 1413:** The *Oxford English Dictionary* notes that "dogger" is another form of the Icelandic *dugga.* Vigfurssion mentions that "thirty English *fiski-duggur* came fishing about Iceland in 1413." The only sketch I have found shows a two-masted boat with a side rudder and no superstructure. Alfred Dudszus, *Das Grosse Buch der Schiffstypen* (Stuttgart: Pietsch Verlag, 2004), 78.

55 **the imperial aims of France:** "The Fur Trade and Eighteenth-Century Imperialism," *William and Mary Quarterly* (Series 3, 40:3, July 1983).

55 **in the Massachusetts area:** Samuel Eliot Morison (ed.), *William Bradford: Of Plymouth Plantation, 1620–1647* (New York: Knopf, 1959), 81, fn 2.

56 **Verrazzano set out in 1524:** Lawrence C. Wroth, *The Voyages of Giovanni da Verrazzano, 1524–1528* (New Haven: Yale University Press, 1970).

56 **natives to be friendly:** John Bakeless, *America as Seen by Its First Explorers: The Eyes of Discovery* (New York: Dover, 1961), 204.

57 **their silent trading:** Similar forms of trading have been noted all over the world among primitive peoples since the time of Herodotus. See William R. Polk, *Neighbors and Strangers* (Chicago: University of Chicago Press, 1997), 132–133.

59 **warned his king in 1564:** D. B. Quinn, "Some Spanish Reactions to Elizabethan Colonial Enterprises," *Transactions of the Royal Historical Society* (London: Royal Historical Society, 1951), 1.

59 **Menéndez was right:** Quinn, "Spanish Reactions," 3. See also John T. McGrath, *The French in Early Florida: In the Eye of the Hurricane* (Gainesville: University Press of Florida, 2000), 117.

60 **stitched together their shirts:** John T. McGrath, *The French,* 82–83, 91. McGrath bases his account on René Goulaine de Laudonnière and on a fifteen-year-old cabin boy, Guillaume Rouffin, who stayed behind and was later captured by the Spaniards.

60 **his account of the new colony:** *Whole and True Discoureryе of Terra Florida* (London, 1562).

60 **the Saint Johns River:** Included among them was Jacques Le Moyne de Morgues who worked as the cartographer and also drew sketches of the area and its people. They are some of our first and most important records. His sketches were reproduced in 1977 by the British Museum in London.

61 **other men's brows:** Quoted by John T. McGrath, *The French,* 109.

61 **put them to the knife:** Matthew J. Connolly, "Four Contemporary Narratives of the Founding of Saint Augustine," *Catholic Historical Review* (60:30, October 1965), 324ff.

61 **Spain's success:** Edward Gaylord Bourne (ed.), *Samuel de Champlain: The Voyages and Explorations of Samuel de Champlain (1604–1616), Narrated by Himself* (New York: A. S. Barnes, 1906; reprint, Dartmouth, Nova Scotia: Brook House, 2000), x.

63 **permeated "British America":** Eric Hinderaker, *Elusive Empires: Constructing Colonialism in the Ohio Valley, 1673–1800* (Cambridge: Cambridge University Press, 1997), 80.

CHAPTER 5: SOCIETY AND WARS IN THE OLD COUNTRIES

64 **less than 1,000 men:** Alan G. R. Smith, *The Government of Elizabethan England* (New York: Norton, 1967), 54; W. T. MacCaffrey, "Place and Patronage in Elizabethan Politics," in S. T. Bindoff et al. (eds.), *Elizabethan Government and Society* (London: University of London Press, 1961), 106–8.

64 **only marginally larger:** Rorer Mettam, *Government and Society in Louis XIV's France* (London: MacMillan, 1977), 16ff.

65 **has commented:** Roy Porter, *English Society in the Eighteenth Century* (London: Penguin, revised edition, 1990), 116.

65 **in time of war:** E. S. De Beer, "The English Revolution" in J. S. Bromley (ed.), *The Rise of Great Britain, Vol. 6, The New Cambridge Modern History* (Cambridge: Cambridge University Press, 1971), 197. France was then thought to enroll at least twenty times that many. David G. Chandler, "The Art of War on Land," in ibid., 741.

66 **policies it handed down:** J. Lough, "France under Louis XIV" in F. L. Carsten (ed.), *The Ascendancy of France,* Vol. 5, *The New Cambridge Modern History* (Cambridge: Cambridge University Press, 1961), 235–236 and Mettam, 16ff.

67 **state-granted benefices:** David Ogg, "Britain after the Restoration," in Carsten (ed.), *The Ascendancy of France,* 306.

67 **would leave for America:** Daniel Defoe writes movingly of his experience as a defeated supporter of Cromwell's government and his fear of the noose. Four of his schoolmates (who had studied under Charles Morton, who was harassed into quitting England and later became vice-president of Harvard) met that fate. He escaped only by almost unbelievable luck. The critic Tom Paulin asserted in the *London Review of Books* (July 19, 2001) in a review of recent books on Defoe that "*Robinson Crusoe* is an epic account of the experience of the English Dissenters under the Restoration." It was a result of this crackdown on the officers of Cromwell's army that my ancestor Robert Pollock, or Polk, escaped to Maryland. See *Polk's Folly* (New York: Doubleday, 2000).

68 **to the American colonies:** Jon Butler, *The Huguenots in America: A Refugee People in New World Society* (Cambridge, Mass.: Harvard University Press, 1983), Part I, "Flight from Terror," 11–68.

68 **than the earth they tread upon:** "General observations for the plantation of New England," May 1629, Winthrop Papers II, 114, quoted in David Hackett Fischer, *Albion's Seed: Four British Folkways in America* (New York: Oxford University Press, 1989), 16.

68 **answered the call:** Abbot Emerson Smith, *Colonists in Bondage: White Servitude and Convict Labor in America, 1607–1776* (New York: Norton, 1971), 47.

68 **of some 21,000 people:** Fischer, *Albion's Seed,* 16.

68 ***Books of Sufferings:*** Laid out in Joseph Besse, *An Abstract of the Sufferings of the People Called Quakers,* published in London in 1733.

68 **to the New World:** Smith, *Colonists* 175, 179.

69 **interpretation of the Gospel:** James Henry, *Sketches of Moravian Life and Character* (Philadelphia: Lippincott, 1859) gives an inside view of the Moravian experience.

70 **died for wante in the strettes 6s.8d.:** Quoted by Henry Kamen, *The Iron Century: Social Change in Europe 1550–1660* (London: Weidenfeld and Nicholson, 1971), 34.

71 **to this realme:** Richard Hakluyt, *The Principall Navigations, Voyages, Traffiques and Discoveries of the English Nation, published as The Voyages* (London: Dent, 1907), 6:61–62.

71 **or rising:** *Observations on Political and Military Affairs* (London, 1671), 145, noted in Lawrence Stone, *The Causes of the English Revolution, 1529–1642* (New York: Harper, 1972), 66–67.

72 **shipped out of England:** Bernard Bailyn puts the figure at 50,000 or "approximately the same number of convicts that would be sent to Australia before 1824." *The Peopling of British North America* (New York: Vintage, 1988), 120–121. Also see Smith, *Colonists,* 149.

74 **nits make lice:** Jonathan Bardon, *A History of Ulster* (Belfast: Blackstaff, 1992), 138.

74 **dabble in revolution:** Elizabeth's spymaster, Sir Francis Walsingham, the patron of Richard Hakluyt and promoter of colonization, maintained about seventy agents in European courts to encourage opposition to Spain. England also sent an expeditionary force of 10,000 soldiers to help the Dutch fight the Spaniards. Smith, 48–49.

74 **John Maynard Keynes wrote:** *A Treatise on Money,* 2:156–157, quoted in Karl de Schweinitz, *The Rise and Fall of British India* (London: Routledge, 1983), 43.

75 **In 1570:** Charlotte M. Gradie, "Spanish Jesuits in Virginia: The Mission that Failed," *The Virginia Magazine of History and Biography* (98:2, April 1988), 131ff. Two French groups also tried but failed to set up outposts in what became South Carolina in 1562 and 1564.

75 **in what is now South Carolina:** J. Leitch Wright, Jr., *Anglo-Spanish Rivalry in North America* (Athens: University of Georgia Press, 1977).

75 **the planting of Jamestown:** In a dispatch of February 23, 1609, quoted in Donald G. Shomette, *Pirates on the Chesapeake* (Centerville, Maryland: Tidewater, 1985), 3–4. This danger had already been pointed out in a dispatch of Ambassador Álvaro de Bazán to King Philip II of March 21/April 9, 1586, reprinted in David Beers Quinn (ed.), *The Roanoke Voyages, 1584–1590* (London: Hakluyt Society, 1955), II: 753. Bazán's agents in London had reported that the English left 500 men and five war vessels on the North Atlantic coast, where, as Quinn comments, "it is reasonably clear that an important function of the North American settlement was to act as an advance base for ships operating against the Indies at large."

75 **pardoned by the king:** Alexander Brown was the first historian to discover that the British actually considered fostering piracy from the North American base. See his *The First Republic in America* (Boston: Houghton Mifflin 1898), 105.

76 **Attacks were certainly expected:** Quinn, *The Roanoke Voyages* II:722.

76 **in 1585:** Rodrigo de Junco to the President of the Council of the Indies, December 1585, reproduced in Quinn, *The Roanoke Voyages* II:746–748.

76 **England in 1588:** Irene A. Wright (ed.), *Further English Voyages to Spanish America, 1583–1594* (London: Hakluyt Society, 1951).

77 **English and Scots colonists:** J. G. Simms, "Conegal in the Ulster Plantation," *Irish Geography* (6:4, 1972), 386.

77 **filth and vermin:** When the Irish revolted in 1641 and massacred several thousand of the colonists, they began a decade-long rebellion that "was crushed by Oliver Cromwell's Puritan army with such ferocious religious zeal" that about one in each three Irish Catholics was killed or died in the accompanying famine; thousands fled to Europe and many others were shipped as slaves to the West Indies or Virginia. About 80 percent of the Irish agricultural lands were confiscated and turned over to Protestants. Kerby A. Miller, *Emigrants and Exiles: Ireland and the Irish Exodus to North America* (New York: Oxford University Press, 1985), 20–21. Miller comments (105) that the Irish word for a person who left Ireland, *deoraí,* means not emigrant but exile, and speculates (137) that perhaps 50,000 to 100,000 Catholics left or were forced to leave Ireland in the 1600s, and more than double that number in the 1700s.

77 **with thin legs:** David Ogg, "Britain after the Restoration," in Carsten (ed.).

78 **during the eighteenth century:** The migrants of one year, 1635, have been exhaustively studied by Alison Games, *Migration and the Origins of the Atlantic World* (Cambridge, Mass.: Harvard University Press, 2001).

CHAPTER 6: THE AFRICAN ROOTS OF AMERICAN BLACKS

79 **want of towns:** *On Poetry* (London n.d. [1733]).

79 **Atlantic Africa:** The term was coined by John K. Thornton. See his entry "The African Background to American Colonization," in Stanley L. Engerman and

Robert E. Gallman (eds.), *The Cambridge Economic History of the United States,* Vol. 1, *The Colonial Era* (Cambridge: Cambridge University Press, 1996), 53.

79 **population was denser:** John Thornton, *Africa and Africans in the Making of the Atlantic World, 1400–1800,* 2nd ed. (Cambridge: Cambridge University Press, 1998), 75. "The average density in seventeenth-century Lower Guinea (roughly the southern half of Ghana, Benin, Togo, and Nigeria) was probably well over thirty people per square kilometer, or well over the average European density of the time. Indeed, the Capuchins who visited the area in 1662 regarded it as so populated it resembled 'a continuous and black anthill' and noted that 'this kingdom of Arda [Allada] and most of this region [Lower Guinea] exceed in number and density [the population] of all other parts of the world.'"

79 **called "Nigritie":** Robert Harms, *The Diligent: A Voyage through the Worlds of the Slave Trade* (New York: Basic Books, 2002), 16–17.

79 **remained profound:** It is noteworthy, for example, that the exhaustive *Encyclopedia of World History* edited by Professor William L. Langer and first published in 1940 does not mention, even in its fifth edition of 1972, the great medieval empire of Mali.

82 **nearly a century would pass:** Leo Africanus, whose real name was Hassan az-Zaiyati, was born in 1485 in Granada, from which his family was expelled in 1492. He studied in the city of Fez in what is now Morocco and traveled throughout Africa. Captured by Christian pirates, he was given as a slave to Pope Leo X, for whom he wrote a memorandum on Africa. An English edition of his work was published by the Hakluyt Society in London in 1896.

82 **buy foodstuffs:** Bruce L. Mouser, *A Slaving Voyage to Africa and Jamaica,* (Bloomington: Indiana University Press, 2002), 71–76. "The chief of their employment[s] seems to be in cultivating rice [,] Making Salt, and tapping their Palm wine tree. This latter appears to be their God. They are a tribe of Bagos, peculiar to them selves, follow their own laws, never make Slaves, or sell any. . . . They have plenty of Stock which may be bought tolerable cheap. . . . The Bagos are very expert in Cultivating rice."

82 **seeds of paradise:** Richard Hakluyt, *Principall Navigations, Voyages, Traffiques and Discoveries of the English Nation,* published as *The Voyages* (London: Dent, 1907), 4:64.

82 **fervent heate thereof:** Hakluyt, *Voyages,* 4:64.

84 **their gold mines:** Harms, *The Diligent,* 135.

84 **before his trip:** P. E. Hair, "Columbus from Guinea to America," *History in Africa* 17 (1990), 113ff.

84 **its own waters:** Thornton, *Africa and Africans,* 36–38.

85 **sugar plantations:** Stephanie E. Smallwood's review of David Eltis, *The Rise of African Slavery in the Americas* in *William and Mary Quarterly* (3rd Series, 58:1, January 2001).

85 **African mercenary soldiers:** John Thornton, "Cannibals, Witches, and Slave Traders in the Atlantic World," *William and Mary Quarterly* (3rd series, 40:2, April 2003), 286.

85 **reside ashore:** Randy Sparks, *The Two Princes of Calabar* (Cambridge, Mass.: Harvard University Press, 2004), 41–42.

85 **twenty-five European forts:** Harms, *The Diligent,* 136.

85 **10 million to 12 million Africans:** Stephen D. Behrendt, *Teacher's Manual* for *The Trans-Atlantic Slave Trade, A Database on CD-ROM,* edited by David Eltis, David Richardson, Stephen D. Behrendt and Herbert S. Klein (Cambridge: Cambridge University Press, 1999). The data constitute essentially a spreadsheet drawn from a statistical study of 27,233 voyages out of an estimated 36,000 actual voyages.

85 **today's Afro-Americans:** David Eltis, *The Rise of African Slavery in the Americas* (Cambridge, Mass.: Harvard University Press, 2000). Eltis believes that without African resistance, the figure might have reached 1 million.

86 **Richard Hakluyt recounts:** *Voyages* 4:39ff.

87 **a full-scale expedition:** The fleet was far grander than many later sent to America by the Spanish, French, and English. It carried "Portuguese priests, masons, carpenters and soldiers, a selection of European cloth and other manufactures, even a printing-press." Roland Oliver and Anthony Atmore, *Medieval Africa 1250–1800* (Cambridge: Cambridge University Press, revised edition, 2003), 170.

87 **first European colony in Africa:** James Duffy, *Portugal in Africa* (Cambridge, Mass.: Harvard University Press, 1962), 38–43. Duffy shows how the relationship deteriorated "under the egotism and greed of the Portuguese [whose] mulatto children became functionaries, agents of the slave-trade, lesser members of the clergy."

87 **to the New World:** Many also went to the sugar-producing islands along the African coast and many ended in Portugal, where by 1550 Lisbon had a population of some 10,000 African slaves. Behrendt, *Teacher's Manual,* 13.

88 **similarly divided:** Commenting on his visit to the African coast, Columbus remarked that in Guinea, that is, the West African coast as he knew it, "there are a thousand different languages." Quoted in P. E. H. Hair. "Columbus from Guinea to America," *History in Africa* 17 (1990), 119. A convenient summary of African linguistics is given in Joseph H. Greenberg, *Language, Culture, and Communication* (Stanford, Calif.: Stanford University Press, 1971), 128ff.

88 **the first millennium A.D.:** Oliver and Atmore, *Medieval Africa,* 79.

89 **nearby villages:** *The Interesting Narrative of the Life of Olaudah Equiano or Gustavus Vassa, the African, Written by Himself* (Leeds, Printed for James Nichols, 1814, reprinted by Henry Louis Gates, Jr., in *The Classic Slave Narratives,* New York: Menton, 1987), 28–29. Doubts about the authenticity of the account have been expressed by Vincent Carretta, "Olaudah Equiano or

Gustavus Vassa? New Light on an Eighteenth Century Question of Identity," *Slavery and Abolition* 20 (1999) and his subsequent book, *Equiano, the African* (Athens: University of Georgia Press, 2005). Also see S. E. Ogude, "Facts into Fiction: Equiano's Narrative Reconsidered," *Research in African Literatures* 13 (1982). But whatever can be said about Equiano, his account is borne out by Randy J. Sparks's study of Ephraim Robin John and Ancona Robin Robin John in his *Two Princes* and by the remarks of the contemporary slave trader Jean Barbot in P. E. H. Hair et al. (eds.), *Barbot on Guinea: The Writings of Jean Barbot on West Africa, 1678–1712* (London: Hakluyt Society, 1992).

89 **a few thousand people:** Thornton, "African Background," 61.

89 **village councils:** Cited in Thornton, "African Background," 64.

90 **to European serfs:** Thornton, *Africa and Africans,* 87–88. A seventeenth-century Italian visitor described them as "slaves in name only."

90 **among their peoples by merchants:** In his essay "The Unity of Later Islamic History," *Journal of World History* 5:4, 892, Marshall G. S. Hodgson speaks of the outward thrust of Islam as carried by merchants; what is interesting about the commercial experience in Mali is that it picked up where the eastern Muslims left off and continued the advance deep into West Africa.

91 **a guardian spirit:** A. B. Ellis, *The Ewe-Speaking Peoples of the Slave Coast of West Africa* (Chicago: Benin Press, reprinted 1965 from an English edition of 1890), 15, 102.

91 **in life and death:** John A. Wilson, *The Burden of Egypt* (Chicago: University of Chicago Press, 1951), 86.

91 **gods permeated all things:** As Henri Frankfort wrote in *Kingship and the Gods: A Study of Ancient Near East Religion as the Integration of Society and Nature* (Chicago: University of Chicago Press, 1948), 30, Ptah, the "Creator of all," put *ka* in all things.

91 **indwelling spirit:** Ellis, *Ewe-speaking Peoples,* 22.

91 **oracular powers:** Wilson, *Burden,* 267–268.

91 **various African peoples:** One of the most fascinating, I think, is the notion of human duality in which the placenta is taken as a doppelgänger. See C. G. Seligmann and Margaret A. Murray, "Note upon an Early Egyptian Standard," *Man* (97, 1911), 165 ff.

91 **inviolate:** See William R. Polk, *Neighbors and Strangers: The Fundamentals of Foreign Affairs* (Chicago: University of Chicago Press, 1997), 238.

91 **called an *edan:*** A. B. Ellis, *The Yoruba-Speaking Peoples of the Slave Coast of West Africa* (Chicago: Benin, reprinted in 1964, original date and publisher not given), 171.

91 **to be molested:** Ellis, *Ewe-Speaking Peoples,* 178–181.

92 **story of Joan of Arc:** The tale of the Angolan Beatriz Kimpa Vita is told by Thornton in "Cannibals," 291–292. For preaching against war (which was the major source of slaves for export) and also for intervening, under a

religious guise, in the choice of succession to rule, she was burned at the stake in 1706.

92 **by eighteenth-century American colonists:** Philip D. Morgan, reviewing Judith A. Carney, *Black Rice: The African Origins of Rice Cultivation in the Americas* (Cambridge, Mass.: Harvard University Press, 2001), in *William and Mary Quarterly* (3rd Series, 53:3, July 2002), 739–742.

92 **of the world:** "Thornton, African Background," 71.

92 **ethnic homeland:** Oliver and Atmore, *Medieval Africa,* 163.

92 **Europe and America:** Thornton, "African Background," 82.

92 **practiced in *Africa*:** Margot Minardi, "The Boston Inoculation Controversy of 1721–1722: An Incident in the History of Race," *William and Mary Quarterly* (3rd Series, 61:1, January 2004), 47.

93 **from their Youth:** Karl Polanyi, *Dahomey and the Slave Trade* (Seattle, Wash.: University of Washington Press, 1966), 36.

93 **staff of experienced officers:** Quoted in Harms, *The Diligent,* 172.

93 **masses of the people:** Polanyi, *Dahomey,* and the following quote is from 54.

94 **terrible wars:** Harms, *The Diligent,* op. cit. 135–136. Since powder deteriorated in the humid climate, fresh supplies had to be imported periodically. By 1730, the Dutch were selling about 180,000 muskets and 20,000 tons of power annually along the Gold Coast and Slave Coast.

94 **trading for firearms:** J. E. Inikori, "The Import of Firearms into West Africa, 1750–1807: A Quantitative Analysis," *Journal of African History* (18, 1997), 339ff, quoted in James A. Rawley, *The Transatlantic Slave Trade* (New York: Norton, 1981), 271.

94 **domesticated animals:** On the origin, spread, and conditions of slavery, see Polk, *Neighbors and Strangers,* 41–43.

95 **legislature, to Africa:** Thornton, *African Background,* 80.

95 **his home on the Gambia River:** He subsequently wrote a letter of thanks, which is included in John W. Blassingame (ed.), *Slave Testimony* (Baton Rouge: Louisiana State University Press, 1997), 5–7.

95 **France or Russia:** On the eve of the French Revolution, perhaps 1 million Frenchmen were serfs, and in Russia at that time the number was far higher. On conditions in Africa, a seventeenth-century Italian visitor is quoted by Thornton (*Africa and Africans,* 87–88) as calling Africans "slaves in name only."

96 **major trade routes:** Thornton, "African Background," 62–63.

96 **Igbo and Ibibio villages:** Sparks, *Two Princes,* 49ff.

97 **the French ship *The Diligent*:** 235ff.

97 **eaten by white cannibals:** Harms, *The Diligent,* 18, 247.

97 **and his owner:** 249–250.

97 **planning a revolt:** Many of course were. David Richardson, "Shipboard Revolts, African Authority, and the Atlantic Slave Trade," *William and Mary*

Quarterly (58:1, January 2001), 69–92. Evidence has now been collected on some 485 "acts of violence" by Africans against slave ships and their crews, including 392 shipboard revolts.

98 **with another captive:** Harms, *The Diligent,* 263.

98 **themselves to death:** To prevent this, a special tool was invented to force-feed them.

98 **died on the journeys:** What is now known is recorded in Eltis et al. (eds.), *Trans-Atlantic Slave Trade,* It has been summarized and illuminated by a series of articles in *William and Mary Quarterly* (58, Issue 1, January 2001), of which some will be cited below; and in Behrendt, *Teacher's Manual.*

CHAPTER 7: EARLY DAYS IN THE COLONIES

101 **aware of their plans:** David Beers Quinn, *The Roanoke Voyages, 1584–1590* (London: Hakluyt Society, 1955), 1:3ff. Quinn analyzes how Richard Hakluyt and others treated the records they decided to publish. In a dispatch to King Philip II, dated March 31–April 9, 1586, Ambassador Álvaro de Bazán reported the English plan.

101 **of the proceeds:** Richard Hakluyt, *The Principall Navigations of the English Nation,* published as *Voyages* (London: Dent, 1907), 6:115–116.

102 **Indians called Tsenacomacoh:** Louis B. Wright and Virginia Freund, *William Strachey's The Historie of Travell into Virginia Britania, 1612* (Cambridge: Cambridge University Press for the Hakluyt Society, 1953), 37.

102 **annoyance of the Spaniards:** Diego Hernández de Quiñones to Philip II, June 12–22, 1585; reprinted in Quinn (ed.), *Roanoke Voyages,* 2:722.

102 **spoke English:** "The Relation of Hernando de Altamirano," reproduced in Quinn (ed.), *Roanoke Voyages,* 2:741.

102 **instrument of empire:** "Sir Walter Ralegh's Indian Interpreters, 1584–1618," *William and Mary Quarterly,* 59, April 202.

102 **Ralph Lane:** "Narrative of the Settlement of Roanoke Island, 1585–1586," in Hakluyt, *Voyages,* 6:150.

103 **rose to the bait:** David Hawke, *The Colonial Experience* (Indianapolis: Bobbs Merrill, 1966), 36.

103 **people had been there:** Quinn, *Roanoke Voyages,* Vol. 2, 810–811.

105 **greatest English merchant:** Hawke, *Colonial Experience,* 67–68.

106 **yielded no return:** John J. McCusker and Russell R. Manard, *The Economy of British America 1607–1789, with Supplementary Bibliography* (Chapel Hill: University of North Carolina Press, 1991), 119.

106 **In 1624:** Lyon G. Tyler (ed.), *Narratives of Early Virginia* (New York: Scribner, 1907), 393–394.

106 **for four months:** Clayton Colman Hall (ed.), *Narratives of Early Maryland 1633–1684* (New York: Scribner, 1910), 92–96.

106 **ventures in Florida:** George Percy, "A Trewe Relacyon of the Procedinges and Ocurrentes of Momente Which Have Hapned in Virginia from the Time Sir Thomas Gates Was Shipwrackte upon the Bermudes An° 1609 untill My Departure Outt of the Country Which was in An° Dni 1612," *Tyler's Quarterly Historical and Genealogical Magazine* (3:4, April 1922), 261.

106 **att the yardes Arme:** Percy, "A Trewe Relacyon," 278–279.

107 **wage war against them:** Irene A. Wright (ed.), *Further English Voyages to Spanish America, 1583–1594* (London: Hakluyt Society, 1951), 187.

107 **George Percy:** Percy, "A Trewe Relacyon," 264. Percy was acting governor of the colony three times during this period. See also Alexander Brown, *The First Republic in America* (Boston: Houghton Mifflin, 1898), 95.

107 **unhappy circumstances:** Smith's account of his adventuresome life has been published in many editions. Of these, I have used *The Adventures and Discovrses of Captain Iohn Smith Fometime Prefident of Virginia and Admiral of New England* (London, Paris, and New York: Caffell, 1883).

107 **view of the wilderness:** "Account of Jamestown and the Early Hardships (1607)," in Louis B. Wright (ed.), *The Elizabethans' America* (Cambridge, Mass.: Harvard University Press, 1965), 167.

108 **As Percy wrote:** Percy, *A Trewe Relacyon,* 170–171.

108 **his own people:** Philip L. Barbour (ed.), *The Jamestown Voyages under the First Charter, 1606–1609* (Cambridge: Cambridge University Press for the Hakluyt Society, 2nd Series, Vols. 136, 137, 1969), 192–193.

109 **wholly devoured it:** Alexander Brown, *The First Republic* (Boston: Houghton Mifflin, 1898), 97.

109 **to maintain themselves:** Norman Egbert McClure (ed.), *The Letters of John Chamberlain,* quoted in Philip Barbour, *The Three Worlds of Captain John Smith* (Cambridge, Mass.: Houghton Mifflin, 1964), 304.

109 **George Percy remembered it:** Percy, *A Trewe Relacyon,* 267.

110 **starved them to deathe:** Percy, *A Trewe Relacyon,* 280.

111 **La Florida:** Edward Gaylord Bourne (ed.), *Samuel de Champlain, Algonquians, Hurons and Iroquois . . . 1603–1616,* trans. of *Les Voyages de la Nouvelle France* (Dartmouth, Nova Scotia: Brook House, 2000), 72–76.

111 **England, Scotland, and Wales:** Abbot Emerson Smith, *Colonists in Bondage: White Servitude and Convict Labor in America 1607–1776* (New York: Norton 1971), 9–10. (Reprint of an earlier edition. The book was mainly written in the 1930s.)

111 **in Massachusetts Bay:** Abbott Lowell Cummings, *The Framed Houses of Massachusetts Bay, 1625–1725* (Cambridge, Mass.: Harvard University Press, 1979), 18.

112 **use a gun:** David Freeman Hawke, *Everyday Life in Early America* (New York: Harper and Row, 1988), 14.

112 **not these settlers' choice:** Harold R. Shurtleff, *The Log Cabin Myth* (Cambridge, Mass.: Harvard University Press, 1939).

112 **dug a cellar:** Americans were escaping the cold, but people living along the Tigris-Euphrates and Persian Gulf who were trying to escape extreme heat built similar semicellars (Arabic, *sirdab*). Independently, both groups had learned that the surrounding earth insulates a room.

112 **before rotting:** A drawing of what one of these house may have looked like is given in Ivor Noël Hume, *Martin's Hundred* (New York: Knopf, 1982), 58–59. A description is given in Pennsylvania Archives, 2nd Series (Harrisburg, 1877), Vol. 5, 182–183.

112 **kept it dry:** James Deetz, *In Small Things Forgotten: An Archaeology of Early American Life* (New York: Anchor, 1996), 20–21. Archaeologists call this form of construction earthfast or post-in-ground. See also Hume, *Martin's Hundred,* 247.

112 **every pane he could find:** Robert J. Brugger, *Maryland: A Middle Temperament 1634–1980* (Baltimore: Johns Hopkins University Press, 1988), 19.

113 **with two servants:** Hume, *Martin's Hundred,* 140–141.

113 **on the frontier:** Deetz, *In Small Things,* 167.

113 **necessary equipment:** Carol Berkin, *First Generations: Women in Colonial America* (New York: Hill and Wang, 1996), 13.

114 **dates from 1721:** Deetz, *In Small Things,* 76–77, 168–169.

114 **touched the land here:** "An Extract of Master Ralph Lanes Letter to Mr. Richard Hakluyt Esquire, and Another Gentleman of the Middle Temple, from Virginia," in Hakluyt, *Voyages,* Vol. 6, 140.

114 **came from Africa:** Darrett B. Rutman and Anita H. Rutman, "Of Agues and Fevers: Malaria in the Early Chesapeake," *William and Mary Quarterly* (3rd Series, 33:1, January 1776).

114 **Grypes of the Gutts:** Brugger, *Maryland,* 23.

114 **before the first birthday:** Louis Green Carr and Lorena S. Walsh, "The Planter's Wife: The Experience of White Women in Seventeenth-Century Maryland," *William and Mary Quarterly* (3rd Series, 34:4, October 1977), 542, 552–553. See also Carol Berkin, *First Generations: Women in Colonial America* (New York: Hill and Wang, 1996), 8. And see Darret B. Rutman, *Small Worlds, Large Questions* (Charlottesville: University Press of Virginia, 1994), 189. Rutman notes that "few of the children of this time and place [late seventeenth-century Virginia] reached their majority without losing at least one parent, while over a third lost both."

115 **iron smelting:** Timber had all but vanished in England by the early seventeenth century and was replaced by coal for heating. By 1600, 200,000 tons were being produced annually and some 300 ships were engaged in hauling it from Newcastle to London. Hawke, *Everyday Life,* 65.

115 **America to England:** Quinn (ed.), *Roanoke Voyages,* Vol. 1, 172.

115 **staves for barrels:** Hawke, *Everyday Life,* 91.

115 **as she could stow:** Samuel Eliot Morison (ed.), *William Bradford: Of Plymouth Plantation, 1620–1647* (New York: Knopf, 1959), 94.

115 **for the Royal Navy:** William Cronon, *Changes in the Land: Indians, Colonists, and the Ecology of New England* (New York: Hill and Wang, 1983), 110.

115 **King James I "blasted":** On his "Counter-Blaste to Tobacco" (1604) see Alden T. Vaughan, "Expulsion of the Savages: English Policy and the Virginia Massacre of 1622," *William and Mary Quarterly* (3rd Series, 35:2, January 1978), 60.

115 **our meat, drinke, cloathing and monies:** Quoted by Lorena S. Walsh, "Slave Life, Slave Society, and Tobacco Production in the Tidewater Chesapeake, 1620–1820," in Ira Berlin and Philip D. Morgan (eds.), *Cultivation and Culture: Labor and the Shaping of Slave Life in the Americas* (Charlottesville: University of Virginia Press, 1993), 170.

116 **a mutuall societie with us:** "A True Report of the Late Discoveries . . . By That Valiant and Worthy Gentleman, Sir Humfrey Gilbert Knight," in Hakluyt, *Voyages,* 49–53.

116 **still alive:** John Rolfe, "Relation of State of Virginia," in William Maxwell (ed.), *Virginia Historical Register* (Richmond, 1848–1853), 1:108–110; quoted in Evarts B. Greene and Virginia D. Harrington, *American Population before the Federal Census of 1790* (New York: Columbia University Press, 1932), 135.

116 **some more, some less:** "The General History of Virginia, The Second Book, Chapter 1," reprinted in Louis B. Wright (ed.), *The Elizabethans' America* (Cambridge, Mass.: Harvard University Press, 1965), 172.

117 **grants had been made:** Marcus Lee Hansen, *The Atlantic Migration 1607–1860* (New York: Harper, 1961), 29.

117 **Queen Elizabeth had specified:** Hakluyt, *Voyages,* 119.

117 **Law of Nations:** Sir George Peckham "A True Report of the Late Discoveries . . ." in Hakluyt, *Voyages,* Vol. 6, 49.

118 **Ralph Lane reported:** "The Second Voyage Made to Virginia by Sir Richard Grinvile for Sir Walter Ralegh, Anno 1585 . . ." in Hakluyt, *Voyages,* Vol. 6, 137–138.

118 **two Indian arrows:** The key features of the charter, actually given to his son, are reproduced in Henry Steele Commager (ed.), *Documents of American History* (New York: Crofts, 1947), 21–22. The token payment was specified in the so-called "Bishop of Durham clause." See also Charles M. Andrews, *The Colonial Period of American History* (New Haven: Yale University Press, 1936), Vol. 2, 279.

118 **Protestants outnumbered Catholics:** Andrews, *Colonial Period,* Vol. 2, 286–289.

119 **not to argue over religion:** "Instructions 13 Novem: 1633 . . ." reproduced in Hall (ed.), *Narratives,* 16.

119 **without molestation:** "A Briefe Relation of the Voyage unto Maryland, 1634," in Hall (ed.), *Narratives,* 40. The subsequent quotations are from the same source.

120 **yeoman-farmer class:** Hall (ed.), *Narratives,* 286–289.

121 **Robert Cushman wrote:** Morison (ed.), *Bradford,* 55.

121 **civil parts of the world:** Bradford was surely one of the most remarkable men of early America. He was born in Yorkshire in 1590 and was a farmer as a young man. After joining the Pilgrims when they moved to Holland, he worked as a weaver and taught himself not only Dutch but also some Latin and Hebrew. He sailed to America on the *Mayflower* and served his colleagues for thirty-three years as their governor and their historian.

121 **the Mayflower Compact:** The text is reproduced in Commager (ed.), *Documents,* 15–16.

121 **to organized society:** Remarkable it was, but not unique, as I point out in Polk, *Neighbors and Strangers,* 311–312. Similar constitutions were known in the ancient world; and probably others were formulated, even if not necessarily written, during the age of exploration.

122 **in the main:** Morison (ed.), *Bradford,* 77. On the settlers' relationship with Tisquantum, the sole survivor of the Patuxet tribe, which was wiped out by disease in 1617, see p. 80. Subsequent information is drawn from Bradford's account.

125 **city upon a hill:** Hawke, *Colonial Experience,* 132–133

CHAPTER 8: "MOTHER ENGLAND" LOSES TOUCH

126 **happened to operate:** Quoted in Jack P. Greene (ed.), *Great Britain and the American Colonies 1606–1763: Documentary History of the United States* (Columbia: University of South Carolina Press, 1970), xi.

126 **an orphant Plantation:** As John Hull later called Massachusetts. Quoted in Bernard Bailyn, *The New England Merchants in the Seventeenth Century* (Cambridge, Mass.: Harvard University Press, 1955), 151.

126 **without royal sanction:** This action was the basis of what came to be called the "Great Charter" of November 1618. See William Waller Hening (ed.), *The Statutes at Large* (Richmond, Va., 1809), Vol. 1, 110–113, reprinted in Greene (ed.), *Great Britain,* 27ff; and David Hawke, *The Colonial Experience* (Indianapolis: Bobbs-Merrill, 1966), 101–102.

127 **House of Commons:** The document authorizing this move has not survived, but it was reaffirmed in an ordinance of July 24, 1621, the text of which is reproduced in Henry Steele Commager (ed.), *Documents of American History* (New York: Crofts, 1947), 13–14.

127 **July 30, 1619:** John Pory, "A Report of the Manner of Proceeding in the General Assembly . . . ," *State Papers Domestic, James I,* 1:45, reprinted in Louis B. Wright (ed.), *The Elizabethans' America* (Cambridge, Mass.: Harvard University Press, 1965), 237–250.

127 **a royal colony:** The proclamation is reprinted in Greene (ed.), Great Britain, 35–37.

127 **reached 50,000:** George Chalmers, *History of the Revolt of the American Colonies* (Boston, 1845), Vol. 1, 29, based on British state papers. Quoted in Evarts B. Greene and Virginia D. Harrington, *American Population before the Federal Census of 1790* (New York: Columbia University Press, 1932), 3.

128 **permissive royal charter:** The text of the charter—dated March 4, 1629—is given in Commager (ed.), *Documents,* 16–18.

129 **to Massachusetts:** Hawke, *Colonial Experience,* 136–137.

130 **a sovereign state:** *Calendar of State Papers, 1660-1668,* §1103, cited in Viola Florence Barnes, *The Dominion of New England: A Study in British Colonial Policy* (New Haven: Yale University Press, 1923; reprint, New York: Frederick Ungar, 1951), 6–7.

131 **considered invading it:** As Edward Randolph reported much later, in 1676, "the whole country [was] complaining of the oppression and usurpation of the magistrates of Boston." The magistrates were, among other things, taxing the inhabitants without their being represented. Text in Michael G. Hall, Lawrence H. Leder, and Michael G. Kammen (eds.), *The Glorious Revolution in America: Documents on the Colonial Crisis of 1689* (New York: Norton, 1964), 19.

131 **United Colonies of New England:** The compact of May 19, 1643, is summarized in Commager (ed.), *Documents,* 26–28.

131 **lawes of England:** Summarized in Commager (ed.), *Documents,* 34.

132 **within its jurisdiction:** Bailyn, *New England,* 120.

133 **its own affairs:** Report by Edward Randolph, quoted in Greene (ed.), *Great Britain,* 69.

133 **Interest of New England:** *Calendar of State Papers, 1675-1676,* §953, 1037, 1067, cited in Barnes, *Dominion,* 13–15.

133 **In his report:** Reprinted in Greene (ed.), *Great Britain* op. cit., 68–70.

133 **but did little:** *Calendar of State Papers, 1677-1680,* §996, cited in Barnes, *Dominion,* 15.

133 **attempted subversion:** *Calendar of State Papers, 1681–1685,* §580, cited in Barnes, *Dominion,* 17–18.

133 **null and void:** *Calendar of State Papers, 1675-1676,* §556, 568, 694, 713 cited in Barnes, *Dominion,* 23.

133 **Dominion of New England:** The king's commission of the governor-designate, Sir Edmund Andros, is given in Greene (ed.), *Great Britain,* 81–84, from Francis N. Thrope (ed.), *Federal and State Constitutions, Colonial Charters, and Their Organic Laws* (Washington, D.C., 1909), Vol. 3, 1863–1867; reproduced in Hall, et al (eds.), *Glorious Revolution,* 25–27.

134 **Puritans to quit England:** "To the Reader . . . Grievances Justly Complained of by the People in New England," Reproduced in Hall, et al, (eds.), *Glorious Revolution,* 30ff.

134 **not having changed them:** "Nathanael Byfield's Account of the Insurrection,

April 29, 1689," reproduced in Hall, et al. (eds.), *Glorious Revolution,* 46–48.

135 **looted at sea:** Although he was in jail, Edmund Randolph presumably had visitors from among the agents and staff he had previously employed; and he wrote a surprisingly detailed letter, on October 8, 1689, describing conditions along the frontier and the seacoast. It is reprinted Hall, et al. (eds.), *Glorious Revolution,* 64–66.

135 **their auncient priviledges:** M. G. Hall (ed.), "The Autobiography of Increase Mather," *Proceedings of the American Antiquarian Society* (71:2, 1962), 335–337, reprinted in Hall, et al. (eds.), *Glorious Revolution,* 75–76.

136 **are absolutely subjected:** Wilcomb E. Washburn, *The Governor and the Rebel: A History of Bacon's Rebellion in Virginia* (New York: Norton, 1972), 20.

137 **muster 1,500 to 2,000 warriors:** Richard White, *The Middle Ground: Indians, Empires, and Republics in the Great Lakes Region, 1650–1815* (Cambridge: Cambridge University Press, 1991), 226.

137 **the king recalled him:** Louis B. Wright, *The Atlantic Frontier: Colonial American Civilization, 1607–1763* (New York: Knopf, 1951), 81–82.

137 **moved over to Maryland:** Leonard Strong, *Babylon's Fall in Maryland . . .* (London, 1655), reprinted in Clayton Colman Hall (ed.), *Narratives of Early Maryland, 1633–1684* (New York: Scribner, 1910), 235.

139 **the local Protestants:** The state records were moved from place to place, first under one side and then under the other, and finally many were destroyed in a fire in the eighteenth century. Leonard Strong's *Babylon's Fall in Maryland* and the *Refutation of Babylon's Fall* by John Langford (also reproduced in Hall [ed.], *Narratives,* 254ff) on behalf of Lord Baltimore are the best available accounts.

139 **wounded or killed:** Letter to Lord Protector Oliver Cromwell from L. Barber, Maryland, April 13, 1655, printed in Hall (ed.), *Narratives,* 262–265.

139 **about twenty to one:** In a population estimated at 25,000. Hall (ed.), *Narratives,* 144.

139 **incoming governor:** Raphael Semmes, *Crime and Punishment in Early Maryland* (Baltimore: Johns Hopkins Press, 1996), 174.

139 **goads driving them:** Michael G. Kammen, "The Causes of the Maryland Revolution of 1689," *Maryland Historical Magazine* (55:4, December 1960), 293ff.

140 **rights to the soil:** In 1716, when the proprietor's son turned Protestant, the colony was restored to the Baltimore family. Charles M. Andrews, *The Colonial Period of American History, Vol. 2: The Settlements* (New Haven: Yale University Press, 1936), 376.

140 **the colony's well-being:** Hawke, *Colonial Experience,* 251.

141 **unless he did:** Letter to Governor Edmond Andros, January 28, 1676; and to Lt. Anthony Brockoles, commander in chief, March 28, 1682, Reprinted in

Michael G. Hall et al., *The Glorious Revolution in America* (New York: Norton, 1972), 94.

141 **Liberties and Privileges:** David S. Lovejoy, "Equality and Empire: The New York Charter of Libertyes, 1683," *William and Mary Quarterly* (3rd Series, 21:4, October 1964).

141 **Schenectady:** Robert Livingston reported to Sir Edmund Andros on April 14, 1690, that 250 French and Indians killed 60 people, abducted 27 and burned the town. Reprinted in Hall, et al. (eds.), *Glorious Revolution,* 89.

142 **convicted and sentenced:** Lawrence H. Leder (ed.), "Records of the Trials of Jacob Leisler and His Associates," *New York History Society Quarterly* (36, 1952), 454. Reprinted in Hall, et al. (eds.) *Glorious Revolution,* 118.

142 **the governor wrote:** Quoted in Greene (ed.), *Great Britain,* xix.

CHAPTER 9: THE GROWTH OF THE COLONIES

144 **to 40,000:** Carl Bridenbaugh, *Cities in Revolt, Urban Life in America, 1743–1776* (New York: Capricorn, 1964), 5, 216.

145 **David Freeman Hawke wrote:** *Everyday Life in Early America* (New York: Harper and Row, 1988), 51–52.

145 **use of nearby timber:** Abbott Lowell Cummings, *The Framed Houses of Massachusetts Bay, 1625–1725* (Cambridge, Mass.: Harvard University Press, 1979), 50–51.

145 **Frederick Jackson Turner mentions:** On the basis of *Massachusetts Province Laws,* Vol. 1, 402. Quoted in "The First Official Frontier of the Massachusetts Bay," *Publications of the Colonial Society of Massachusetts,* (17, April 1914), 250–271; reprinted in *The Frontier in American History* (New York: Holt, Rinehart, and Winston, 1962).

146 **roof beam:** I owe this insight to the great German-American architect Walter Gropius, who meticulously examined the house with me.

146 **house painted:** See Alice Morse Earle, *Home Life in Colonial Days* (New York: Grosset and Dunlap, 1898), 27. Earle remarks, "Painters do not appear in any of the early lists of workmen."

146 **"half-faced camps":** Earle, *Home Life,* 3–5. Houses, being made of unseasoned wood, rarely lasted a decade. On Maryland see Gloria Main, *Tobacco Colony* (Princeton: Princeton University Press, 1982), 153.

146 **from the weather:** *Voyages dans l'Amérique Septentrionale* (Paris: Chez Prault, Imprimeur du Roi, 1786). Revised trans. by Howard C. Rice, Jr., *Travels in North America in the Years 1780, 1781, and 1782.* (Chapel Hill: University of North Carolina Press, 1963), 391.

146 **death in a City:** Letter to Susanna Wright. Leonard W. Labaree (ed.), *The Papers of Benjamin Franklin* (New Haven: Yale University Press), Vol. 4, 336.

146 **furniture-making:** Luke Beckerdite, "An Identity Crisis: Philadelphia and

Baltimore Furniture Styles of the Mid Eighteenth Century," in Catherine E. Hutchins (ed.), *Shaping a National Culture: The Philadelphia Experience, 1750–1800* (Winterthur, Del.: Henry Francis du Pont Winterthur Museum, 1994), 243ff.

147 **at least twenty cords:** Hawke, *Everyday Life* 55. See also Carl Bridenbaugh, *Cities* op. cit., 26. Bridenbaugh reports that in Manhattan "a medium-sized household consumed fifty or more cords of 'good walnut wood' annually."

147 **from city to city:** Characteristically, he did not patent it, writing, "That as we enjoy great Advantages from the Inventions of others, we should be glad of an Opportunity to serve others by any Invention of ours, and this we should do freely and generously." From *The Autobiography*, reprinted in Labaree, *Franklin*, Vol. 2, 419.

147 **a second story:** Alice Elaine Mathews, *Society in Revolutionary North Carolina* (Raleigh: North Carolina Department of Cultural Resources, 1976), 51–52.

148 **as late as the 1840s:** Family papers from my grandmother, who grew up there; and Herschel Gower, "Belle Meade" *Tennessee Historical Quarterly*, (22, September 1963), 4–5.

148 **his more intellectual activities:** Clement Eaton, *A History of the Old South* (New York: Macmillan, 1969), 59.

148 **known as a "necessary":** D. A. Tompkins, *History of Mecklenburg County and the City of Charlotte from 1740 to 1903* (Charlotte: Observer, 1903), 23.

148 **in his diary:** *The Diaries of George Washington* (Boston: Houghton Mifflin, 1925), CD-ROM edition (Oakman, Ala.: H-Bar Enterprises).

148 **May 9, 1759:** Labaree (ed.), *Franklin*, Vol. 8, 342ff.

149 **in his journal in 1774:** Edward Miles Riley (ed.): *The Journal of John Harrower, an Indentured Servant in the Colony of Virginia 1773–1776* (New York: Holt, Rinehart, and Winston for Colonial Williamsburg, 1963), 56.

149 **on the frontier:** Richard J. Hooker (ed.), *Charles Woodmason: The Carolina Backcountry on the Eve of the Revolution* (Chapel Hill: University of North Carolina Press, 1953), passim.

149 **taller than a British soldier:** Charles Patrick Neimeyer, *America Goes to War: A Social History of the Continental Army* (New York: New York University Press, 1996), 39.

150 **4.5 million pounds:** J. H. Plumb, *England in the Eighteenth Century (1714–1815)* (Harmondsworth: Penguin, 1969 edition), 126. See also Alice Hanson Jones, *Wealth of a Nation to Be: The American Colonies on the Eve of the Revolution* (New York: Columbia University Press, 1980), 135, who has estimated the total foreign debt of Americans at 6 million pounds.

150 **mercantile houses in London:** Quoted in Arthur M. Schlesinger, *Prelude to Independence: The Newspaper War on Britain 1764–1776* (New York: Vintage, 1965), 9.

150 **gentry of Virginia into rebels:** Cited in Charles A. Beard, *Economic Origins of*

Jeffersonian Democracy (New York, 1915), 297–298; quoted in Schlesinger, *Prelude,* 9.

150 **for half a century:** Eaton, *Old South,* 17.

150 **most important market, London:** Under the pseudonym "A Lover of Mankind," *A Discourse Concerning the Plague with Some Preservatives against It* (London, 1721), reprinted in Maude Woodfin (ed.), *Another Secret Diary of William Byrd of Westover* (Richmond, Va.: Dietz, 1942), 439.

151 **foxes from England:** Eaton, *Old South,* 57.

151 **subscription of the residents:** *Autobiography* reprinted in *Writings* (New York: The Library of America, 1987), 1446ff.

151 **he had undergone:** Quotation from Mather in Margot Minardi, "The Boston Inoculation Controversy of 1721–1722: An Incident in the History of Race," *William and Mary Quarterly* (3rd Series, 61:1, January 2004), 47.

152 **it was widely adopted:** Milbry C. Polk and Mary Tiegreen, *Women of Discovery* (New York: Random House, 2001), 177–179.

152 **contracted smallpox:** Bridenbaugh, *Cities,* 129.

152 **few property rights:** Carol Berkin, *First Generations: Women in Colonial America* (New York: Hill and Wang, 1997), 14ff.

152 **Charles Woodmason remarked:** Hooker (ed.), *Woodmason,* 39.

152 **a child a year:** This is the impression one gets from *William Byrd's Diary,* Woodfin (ed.), and it is borne out in entries in the Polk family Bible and many other sources.

153 **Charles Woodmason reported:** Hooker (ed.), *Woodmason,* 38.

153 **read law there:** Carl and Jessica Bridenbaugh, *Rebels and Gentlemen: Philadelphia in the Age of Franklin* (New York: Oxford University Press, 1965), 55–56, 179ff. A more general account of cultural life in Philadelphia is Nina Reid-Maroney, *Philadelphia's Enlightenment, 1740–1800: Kingdom of Christ, Empire of Reason* (Westport, Conn.: Greenwood, 2001).

153 **from North Carolina:** Alan D. Watson, *Society in Colonial North Carolina* (Raleigh: North Carolina Department of Cultural Resources, 1996), 77. See also Charles J. Stillé, *The Life and Times of John Dickinson, 1732–1808, Being Volume XIII, Memoirs of the Historical Society* (Philadelphia: Historical Society of Pennsylvania, 1891), Vol. 1, 26. Stillé's numbers do not quite match those of Alan Watson.

153 **also widely read:** Schlesinger, *Prelude,* 41.

153 **then as today:** Louis B. Wright, *The Cultural Life of the American Colonies, 1607–1763* (New York: Harper and Row, Torchbook edition, 1962), 128ff.

154 **in Everyman's Hands:** Bridenbaugh, *Cities,* 387.

154 **captured public interest:** Schlesinger, *Prelude,* 51ff.

154 **committees in conversation:** In his *Itinerarium,* quoted in Carl and Jessica Bridenbaugh, *Rebels and Gentlemen,* 22.

155 **their jakes on our tables!:** Labaree (ed.), *Franklin,* Vol. 8, 351, also see

Richard Hofstadter, *America at 1750: A Social Portrait* (New York: Vintage, 1973), 47–49.

155 **the mixture you have made:** Labaree (ed.), *Franklin,* Vol. 8, 351.

156 **Reverend William Smith, wrote:** Quoted in Carl and Jessica Bridenbaugh, *Rebels and Gentlemen,* 16.

156 **sugarcane at New Smyrna:** Esmond Wright, "American Independence in its American Context . . ." in A. Goodwin (ed.), *The New Cambridge Modern History,* Vol. 8, *The American and French Revolutions* (Cambridge: Cambridge University Press, 1971), 510.

156 **even between cities:** Carl Bridenbaugh, *Cities,* argues that "no concept about the last thirty-five years of the colonial period is more demonstrably erroneous than the one that the colonies were isolated one from another in thought and in deed, that travel by land was infrequent" (55), but he furnishes various examples to demonstrate the contrary. Certainly contemporary accounts are full of the perils of travel, and this remained true long after the Revolution.

156 **neither bridges nor ferries:** Wright, *American Independence,* 516.

157 **Liverpool to Boston:** Hawke, *Everyday Life,* 148.

157 **at how primitive these roads were:** Marquis de Chastellux, *Travels,* 83.

157 **often took five to six weeks:** Wesley Everett Rich, *The History of the United States Post Office to the Year 1829* (Cambridge, Mass.: Harvard University Press, 1924), 30–36.

158 **more frequent:** Schlesinger, *Prelude,* 6–7.

158 **additional 100 miles:** Rich, *Post Office,* 40.

158 **twopence a pound:** consistent figures do not exist for any one period but can be assembled from the following: for the British soldier on the eve of the Revolution, Edward P. Curtis, *The Organization of the British Army in the American Revolution* (New Haven: Yale University Press, 1926), 22; for the price of bread in New York, *Minutes of Common Council* noted in Michael G. Hall, Lawrence H. Leder, and Michael G. Kammen (eds.), *The Glorious Revolution in America* (New York: Norton, 1964), 86. For English wages and prices see George Rudé, *Wilkes and Liberty: A Social Study of 1763–1774* (London: Oxford University Press, 1962), 8, and Phyllis Deane, *The First Industrial Revolution* (Cambridge: Cambridge University Press, 1965), 8–9, and for comparison in America, John J. McCusker and Russell R. Menard, *The Economy of British America, 1607–1789* (Chapel Hill: University of North Carolina Press, 1991), 247.

158 **secrecy in their discussions:** John Adams's letter of September 29, 1774, to William Tudor, quoted in Schlesinger, *Prelude,* 208.

158 **duty of every citizen:** Pauline Maier, *From Resistance to Revolution: Colonial Radicals and the Development of American Opposition to Britain, 1765–1776* (London: Routledge and Kegan Paul, 1973), 4–5; and Bridenbaugh, *Cities,* 299.

159 **obliged to respond:** Benjamin Franklin gives an example in the *Pennsylvania Gazette* of April 11, 1751, reprinted in *Writings,* (New York: The Library of America, 1987), 357–359.

159 **mobs:** For example, food riots in Somerset and Wiltshire in 1766–1767 were put down by some 3,000 troops. See also Roy Porter, *English Society in the Eighteenth Century* (London: Penguin, rev. edition, 1991), 100–103. Porter notes that "disorder pockmarked Georgian England."

159 **in the 1770s:** Pauline Maier, *Resistance,* 4–5; and Dirk Hoerder, *Crowd Action in Revolutionary Massachusetts 1765–1780* (New York: Academic, 1977), 49–56.

159 **right and proper:** Eric Foner, *Tom Paine and Revolutionary America* (New York: Oxford University Press, 1976), 53–56. Such mobs were often egged on or actually led by members of the local gentry.

159 **Virginia in 1619:** Letter of September 30, 1619, reprinted in Louis B. Wright (ed.), *The Elizabethans' America* (Cambridge, Mass.: Harvard University Press, 1965), 253.

160 **for stealing one:** Ivor Noël Hume, *Martin's Hundred* (New York: Knopf, 1982), 147–148.

161 **distinct, and so segregatable, black slaves:** Abbot Emerson Smith, *Colonists in Bondage: White Servitude and Convict Labor in America 1607–1776* (New York: Norton 1971), 27–29.

161 **Dutch ship in 1619:** One took the name Anthony Johnson. The man who bought him allowed him to farm, get married, and finally become free. Johnson acquired by headright 250 acres on the eastern shore of Maryland which he and his children farmed using black slaves. Ira Berlin, *Many Thousands Gone: The First Two Centuries of Slavery in North America,* (Cambridge, Mass.: Harvard University Press 1998), 29–31.

161 **in English-controlled Virginia:** *Virginia Magazine of History and Biography,* (7), 364–367; cited in Evarts B. Greene and Virginia D. Harrington, *American Population before the Federal Census of 1790* (New York: Columbia University Press, 1932), 144.

161 **Smith wrote:** Smith, *Colonists,* 113ff.

161 **sixty-eight captives:** Smith, *Colonists,* 182.

162 **described the system:** Smith, *Colonists,* 69ff and 83.

CHAPTER 10: BLACKS IN AMERICA

164 **like a domesticated animal:** Mario Liverani, "The Ideology of the Assyrian Empire," in Mogen Trolle Larsen, *Power and Propaganda: A Symposium on Ancient Empires* (Copenhagen: Akademisk Forlag, 1979), 311.

164 **"captive alien":** Middle Iranian, *dast-grab;* Latin, *captus;* Greek, *doulos;* etc. Keith Hopkins, *Conquerors and Slaves* (Cambridge: Cambridge University

Press, 1978), esp. 76–77; Joseph Vogt, *Ancient Slavery and the Ideal of Man* (Oxford: Blackwell, 1974), 39ff; J. P. V. D. Balsdon, *Romans and Aliens* (London: Duckworth, 1979), 79ff; Orlando Patterson, *Slavery and Social Death* (Cambridge, Mass.: Harvard University Press, 1982), 345ff; K. R. Bradley, *Slaves and Masters in the Roman Empire* (Oxford: Oxford University Press, 1987), 113ff; C. I. Beckworth, *The Tibetan Empire* (Princeton: Princeton University Press, 1993), 103 on *lu*.

164 **destined for slavery:** I have discussed these and other issues relating to "strangers" in *Neighbors and Strangers: The Fundamentals of Foreign Affairs* (Chicago: University of Chicago Press, 1997), 296–297.

164 **close to apes:** That the black was closer to the animal, Jefferson astonishingly and falsely asserts by "the preference of the Oran-ootan for the black women over those of his own species." William Peden (ed.), *Thomas Jefferson, Notes on the State of Virginia* (Chapel Hill: University of North Carolina Press), query 14, 138–139.

164 **wool:** He instructed his steward to get "a strong horn comb and [directed] him to keep his head well combed, that the hair, or wool may grow long." W. D. Conway (ed.), *George Washington and Mount Vernon* (Brooklyn, N.Y.: Long Island Historical Society Memoirs, 1889) Vol. 4, 244; quoted in Michael Mullin, *American Negro Slavery: A Documentary History* (Columbia: University of South Carolina Press, 1976), 104.

164 **apply to a slave:** *State v. Tom*, noted in Bryce R. Holt, *The Supreme Court of North Carolina and Slavery* (New York: AMS, 1970, from the 1927 edition), 29.

164 **set them aside:** William Peden (ed.), *Jefferson*, query 14, 139.

165 **the chyle:** "A milky fluid containing emulsified fat and other products of digestion, formed from the chyme in the small intestine and conveyed by the lacteals and the thoracic duct to the veins." *Random House Webster's Unabridged Dictionary*.

165 **pervading darkness:** Quotation in Kenneth M. Stampp, *The Peculiar Institution: Slavery in the Ante-Bellum South* (New York: Vintage, 1989), 8. Note, however, that most of Stampp's book deals with the nineteenth century.

165 **becoming a "zombie":** Viktor Frankl, *From Death-Camp to Existentialism* (Boston: Beacon, 1959); Bruno Bettelheim, *The Informed Heart* (London: Thames and Hudson, 1961); Terrence des Pres, *The Survivor: An Anatomy of Life in the Death Camps* (New York: Oxford University Press, 1976).

165 **one description:** See the engraving from a daguerreotype of 1860, published in *Harper's Weekly*, reproduced in Daniel P. Mannix with Malcolm Cowley, *Black Cargoes* (New York: Viking, 1962), opposite 147.

166 **died on the voyage:** Herbert S. Klein, Stanley L. Engerman, Robin Haines, and Ralph Schlomowitz, "Transoceanic Mortality: The Slave Trade in Comparative Perspective," *William and Mary Quarterly* (3rd Series, 58:1, January 2001), 93–117.

166 **thrown overboard:** Daniel P. Mannix with Malcolm Cowley, *Black Cargoes* (New York: Viking, 1962), 126; and James A. Rawley, *The Transatlantic Slave Trade* (New York: Norton, 1981), 298–299. When the case was appealed, the court held that there was a "higher law" that required humanity toward the slaves, but except for saving this particular insurance company money, its decision seems to have had scant effect.

166 **sick and thin:** I. A. Wright (ed.), *Spanish Documents Concerning English Voyages to the Caribbean, 1527–1568* (London: Hakluyt Society, 1929), quoted in Kenneth R. Andrews, *Trade, Plunder, and Settlement: Maritime Enterprise and the Genesis of the British Empire, 1480–1630* (Cambridge: Cambridge University Press, 1984), 125.

166 **overwhelm the crews:** David Richardson, "Shipboard Revolts, African Authority, and the Atlantic Slave Trade," *William and Mary Quarterly* (3rd Series, 58:1, January 2001), 69–92.

166 **from the middle passage:** David Eltis, *The Rise of African Slavery in the Americas* (Cambridge: Cambridge University Press, 2000), 160.

166 **are now known:** Richardson, "Shipboard Revolts."

166 **Olaudah Equiano:** Henry Louis Gates, Jr., *The Life of Olaudah Equiano: The Classic Slave Narratives* (New York: Mentor, 1987).

167 **into oil and eaten:** John Thornton, *Africa and Africans in the Making of the Atlantic World, 1400–1680* (Cambridge: Cambridge University Press, 1992), 161. Thornton comments, "This was a near universal opinion."

167 **would slavery be legalized:** Clement Eaton, *A History of the Old South* (New York: MacMillan, 1969), 30.

167 **in New Netherland:** William Stuart, "Negro Slavery in New Jersey and New York," *Americana Illustrated* (16:4, October 1922), 348; cited in David Steven Cohen, *The Ramapo Mountain People* (New Brunswick, N.J.: Rutgers University Press, 1974), 25–26.

168 **among America's foremost slavers:** Laura E. Wilkes, *Missing Pages in American History* (Washington D.C.: Press of R. L. Pendelton, 1919), 9, 15.

168 **trained to fight:** Thornton, *Africa and Africans,* 149–150.

168 **to the Lords of Trade:** Wilkes, *Missing Pages,* 8.

169 **boycott of British goods:** Walter Minchinton, Celia King, and Peter Waite (eds.), *Virginia Slave-Trade Statistics 1698–1775* (Richmond: Virginia State Library, 1984), xiiff.

169 **British American colonies:** Of each 100, fifty-five were sent to Spanish and Portuguese colonies and thirty-four went to the Caribbean. Robert William Fogel and Stanley L. Engerman, *Time on the Cross: The Economics of American Negro Slavery* (Boston: Little Brown, 1974), 14 (table of distribution).

169 **Shelter Island:** John Noble Wilford, "American Slavery in the North," *International Herald Tribune* (August 2, 1999), 9.

169 **one of the few narratives of this period:** *History of Mary Prince, A West Indian*

Slave, Related by Herself (London, 1831); reprinted in Gates, *Slave Narratives,* 197.

171 **paint, powder, or lead dust:** One job of ships' "surgeons" was to prepared the slaves for market, "to heal as well as hide diseases. Black skins were made to glow with oil, wounds closed, and scars concealed with ointments. Mercury and other drugs employed in these artifices would later cause a disease to erupt with even greater virulence." Rawley, *Transatlantic,* 297.

171 **their African origin:** Judith A. Carney, *Black Rice: The African Origins of Rice Cultivation in the Americas* (Cambridge, Mass.: Harvard University Press, 2001). Reviewed by Philip D. Morgan in *William and Mary Quarterly* (59:3, July 2002), 740.

172 **a tiny minority:** Ira Berlin, *Many Thousands Gone: The First Two Centuries of Slavery in North America,* (Cambridge, Mass.: Harvard University Press, 1998), 136–137.

172 **Frederick Douglass remembered:** *Narrative of the Life of Frederick Douglass, an American Slave, Written by Himself* (Boston: Anti-Slavery Office, 1845), reprinted in Gates, *Slave Narratives.* Quotations are from pp. 260, 264, and 271.

172 **round his Loins:** Quoted in Philip D. Morgan, *Slave Counterpoint: Black Culture in the Eighteenth-Century Chesapeake and Lowcountry* (Chapel Hill: University of North Carolina Press, 1998), 133.

172 **Negro cloth:** Stampp, *Peculiar Institution,* 288 ff.

173 **Mary Prince:** *History of Mary Prince, A West Indian Slave, Related by Herself* (London, 1831); reprinted in Gates, *Slave Narratives* 191, 197, 198.

173 **least over time:** Mullin (ed.), *Negro Slavery,* 97.

173 **". . . was often intimate."** Rhys Isaac, *Landon Carter's Uneasy Kingdom,* (Oxford: Oxford University Press, 2004), chapter 13, "Landon and Nassaw." See also the diary of William Byrd. Louis B. Wright and Marion Tinling (eds.) *The Great American Gentleman William Byrd of Westover in Virginia* (New York: Capricorn, 1963), passim; and Maude H. Woodfin (ed.), *Another Secret Diary of William Byrd of Westover* (Richmond, Va.: Dietz, 1942), passim.

174 **comfortable appearance:** Quoted from J. H. Easterby (ed.), *South Carolina Rice Plantation,* 347 in Stampp, *Peculiar Institution,* 289–290.

174 **begging for mercy:** "An Account of the Life of Mr. David George, from Sierra Leone in Africa. . . ." *Baptist Annual Register, Vol. 1 (1790–1793),* 473; reprinted in Morgan, *Slave Counterpoint,* 392.

175 **in such circumstances:** Peden (ed.), *Jefferson,* 162–163, and Isaac, *Landon Carter,* 230–232.

175 **Sally Hemings:** Alexander O. Boulton, "The Monticello Mystery—Case Continued," *William and Mary Quarterly,* (3rd Series, 58:4, October 2001), 1039–1046. As Boulton comments in this review essay, "The story of the

Jefferson-Hemings relationship is part of an ongoing national conversation on issues of race, sexuality, and American culture. . . . Have we excluded discussions of race and sexuality in an effort to make our historical memory conform to a Jeffersonian rhetoric of equality?"

175 **a sexual union:** Walter Clark (ed.), *The State Records of North Carolina* (Winston: State of North Carolina, 1895–1906), Vol. 23, 62–66, quoted in Marvin L. Michael Kay and Lorin Lee Cary, *Slavery in North Carolina, 1748–1775* (Chapel Hill: University of North Carolina Press, 1995), 67.

175 **followed suit in 1691:** Eaton, *Old South,* 30. See also David D. Smits, " 'Abominable Mixture,' " *The Virginia Magazine of History and Biography* (95:2, 1987), 157–158.

175 **this dominion forever:** William Waller Hening (ed.), *The Statutes at Large, being a Collection of all the Laws of Virginia, from the First Session of the Legislature in the year 1619* (Philadelphia, 1823), Vol. 3, 86–88.

175 **each White man or woman:** Walter Clark (ed.), *The State Records,* Vol. 33, 20, quoted in Jeffrey J. Crow, *The Black Experience in Revolutionary North Carolina* (Raleigh: Division of Archives and History, 1996), 20.

175 **of this Province:** Quoted in Crowe, *Black Experience,* 21.

175 **about the matter:** Quoted in Crowe, *Black Experience,* 30.

176 **by dark-skinned men:** Nancy L. Paxton, "Mobilizing Chivalry: Rape in British Novels about the Indian Uprising of 1857," *Victorian Studies* (Fall 1992), 5ff. Paxton notes that the theme of Englishwomen being raped by dark Indian men figured in some eighty English novels.

176 **and a white mother:** Martha Hodes, *White Women, Black Men: Illicit Sex in the Nineteenth-Century South* (New Haven: Yale University Press, 1997); and Ira Berlin, *Many Thousands Gone,* 44–45.

176 **produced mulatto babies:** Mullin (ed.) *Negro Slavery,* 101.

176 **the eighteenth century:** W. Jeffrey Bolster, *Black Jacks: African American Seamen in the Age of Sail* (Cambridge, Mass.: Harvard University Press, 1997), 7. Some free blacks even served on slave ships (58–59).

177 **land to farm:** Alexander C. Flick (ed.), *History of the State of New York* (New York: Columbia University Press, 1933–1937), Vol. 1, 342; quoted in Cohen, *Ramapo Mountain People,* 25–26.

177 **a "run-away":** His advertisement is given in Mullin (ed.), *Negro Slavery,* 112. John Hope Franklin's book *Runaway Slaves* (Oxford: Oxford University Press, 1999) deals with the nineteenth century, when a large proportion of blacks had acquired language skills and other knowledge that enabled them to move more easily.

177 **act of 1691:** Hening (ed.), *Statutes,* Vol. 3, 86–88.

178 **good bound Hat:** *Virginia Gazette,* September 12, 1771; reproduced in Mullin (ed.), *Negro Slavery,* 82–83.

178 **a free colony:** Robert Harms, *The Diligent* (New York: Basic Books, 2002), 279.

178 **quite powerful:** Review by E. J. Hobsbawm of Richard Price's *Alabi's World* in *New York Review of Books* (December 6, 1990).

178 **Norfolk, Virginia:** Kay and Cary, *Slavery,* 98, 99.

179 **the job for them:** Berlin, *Many Thousands Gone,* 67.

179 **still exists:** Cohen, *Ramapo Mountain People.* Among the strains of migrants are said to be Tuscarora Indians from South Carolina, slaves manumitted by the Dutch in the middle of the seventeenth century, and blacks who escaped from the South.

179 **if they should:** Quoted in Mullin (ed.), *Negro Slavery,* 83.

179 **sheltered runaway blacks:** Verner W. Crane, *The Southern Frontier, 1670–1732* (Ann Arbor: University of Michigan Press, 1932), 184, 247. This practice gave rise to a new group of mulattoes, a mixture of black and Indian.

180 **with the two ears:** Thomas Cooper and David J. McCord (eds.), *The Statutes at Large of South Carolina* (Columbia, 1836–1841), Vol. 7, 415; noted in A. Leon Higginbotham, Jr., *In the Matter of Color: Race and the American Legal Process—The Colonial Period* (New York: Oxford University Press, 1978), 198. To understand the specification of "two ears" consider that scalping was an eighteenth-century means of making a "body count." To ensure that a single scalp had not been divided so as to earn a double bounty, each scalp had to be entire, with both ears.

180 **two Drums beating:** As described in a letter to *Gentleman's Magazine* (London) dated October 2, 1740; quoted in Mullin (ed.), *Negro Slavery,* 84–86.

181 **harsh set of laws:** Higginbotham, *Color,* 38ff. See also Theodore Brantner Wilson, *The Black Codes of the South* (Tuscaloosa: University of Alabama Press, 1965).

181 **teaching him to write:** Higginbotham, *Color,* 198. This act and subsequent acts formed the basis for the code in Georgia as well. See Ulrich B. Phillips, *American Negro Slavery* (Baton Rouge: Louisiana State University Press, 1966, from the 1918 edition), 492.

182 **into submission:** Kay and Cary, *Slavery,* 333–334, fn 31.

182 **the previous year:** Kay and Cary, *Slavery,* 94.

182 **slaves are executed:** Holt, *Supreme Court,* 7.

182 **legislature in 1669:** Hening, *Statutes,* 36.

182 **which specified:** Hening, *Statutes,* 55.

182 **a first offense:** Holt, *Supreme Court,* 12.

183 **fine of 100 pounds:** Cooper and McCord, *Statutes,* 198.

183 **candidates for freedom:** John Hope Franklin, *From Slavery to Freedom: A History of Negro Americans,* 3rd ed. (New York: Vintage, 1969), 86.

183 **followed by five other colonies:** Albert J. Raboteau, *Slave Religion: The "Invisible Institution" in the Antebellum South* (New York: Oxford University Press, 1978), 99.

183 **of our assembly:** Ministers of the South Carolina Society for the Propagation of

the Gospel to Mr. Gideon Johnston, March 4, 1713; reprinted in Mullin (ed.), *Negro Slavery,* 53.

183 **attendance at church:** Walter Clark (ed.), *The State Records of North Carolina* (Winston and Goldsboro, 1895–1907), Vol. 33, 20; quoted in Crowe, *Black Experience,* 20.

183 **each separate plantation:** Particularly as rice became a major plantation crop, larger groups of black slaves were brought together. By about 1720, three out of four black slaves lived alongside at least ten others; as the trend toward larger plantations continued, the size of slave populations grew apace. By the middle of the eighteenth century, these groups could be described as small societies approximating the kind that slaves or their ancestors had known in Africa.

184 **whites were doing:** Sylvia R. Frey, *Water from the Rock: Black Resistance in a Revolutionary Age* (Princeton: Princeton University Press, 1991), 23–25.

184 **no one knows:** W. E. B. Du Bois was a pioneer in at least asking such questions in *The Souls of Black Folks* and *The Negro Church* (both 1903). See also Raboteau, *Slave Religion,* 96ff.

184 **rebellious colonists:** Benjamin Quarles, "Lord Dunmore as Liberator," *William and Mary Quarterly* (3rd Series, 15:4, October 1958), 494ff. Adapted and separately published: Benjamin Quarles, *The Negro in the American Revolution* (Chapel Hill: University of North Carolina Press, 1996, from the 1961 edition), 19ff.

185 **Herbert Aptheker has estimated:** *American Negro Slave Revolts, 1526–1860* (New York: International, 1939). See also David Brion Davis, review of Robin Blackburn's *The Making of New World Slavery* (London: Verso, 1998) in *New York Review of Books,* June 11, 1998; and Karen Ordahl Kupperman: "The Founding Years of Virginia—and the United States," *Virginia Magazine of History and Biography* (104:1), 155ff.

185 **a very wrong foundation:** Kenneth Coleman, *The American Revolution in Georgia, 1763–1789* (Athens: University of Georgia Press, 1958), 45–46.

185 **run for freedom:** Their generally sad fate at the hands of the British falls outside the scope of this narrative. Only a few achieved freedom. Many ended by exchanging a slavery in North America for a slavery in the West Indies, trading, as it were the tillage of cotton or rice for sugar.

CHAPTER 11: WHITES, INDIANS, AND LAND

186 **the Powhatan in 1609:** Cited by J. Frederick Fausz, "Merging and Emerging Worlds," in Lois Green Carr, Philip D. Morgan, and Jean B. Russo (eds.), *Colonial Chesapeake Society* (Chapel Hill: University of North Carolina Press, 1988), 51.

186 **among the Pueblo people:** Because the areas now composing the states of Texas and New Mexico and the region farther west did not have a major effect

upon American history until well after the period described in this book, I have not dealt with them here. For the revolts of the Pueblo people, see David J. Weber, *The Spanish Frontier in North America* (New Haven: Yale University Press, 1992), 132ff.

188 **and possess theirs:** Hugh Talmage Lefler (ed.), *A New Voyage to Carolina . . . by John Lawson* (Chapel Hill: University of North Carolina Press, 1967), 243.

189 **will make many adventures upon you:** *The Original Writings and Correspondence of the Two Richard Hakluyts* (London: Hakluyt Society, 2nd series, 1935), Vol. 76, 495; quoted by David D. Smits, "Abominable Mixture: Toward the Repudiation of Anglo-Indian Intermarriage in Seventeenth Century Virginia," *Virginia Magazine of History and Biography* (95:2, 1987), 164. Governor Argoll issued an edict to this effect on May 28, 1618: "No trade with ye perfidious Savages nor familiarity lest they discover our weakness."

189 **scarce fifty remained:** Samuel Eliot Morison (ed.), *William Bradford of Plymouth Plantation, 1620–1647* (New York: Knopf, 1959), 77.

189 **As Smith wrote:** *The Adventvres and Discovrses of Captain Iohn Smith, Fometime Prefident of Virginia, and Admiral of New England, Newly Ordered by Iohn Ashton* (London: Caffell, 1883), 281–282.

190 **Travelers in the Carolinas:** Colin G. Calloway, *New Worlds for All: Indians, Europeans, and the Remaking of Early America* (Baltimore: Johns Hopkins University Press, 1997), 33ff.

190 **Indian people were gone:** Kathleen A. Deagan, "Spanish-Indian Interaction in Sixteenth-Century Florida and Hispaniola," in William W. Fitzhugh (ed.), *Cultures in Contact* (Washington D.C.: Smithsonian Institution Press, 1985), 288.

190 **A Catholic priest in Louisiana speculated:** Father Le Maire in 1718, in *Correspondance Générale: Louisiane Archives des Colonies,* Série C13A, Archives Nationales de Paris. Quoted in Gail Alexander Buzhardt and Margaret Hawthorne, *Rencontres sur le Mississippi, 1682–1763* (Jackson: University Press of Mississippi, 1993), 101.

190 **Thomas Harriot reported in 1587:** "A Briefe and True Report of the New Found Land of Virginia . . ." in Hakluyt, *The Principall Navigations of the English Nation,* reprinted as *Voyages* (London: Dent, 1907), Vol. 6, 191.

190 **recorded in 1634:** Morison (ed.), *Bradford,* 270–271.

191 **a factor of twenty to one:** Daniel K. Richter, "War and Culture: The Iroquois Experience," *William and Mary Quarterly* (3rd Series, 40:4, October 1983), 537, fn 34. He cites Russell Thornton, "American Indian Historical Demography: A Review Essay with Suggestions for Future Research," *American Indian Culture and Research Journal* (3:1, 1979), 69–74.

191 **nineteen tribes in 1670:** Wilcomb E. Washburn, *The Governor and the Rebel: A History of Bacon's Rebellion in Virginia* (New York: Norton, 1957), 19–20.

191 **to less than 3,000:** Francis Jennings, *Empire of Fortune: Crowns, Colonies, and Tribes in the Seven Years' War in America* (New York: Norton, 1988), 446, fn 23.

191 **the Catawbas:** As John Lawson noted in 1701. James H. Merrell, "The Indians' New World: The Catawba Experience," *William and Mary Quarterly* (3rd Series, 41:4, October 1984), 548.

191 **a hollow peninsula:** "That Demonic Game," *The Americas* (35, July 1978), quoted in Weber, *Spanish Frontier,* 118.

191 **had to have this weapon also:** Francis Jennings, *The Ambiguous Iroquois Empire* (New York: Norton, 1984), 80–81; and Daniel K. Richter, *The Ordeal of the Longhouse: The Peoples of the Iroquois League in the Era of European Colonization* (Chapel Hill: University of North Carolina Press, 1992), 63–64.

192 **without all redemption:** John Pory "A Report of the Manner of Proceeding in the General Assembly Covented at James City . . . July 30–August 4, 1619," in Louis B. Wright (ed.), *The Elizabethans' America* (Cambridge, Mass.: Harvard University Press, 1965), 245.

192 **violated the restrictions:** Helen C. Rountree and Thomas E. Davidson, *Eastern Shore Indians of Virginia and Maryland* (Charlottesville: University Press of Virginia, 1997), 136.

192 **Father Andrew White, S.J., in 1634:** "A Brief Relation of the Voyage unto Maryland," in Clayton Colman Hall (ed.), *Narratives of Early Maryland, 1633–1684* (New York: Scribner, 1910), 44.

192 **long been debated:** Peter C. Mancall, *Deadly Medicine: Indians and Alcohol in Early America* (Ithaca, N.Y.: Cornell University Press, 1995), 1ff.

192 **they were all type O:** Luigi Luca Cavalli-Sforza, *Genes, Peoples, and Languages* (New York: Farrar, Straus, and Giroux, 2000), 14; and Luigi Luca Cavalli-Sforza and Francesco Cavalli-Sforza, *The Great Human Diasporas* (London: Perseus, 1995), 109. See also Charles A. Hanna, *The Wilderness Trail* (Lewisburg, Pa.: Wennawoods, 1995, reprinted from the 1911 edition), Vol. 1, 163, 274, 275.

193 **their predicament:** "Narratives of the Late Massacres," in *Writings* (New York: Library of America, 1987), 553.

193 **Indians would destroy themselves:** Johnson's sympathetic biographer, James Thomas Flexner, in *Lord of the Mohawks* (Boston: Little, Brown, revised edition, 1979) preferred to skip over this aspect of his relationship with the Indians. It was picked up in Fred Anderson, *Crucible of War: The Seven Years' War and the Fate of Empire in British North America, 1754–1766* (New York: Knopf, 2000), 818–819, fn 7.

193 **Ensign Thomas Savage:** Rountree and Davidson, *Eastern Shore,* 50.

194 **25,000 tons of rum:** John J. McCusker and Russell R. Menard, *The Economy of British America, 1607–1789, with Supplementary Bibliography* (Chapel Hill: University of North Carolina Press, 1991), 290; and John L. Bullion, *A Great*

and Necessary Measure: George Grenville and the Genesis of the Stamp Act, 1763–1765 (Columbia: University of Missouri Press, 1982), 39.

194 **war chief of the Lower Cherokee:** Quoted in David H. Corkran, *The Cherokee Frontier* (Norman: University of Oklahoma Press, 1962), 14.

194 **ax-maker:** Richter, *Longhouse,* 75.

194 **wilder Germans at bay:** I have discussed this strategy in *Neighbors and Strangers: The Fundamentals of Foreign Affairs* (Chicago: University of Chicago Press, 1997), 129ff.

194 **concluded with us:** Frederic W. Gleach, *Powhatan's World and Colonial Virginia: A Conflict of Cultures* (Lincoln: University of Nebraska Press, 1997), 188–189; and E. Lawrence Lee, *Indian Wars in North Carolina, 1663–1763* (Raleigh: Carolina Charter Tercentenary Commission, 1963), 49.

194 **enslaved or massacred:** J. Leitch Wright, Jr., *The Only Land They Knew: The Tragic History of the American Indians in the Old South* (New York: Free Press, 1981), 138–139.

195 **to plant and fortefye:** Quoted in Alden T. Vaughan, "Expulsion of the Savages: English Policy and the Virginia Massacre of 1622," *William and Mary Quarterly* (3rd Series, 35:1, January 1978), 63.

195 **related in 1610:** George Percy, "A Trewe Relacyon of the Procedinges and Ocurrentes of Momente Which Have Hapned in Virginia from the Time Sir Thomas Gates Was Shipwrackte upon the Bermudes An° 1609 untill My Departure Outt of the Country Which Was in An° Dni 1612," *Tyler's Quarterly Historical and Genealogical Magazine* (3:4), 259ff.

196 **their new settlements:** Gleach, *Powhatan's World,* 148ff.

196 **the greatest labor:** Report of Edward Waterhouse in 1622, in Susan M. Kingsbury (ed.), *The Records of the Virginia Company of London* (Washington, D.C.: U.S. Government Printing Office, 1933), Vol. 3; cited in Gleach, *Powhatan's World,* 159.

197 **part of their heads:** Quoted in Carl Bridenbaugh, "Opechancanough: A Native American Patriot," in *Early Americans* (New York: Oxford University Press, 1981), 45.

197 **without special permission:** Philip Alexander Bruce, *Institutional History of Virginia* (New York: Putnam, 1910), 72.

197 **upon paine of death:** John D. Cushing (ed.), *Colonial Laws of Virginia, 1619–1660* (New York, 1915), 52, 123; quoted in Smits, *Abominable Mixture,* 186ff.

198 **they could find:** Alfred A. Cave, *The Pequot War* (Amherst: University of Massachusetts Press, 1996), 69ff; John Grenier, *The First Way of War* (Cambridge: University of Cambridge Press, 2005), 21ff.

198 **saying in 1675:** Colin G. Calloway (ed.), *The World Turned Upside Down: Indian Voices from Early America* (Boston: St. Martin's, 1994), 21.

198 **Deare in a Forrest:** Douglas Edward Leach, *Flintlock and Tomahawk: New*

England in King Philip's War (New York: Macmillan, 1958, reprinted Hyannis, Mass.: Parnassus Imprints, 1995), 6. An alternative interpretation of the war is offered in James D. Drake, *King Philip's War: Civil War in New England, 1675–1676* (Amherst: University of Massachusetts Press, 1999). Drake argues that the societies were so mixed that the war was not so much a conflict between whites and Indians as a sort of civil war in which the Indians fought on both sides. That is, the whites were able to recruit Indian allies. This was certainly true there and nearly everywhere else in colonial America. But, of course, the whites were not so divided. I find Leach's basic notion persuasive: that either the Indians or the whites had to prevail and that they could not long coexist.

198 **in just a few hours:** E. Lawrence Lee, *Indian Wars in North Carolina, 1663–1763* (Raleigh: Carolina Charter Tercentenary Commission, 1963), 28ff.

199 **tells the story:** Bridenbaugh, *Early Americans,* 1ff.

200 **incorrigible liars:** Eric Hinderaker, *Elusive Empires: Constructing Colonialism in the Ohio Valley 1673–1800* (Cambridge: Cambridge University Press, 1997), 36–38.

201 **been carried off:** Wright, *The Only Land,* 140.

201 **living in sixty villages:** Hinderaker, *Elusive Empires,* 11–16.

202 **a leading Indian:** Quoted in Hinderaker, *Elusive Empires,* 13.

202 **a long-standing practice:** Quoted in Richter, "War and Culture," 531. Even in success, as Richter has pointed out (*Longhouse,* 65–66), there was at least a cultural price: over time, immigrants came to constitute a majority of the societies into which they had been absorbed. A French Jesuit missionary estimated that by 1660 more than half of the Iroquois were recent immigrants—adoptees—of foreign extraction. Later, the number may have risen to two-thirds or more.

203 **ancestral spirits:** Karl Polanyi, *Dahomey and the Slave Trade* (Seattle: University of Washington Press, 1966), 34.

203 **ate the captives:** This practice was vividly described by Samuel de Champlain, in *The Voyages and Explorations of Samuel des Champlain (1604–1616). Narrated by Himself,* trans. Annie Nettleton Bourne. Edited with introduction and notes by Edward Gaylord Bourne (New York: Barnes & Company, 1906; reprint, Dartmouth, Nova Scotia: Brook House, 2000), 104–109.

203 **founded Detroit in 1701:** Hinderaker, *Elusive Empires,* 46–49.

203 **150 French residents:** Olive Patricia Dickason, "From 'One Nation' in the Northeast to 'New Nation' in the Northwest," in Jacqueline Peterson and Jennifer S. H. Brown (eds.), *The New Peoples* (Winnipeg: University of Manitoba Press, 1985).

204 **extension of kinship:** Polk, *Neighbors and Strangers,* 219ff.

204 **diplomats or negotiators:** Richter, *Longhouse,* 28ff.

205 **Christianitie for their soules:** Quoted in James Axtell, *The Invasion Within:*

The Contest of Cultures in Colonial North America (New York: Oxford University Press, 1985), 133.

206 **". . . captives into slavery"** James Savae (ed.), John Winthrop, The History of New England from 1630 to 1649 (Boston: Little Brown, 1853), I:278–279.

CHAPTER 12: THE FRENCH AND INDIAN WAR

210 **was usually ineffective:** Ian K. Steele, *Warpaths: Invasion of North America* (New York: Oxford University Press, 1994), 195ff.

210 **practically nothing:** Observation by Archibald Kennedy, a member of the New York Council, in *Serious Advice to the Inhabitants of the Northern Colonies on the Present Situation of Affairs* (New York, 1755), 5–6; quoted in Alan Rogers, *Empire and Liberty: American Resistance to British Authority, 1755–1763* (Berkeley: University of California Press, 1974), 49.

211 **had not been well founded:** John Bigelow (ed.), *The Life of Benjamin Franklin Written by Himself* (Oxford: Geoffrey Cumberlege, 1924, reprinted from the 1868 edition), 189–190.

211 **". . . make any impression"** Bigelow (ed.), *Benjamin Franklin,* 187–188. ". . . deal with them another time." Fred Anderson, *Crucible of War: The Seven Years' War and the Fate of Empire in British North America, 1754–1766* (New York: Knopf, 2000), 95 ff.

211 **without cavalry in open country:** Edward P. Hamilton: *Adventure in the Wilderness: The American Journals of Louis-Antoine de Bougainville, 1756–1760* (Norman: University of Oklahoma Press, 1964), 148–149.

212 **Indian methods of fighting:** Quoted in John Shy, *Toward Lexington: The Role of the British Army in the Coming of the American Revolution* (Princeton: Princeton University Press, 1965), 129.

212 **ceremonial costume and parade:** "Fighting 'Fire' with Firearms: The Anglo-Powhatan Arms Race in Early Virginia," *American Indian Culture and Research Journal* (3:4, 1979), 34.

212 **regulars out from England:** Alan Rogers, *Empire and Liberty: American Resistance to British Authority, 1755–1763* (Berkeley: University of California Press, 1974), 41–47.

213 **officers of junior rank:** Stipulated in the Rules and Articles of War. See Fred Anderson, *A People's Army: Massachusetts Soldiers and Society in the Seven Years' War* (Chapel Hill: University of North Carolina Press, 1984; reprint, Norton, 1985), 169.

213 **commission in the regular British army:** Anderson, *A People's Army,* 758; and John Shy, *Toward Lexington,* 393. Anderson comments that in the reorganization of Virginia's militia, Washington would be demoted to captain and "would have lost more status and honor than a proud Virginia gentleman could afford, and he could never have attracted Braddock's attention."

213 **daily insults:** Fred Anderson quotes a number of such reports. Anderson, *A People's Army,* 113ff.

213 **Braddock could muster only eight:** Anderson, *Crucible of War,* 96, 98.

213 **both should be destroyed:** *Works* (London, new edition, 1806), Vol. 11, 340.

213 **Sir William Johnson, proposed:** Letter to General Thomas Gage, December 26, 1763, quoted in Francis Jennings, *Empire of Fortune: Crowns, Colonies, and Tribes in the Seven Years' War in America* (New York: Norton, 1988), 438.

214 **Using the British Army's code:** Sydney Bradford, "Discipline in the Morristown Winter Encampments," *Proceedings of the New Jersey Historical Society* (80, January 1962), 1ff.

215 **officers and all:** Letter to Lord George Sackville, August 7, 1755, in Bekcles Willson (ed.), *The Life and Letters of James Wolfe* (London, 1909), 392.

215 **impatient of war:** Quoted in Piers Mackesy, *The War for America, 1775–1788* (Cambridge, Mass.: Harvard University Press, 1965), 30.

215 **a distant memory:** As Lord Jeffery Amherst found of New England militiamen in 1759. Anderson, *A People's Army,* 75.

215 **heat a house for the winter:** My calculations are based on Charles Knowles Bolton, *The Private Soldier under Washington* (New York: Scribner, 1902; reprint, Williamstown, Mass.: Corner House, 1976), 107.

215 **only three arms factories:** D. A. Tompkins, *History of Mecklenburg County and the City of Charlotte from 1740 to 1903* (Charlotte: Observer, 1903), 43.

215 **even knew how to use guns:** As Edmund S. Morgan notes in a review of Michael A. Bellesiles, *Arming America: The Origins of a National Gun Culture* (New York: Knopf, 2000), in probate records in the period 1765–1790 only about 14 percent of the inventories list guns, and half of those guns are described as not working. *New York Review of Books,* October 19, 2000.

215 **neglected or perverted:** "Address on Slavery," quoted in Philip D. Morgan, *Slave Counterpoint: Black Culture in the Eighteenth-Century Chesapeake and Lowcountry* (Chapel Hill: University of North Carolina Press, 1998), 388.

215 **French and Indian War:** Anderson, *Crucible,* 773, citing an unpublished paper by Thomas Purvis.

215 **40,000 men:** Edward P. Curtis, *The Organization of the British Army in the American Revolution* (New Haven: Yale University Press, 1926), 1–3.

216 **75 percent of government expenditure:** John Rule, *The Vital Century, 1714–1815* (London: Longman, 1992), 276.

216 **died or were listed as missing:** Eric Robson, "The Armed Forces and the Art of War," in J. O. Lindsay (ed.), *The New Cambridge Modern History,* Vol. 7, *The Old Regime, 1713–1763* (Cambridge: Cambridge University Press, 1971), 184.

216 **England's "military-industrial complex":** Christopher Lloyd, "Armed Forces and the Art of War," in A. Goodwin (ed.), *The New Cambridge Modern History,* Vol. 8, *The American and French Revolutions, 1763–1793* (Cambridge: Cambridge University Press, 1971), 174–177.

216 **exported to England:** John J. McCusker and Russell R. Menard, *The Economy of British America, 1607–1789, with Supplementary Bibliography* (Chapel Hill: University of North Carolina Press, 1991), 321.

216 **as Britain then had:** Roy Porter, *English Society in the Eighteenth Century* (London: Penguin, revised edition, 1991).

217 **Benjamin Franklin commented:** *Pennsylvania Gazette,* August 30, 1744; reprinted in Benjamin Labaree, *The Papers of Benjamin Franklin* (New Haven: Yale University Press), Vol. 3, 54. Franklin gives a contemporary account of the Louisbourg campaign in the following pages.

217 **continued to resist:** Richard White, *The Middle Ground: Indians, Empires, and Republics in the Great Lakes Region, 1650–1815* (Cambridge: Cambridge University Press, 1991), 287ff.

217 **1762 at Detroit:** White, *The Middle Ground,* 279.

217 **the British had built forts:** Howard H. Peckham, *Pontiac and the Indian Uprising* (Chicago: University of Chicago Press, 1947, Phoenix edition, 1961), 96.

218 **immediately be put to death:** Quoted in Peckham, *Pontiac,* 226.

218 **extirpate this execrable race:** Letter of July 16, 1763. See Bernard Knollenberg "General Amherst and Germ Warfare," *Mississippi Valley Historical Review* (41, 1954–1955).

218 **Jon Butler has commented:** *Becoming America: The Revolution before 1776* (Cambridge, Mass.: Harvard University Press, 2000), 129.

CHAPTER 13: FROM INTERNATIONAL WAR TO COLONIAL WAR

219 **would not be permitted:** "By the King. A Proclamation. George, R." October 7, 1763, from Annual Register for 1763, reprinted in Samuel Eliot Morison (ed.), *Sources and Documents Illustrating the American Revolution 1764–1788* 2nd ed. (New York: Oxford University Press, 1929), 1ff.

220 **its annual budget:** Peter D. G. Thomas, "The Cost of the British Army in North America, 1763–1775," *William and Mary Quarterly* (3rd Series, 45, July 1988); and John Shy, *Toward Lexington: The Role of the British Army in the Coming of the American Revolution* (Princeton: Princeton University Press, 1965), 240–241.

220 **unsurveyed lands:** *History of Cumberland and Adams Counties, Pennsylvania* (Chicago: Warner, Burn, 1886), 8.

220 **tomahawk right:** Also known as "pipe and tomahawk rights." Tomahawk rights involved simply blazing or slashing trees around the boundaries of a claimed piece of land. As Harry Foreman wrote in 1966, "This procedure is still practiced by surveyors through wooded or mountainous areas [of Pennsylvania] to this day." Paper given at Kittochtinny Historical Society, *Collected Papers,* Kittochtinny, Pennsylvania, June 30, 1966, 137. One (probably typical) toma-

hawk claim is described in *Colonial Records of North Carolina,* Vol. 7, *1765–1768* (Raleigh: Josephus Daniels, 1890), 825: "four Trees standing in a square Form marked with Notches and Blases, and on one of them the Letters G. R. These Trees were about five or six hundred yards to the Eastward of Cold Water Creek, and terminates upon the old Western Indian Path, upon the Eastern Bank of Cold Water, on a large (Gum it is thought to be) Tree the Letters W. C. (for William Churton) 1756 is marked."

220 **dispose of tribal land:** From the first encounters in the sixteenth century, whites treated "chiefs" as the equivalent of European monarchs who ruled absolutely. But, as Richard White pointed out regarding the peoples involved in a key set of negotiations, the Fort Stanwix negotiations, "chiefs can have influence, but they do not have authority." *The Middle Ground: Indians, Empires, and Republics in the Great Lakes Region, 1650–1815* (Cambridge: Cambridge University Press, 1991), 206.

220 **for murdering an Indian:** Jack M. Sosin, *Whitehall and the Wilderness: The Middle West in British Colonial Policy, 1760–1775* (Lincoln: University of Nebraska Press, 1961), 108–109.

221 **John Stuart:** J. Russell Snapp, *John Stuart and the Struggle for Empire on the Southern Frontier* (Baton Rouge: Louisiana State University Press, 1996), 42.

221 **royal governors of the colonies:** Esmond Wright, "American Independence in Its American Context," in A. Goodwin (ed.), *The New Cambridge Modern History,* Vol. 8, *The American and French Revolutions, 1763–1793* (Cambridge: Cambridge University Press, 1971), 523.

221 **vast stretches of Indian lands:** Sosin, *Whitehall,* 105; and Francis Jennings, *Empire of Fortune: Crowns, Colonies, and Tribes in the Seven Years' War in America* (New York: Norton, 1988), 440.

221 **For personal gain, Johnson gave:** James Thomas Flexner, *Lord of the Mohawks: A Biography of Sir William Johnson* (Boston: Houghton Mifflin, 1979), 328.

221 **in the royal proclamation:** Sosin, *Whitehall,* 176.

221 **illegal land-grabbing:** Using Sir William Johnson as the "cutout" for these acquisitions, as Johnson revealed in his papers: *Johnson Papers,* Vol. 14, 214; cited in Jennings, *Empire of Fortune,* 451.

221 **in someone else's pocket:** Roy Porter, *English Society in the Eighteenth Century* (London: Penguin, revised edition, 1991), 61.

221 **people greedy for land:** Eric Hinderaker, *Elusive Empires: Constructing Colonialism in the Ohio Valley 1673–1800* (Cambridge: Cambridge University Press, 1992), 13–14.

222 **lately much solicited:** Letter to the Board of Trade, November 6, 1747; in Louis Knott Koontz, *Robert Dinwiddie: His Career in American Colonial Government and Westward Expansion* (Glendale, Calif.: Arthur H. Clark, 1941), 157.

223 **officials he had suborned:** A partial list is given in Kenneth P. Bailey, *The Ohio Company of Virginia and the Westward Movement, 1748–1792* (Glendale, Calif.: Arthur Clark, 1939), 239.

223 **mostly unnamed "associates":** Koontz, *Dinwiddie,* 157.

223 **the Half-King:** Use of that title is an echo of the assertion from the early days in Jamestown that tribal elders were the equivalent of European monarchs and thus had the right to dispose of tribal lands. "Half-king" is roughly equivalent to "viceroy." He was a leader of his own people but not completely independent of his Iroquois suzerain.

223 **had no experience:** Quoted in Hinderaker, *Elusive Empires,* 140.

223 **his definition of diplomacy:** This is covered in Washington's report, which was printed immediately upon his return by order of Governor Dinwiddie. It is available in Benson J. Lossing (ed.), *The Diary of George Washington, from 1789 to 1791 . . . Together with His Journal of a Tour to the Ohio in 1753* (New York: Charles R. Richardson, 1860; reprint, Williamstown, Mass.: Corner House, 1978).

223 **his diary:** I have used the CD-ROM reproduction of the diaries for 1748–1799 (Oakman, Ala.: H-Bar Enterprises, 1996).

223 **since we left Venango:** Washington describes this as "an old Indian town, situated at the mouth of French Creek, on the Ohio." Lossing, *Diary,* 229.

224 **French-inspired Indian rapine:** Koontz, *Dinwiddie,* 258.

224 **lands on the Ohio:** Koontz, *Dinwiddie,* 282. George Washington would ultimately buy out his colleagues and lay claim to the whole parcel.

224 **Leading the charge:** Washington's own record of the event was lost with all his papers when disaster struck his little force. The French ultimately published a version, which Washington repudiated.

224 **first shot in the French and Indian War:** Koontz, *Dinwiddie,* 241.

225 **high wartime taxes:** John Rule, *The Vital Century: England's Developing Economy, 1714–1815* (London: Longmans, 1992), 289–291.

227 **burden too heavy to tolerate:** C. L. Hunter, *Sketches of Western North Carolina, Historical and Biographical* (Baltimore: Regional Publishing, reprint, 1970), 15–20.

227 **Between 1765 and 1771:** William L. Saunders (ed.), *Colonial Records of North Carolina, Published under the Supervision of the Trustees of the Public Libraries, by Order of the General Assembly* (Raleigh: Josephus Daniels, Printer to the State, 1890), Vol. 7, 12ff; and Vol. 8.

227 **8,000 taxpayers:** Marvin L. Michael Kay, "The North Carolina Regulation, 1766–1776: A Class Conflict," in Alfred F. Young (ed.), *The American Revolution: Explorations in the History of American Radicalism* (DeKalb: Northern Illinois University Press, 1976), 73; and Marjoleine Kars, *Breaking Loose Together: The Regulator Rebellion in Pre-Revolutionary North Carolina* (Chapel Hill: University of North Carolina Press, 2002). These studies emphasize the religious motivations of the Regulators.

228 **remedies for their grievances:** "Regulator Advertisement Number 1," issued in August 1766, called for an end to excessive taxes, high rents, and extortionate fees, all local issues. Blackwell P. Robinson, *The Five Royal Governors of North Carolina, 1729–1775* (Raleigh: State Department of Archives and History, 1968), 51–52.

228 **relatively prosperous settlers:** Alice Elaine Mathews, *Society in Revolutionary North Carolina* (Raleigh: North Carolina Department of Cultural Resources, 1976), 40–42.

228 **collected by them:** William S. Powell, *The War of the Regulation and the Battle of Alamance, May 16, 1771* (Raleigh: North Carolina Department of Cultural Resources, 1976), 7.

229 **if called upon:** Francis L. Hawks, David L. Swain, and William A. Graham, *Revolutionary History of North Carolina* (Raleigh: William D. Cook, 1853), 63.

229 **were forgiven:** A huge price—100 pounds and 1,000 acres of land—was placed on Harmon's head if he was taken dead or alive. Paul David Nelson, *William Tryon and the Course of Empire* (Chapel Hill: University of North Carolina Press, 1990), 86.

229 **in the larger cause:** Robinson, *Royal Governors,* 64. Robinson comments that of 883 known Regulators, 289 fought on the rebels' side and only thirty-four on the Loyalist side. It is not known how the remaining 560 aligned themselves.

CHAPTER 14: PRODUCTION AND COMMERCE

230 **by Richard Hakluyt:** "The First Voyage Made to the Coast of America, with Two Barks, Where in Were Captaines M. Philip Amadas, and M. Arthur Barlowe, Who Discovered Part of the Countrey Now Called Virginia, Anno 1584. Written by One of the Said Captaines, and Sent to Sir Walter Ralegh knight, at whose charge and direction, the said voyage was set forth"; and "A Briefe and True Report of the Commodities as Well Marchantable as Others, Which Are to Be Found and Raised in the Countrey of Virginia, Written by M. Thomas Harriot: Together with Master Ralph Lane His Approbation Thereof in All Points." Richard Hakluyt, *Principall Navigations, Voyages, Traffiques and Discoveries of the English Nation,* reprinted as *Voyages* (London: Dent, 1907), Vol. 6, 127, 169ff.

231 **sea to sea:** Bernard Bailyn, *The New England Merchants in the Seventeenth Century* (Cambridge, Mass.: Harvard University Press, 1955), 3–5.

232 **could be swallowed:** John Bakeless, *America as Seen by Its First Explorers* (New York: Dover, 1961), 189, 212.

232 **wryly pointed out:** John J. McCusker and Russell R. Menard, *The Economy of British America, 1607–1789, with Supplementary Bibliography* (Chapel Hill: University of North Carolina Press, 1991), 99.

232 **120-ton *Desire*:** The best general account I have seen is William A. Baker, *Colonial Vessels: Some Seventeenth-Century Sailing Craft* (Barre, Mass.: Barre Publishing, 1962).

232 **the 300-ton *Mary Ann*:** John Winthrop, *The History of New England from 1630 to 1649* (Boston: Little, Brown, new edition, 1853), Vol. 2, 37.

232 **in 1713 at Gloucester:** Björn Landström, *The Ship: An Illustrated History* (Garden City, N.Y.: Doubleday, n.d. [1962]), 174–175; and Jack Coggins, *Ships and Seamen of the American Revolution* (Harrisburg, Pa.: Promontory, 1969), 52–53.

232 **mostly to England:** Jacob Price, quoted in John J. McCusker and Russell R. Menard, *Economy,* 82, 321.

233 **dangerous design this spring:** Quoted in James Deetz, *In Small Things Forgotten: An Archaeology of Early American Life* (New York: Anchor, 1996), 47.

233 **from West Africa:** Judith A. Carney, *Black Rice: The African Origins of Rice Cultivation in the Americas,* (Cambridge, Mass.: Harvard University Press, 2001). A description of how rice was grown by the Baga of the Windward Coast is presented in the log of the *Sandown,* edited by Bruce L. Mouser as *A Slaving Voyage to Africa and Jamaica* (Bloomington: Indiana University Press, 2002), 71. A drawing of an African rice paddy appears on p. 76.

233 **made slavery extremely profitable:** Clement Eaton, *A History of the Old South,* 2nd ed. (New York: Macmillan, 1969), 19–20. Prices of slaves varied but fell into a range of 25 to 50 pounds during most of the eighteenth century. See Alice Hanson Jones, *Wealth of a Nation To Be: The American Colonies on the Eve of the Revolution* (New York: Columbia University Press, 1980), 113; and Hugh Thomas, *The Slave Trade,* (New York: Simon and Schuster, 1997), 402–403.

234 **their economy came to depend:** Judith A. Carney, *Black Rice.* Consequently, plantation owners in Carolina tended to purchase more slaves from the Bight of Biafra than did the tobacco growers farther north. Stephen D. Behrendt, *Teachers' Manual for The Trans-Atlantic Slave Trade: A Database on CD-Rom, Edited by David Eltis, David Richardson, Stephen D. Behrendt and Herbert S. Klein* (Cambridge: Cambridge University Press, 1999), 9.

235 **the going rate in England:** About four shillings sixpence daily. John L. Bullion, *A Great and Necessary Measure: George Grenville and the Genesis of the Stamp Act, 1763–1765* (Columbia: University of Missouri Press, 1982), 32.

235 **virtually unusable:** As a merchant in Philadelphia wrote to Benjamin Franklin on June 19, 1765. Leonard Labaree (ed.), *The Papers of Benjamin Franklin* (New Haven: Yale University Press, 1968), Vol. 12, 185.

235 **for his neighbors:** Samuel Eliot Morison, *Builders of the Bay Colony* (Boston: Houghton Mifflin, 1955), 138, 153–154.

236 **late eighteenth century:** Barbara L. Solow, "The Transatlantic Slave Trade: A New Census," *William and Mary Quarterly* (58, January 1, 2001).

236 **about half a mile long:** McCusker and Menard, *Economy,* 315 fn.

236 **seventy-three distilleries:** Bullion, *Measure,* 39.

236 **about 155:** McCusker and Menard, *Economy,* 290.

236 **Adam Smith wrote:** *An Inquiry into the Nature and Causes of the Wealth of Nations* (London, 1776; reprint, New York: Modern Library, 1937), 543–548.

237 **non-British destinations:** Bailyn, *Merchants,* 83.

237 **triangular trade:** The concept was based on the historian George Mason's study of a voyage made in 1752–1753 by the brigantine *Sanderson.* But it is clearly a simplification of the way ships actually sailed, as is pointed out in James A. Rawley, *The Trans-Atlantic Slave Trade* (New York: Norton, 1981), 344–345. Actually, the first such voyage was probably made in 1644 and financed by a group of Bostonian merchants. See Bernard Bailyn, *Merchants,* 84.

237 **in ballast:** J. E. Merritt, "The Triangular Trade," *Business History* (3:1, 1960); quoted in James A. Rawley, *Trans-Atlantic,* 260–261.

237 **The French:** Robert Harms, *The Diligent: A Voyage through the Worlds of the Slave Trade* (New York: Basic Books, 2002), 81–82, 100.

238 **market of the mother country:** The commodities originally enumerated in 12 Car. II, c. 18, §18, were sugar, tobacco-cotton-wool, indigo, ginger, fustic and other dyeing woods. "The rest are called *non-enumerated;* and may be exported directly to other countries, provided it is in British or Plantation ships. Of which the owners and three-fourths of the mariners are British subjects." Smith, *Wealth,* 543–544. The main provisions of the act are quoted in Samuel Eliot Morison (ed.), *Sources and Documents Illustrating the American Revolution, 1764–1788* 2nd ed. (New York: Oxford University Press, 1929, reissued 1965), 74ff.

239 **"Britain was at war":** Carl Bridenbaugh, *Cities in Revolt: Urban Life in America 1743–1776* (New York: Capricorn, 1964), 64–67.

239 **wrote to Benjamin Franklin:** Labaree (ed.), *Franklin,* Vol. 12, 183–189. See also John Dickinson, *Letters from a Farmer in Pennsylvania* (Indianapolis: Liberty Fund, 1999), 25. Dickinson expressed the anger colonists felt.

239 **any foreign market:** Forbidden in the Woolen Act of 1699, 10 Wm. III, c. 16; and by the Hat Act of 1732, 5 Geo. II, c. 22. This was done deliberately "for the better encouraging the making Hats in *Great Britain.*" E. L. Jones, "The European Background," in Stanley L. Engerman and Robert E. Gallman (eds.), *The Cambridge Economic History of the United States,* Vol. 1, *The Colonial Era* (Cambridge: Cambridge University Press, 1996), 124. These and similar laws were satirized by Franklin as "an Edict by the King of Prussia" in *Writings,* (New York: Library of America, 1987), 698–703.

239 **for overseas commerce:** Very little information exists on the coastal trade; an attempt to reconstruct it is offered in James F. Shepherd and Samuel Williamson, "The Coastal Trade of the British North American Colonies, 1768-1772," *Journal of Economic History* (32:4, December 1972). These authors conclude that the trade was small but essential.

239 **to Britain's enemies:** Bridenbaugh, *Cities,* 65.

240 **". . . against him for smuggling":** J. H. Plumb, *England in the Eighteenth Century, 1714–1815* (Harmondsworth: Penguin, 1969 edition), 126.

240 **made profiteer John Hancock a patriot:** A. M. Schlesinger, *The Colonial Merchants and the American Revolution, 1763–1776* (New York, 1918); noted in Esmond Wright, "American Independence in its American Context: Social and Political Aspects: Western Expansion," in A. Goodwin (ed.), *The New Cambridge Modern History* (Cambridge: Cambridge University Press, 1971), Vol. 8, 522.

240 **subsidized those who did:** McCusker and Menard, *Economy,* 95; Bailyn, *Merchants,* 20–21.

240 **already settled from leaving:** As Massachusetts did from 1645 on. *Massachusetts Province Laws,* Vol. 1, 402, quoted in Frederick Jackson Turner, "The First Official Frontier of the Massachusetts Bay," *Publications of the Colonial Society of Massachusetts* (17, April 1914), 250–271; reprinted in *The Frontier in American History* (New York: Holt, Rinehart, and Winston, 1962). A few years later, between 1669 and 1675, various new regulations were promulgated that affected the inland towns. On March 12, 1694, Massachusetts listed a further group of towns whose inhabitants were forbidden to leave. Anyone who did leave would lose his title to land or, if not a landowner, be imprisoned. Still more towns were added in 1699–1700, "tho' they be not frontiers as those towns first named, yet lye more open than many others to an attack of an Enemy."

240 **money was in short supply:** Jones, *Wealth,* 132–133.

240 **set downe from time to time:** Bailyn, *Merchants,* 48.

240 **wheat at 6 shillings:** Winthrop, *History,* Vol. 2, 8.

240 **warehouses into "money":** Information in this paragraph is drawn from McCusker and Menard, *Economy,* 338, Chap. 3 and p. 337.

241 **no longer legal tender:** Labaree (ed.), *Franklin,* Vol. 11, 238.

241 **Adam Smith noted in 1776:** Smith, *Wealth,* 548–549.

241 **2,508 tons:** Theodore Thayer, *Colonial and Revolutionary Morris County* (Morristown, N.J.: Morris County Heritage Commission, 1975), 65.

242 ***Letters from a Farmer in Pennsylvania:*** Printed in Forrest McDonald (ed.), *Empire and Nation* (Indianapolis: Liberty Fund, 1999), 25–26.

242 **fit for use:** Labaree (ed.), Franklin, Vol. 12, 182–183.

242 **about fourteen cords:** Bernard Bailyn, *Voyagers to the West* (New York: Knopf, 1986). Bailyn, focusing mainly on the last five years or so before the Revolution, summarizes specialized studies on iron production including Theodore W. Kury, "Iron and Settlement . . ." in H. J. Walker and W. G. Haag (eds.), *Man and Cultural Heritage* (Baton Rouge: State University of Louisiana Press, 1974); and Ribert J. Sim and Harvey B. Weiss, *Charcoal-Burning in New Jersey* (Trenton: New Jersey Agricultural Society, 1949–1955), 245–254.

242 **Adam Smith summed it up:** Smith, *Wealth,* 547.

244 **Scranton, Pennsylvania:** The Oxford furnace has been well documented from papers deposited in the Library of Congress by a member of the Shippen family of Philadelphia. These are drawn upon in A. G. Yount, "The Old Oxford Furnace" (Belvidere and Washington, N.J.: Hicks, 1975). I have supplemented this source with papers remaining from my mother's ancestors, the Henrys and Scrantons, who were at various times part owners.

245 **world's iron production:** McCusker and Menard, *Economy, 326.*

245 **New Jersey highlands:** Bailyn, *Voyagers,* 249–252.

245 **an iron panel at the back:** The earliest surviving panels date from the reign (and often bear the coat of arms) of King George II (r. 1727–1760).

245 **"durable" product:** Franklin described and publicized his fireplace in a pamphlet, *An Account of the New-Invented Pennsylvania Fire Places: Wherein Their Construction & Manner of Operation Is Particularly Explained; Their Advantages above Every Other Method of Warming Rooms Demonstrated; and all Objections That Have Been Raised against the Use of Them Answered & Obviated, &c.* Discussed in *The Autobiography,* Part 3, written much later, in August 1788, and reprinted in *Writings* (New York: Library of America, 1987), 1417–1418. By that time, wood was already in such short supply and in inconvenient locations that one selling point of the stove was, as Franklin said, "a great Saving of Wood to the Inhabitants."

246 **sustained the Revolutionary army:** There is no satisfactory modern biography of Henry. The best biography available is Heerbert H. Beck, "William Henry, Patriot, Master Gunsmith, Progenitor of the Steamboat," *Transactions of the Moravian Historical Society* (16, Part 2, 1955), 69ff. Less satisfactory is Francis Jordan, Jr., *The Life of William Henry, 1729–1786* (Lancaster, Pa.: New Era, 1910).

246 **at 110 million pounds:** Jones, *Wealth,* 50–51.

CHAPTER 15: REPRESENTATION AND TAXATION

247 **no satisfactory accounting existed:** The figures were notoriously inaccurate and would be for many years. See J. H. Plumb, *England in the Eighteenth Century (1714–1815)* (Harmondsworth: Penguin, 1969), 27. See also Piers Mackesy, *The War for America, 1775–1783* (London: Longmans, 1964), 6–7. Mackesy notes that the British government had no audit procedures to control expenditure; departments simply ran up debt, and afterward there was little or no accountability. See also John Brewer, *The Sinews of Power: War, Money, and the English State, 1688–1783* (New York: Knopf, 1981), discussed in Fred Anderson, *Crucible of War: The Seven Years' War and the Fate of Empire in British North America, 1754–1766* (New York: Knopf, 2000), 811, fn5.

247 **the government's yearly revenues:** R. R. Mitchell, *European Historical Statistics, 1750–1970* (London: Macmillan, 1978), 370.

247 **had almost doubled:** John L. Bullion, *A Great and Necessary Measure: George Grenville and the Genesis of the Stamp Act, 1763–1765* (Columbia: University of Missouri Press, 1982), 18–19.

248 **expenditures were 17,993,000 pounds:** Roy Porter, *English Society in the Eighteenth Century,* revised ed. (London: Penguin Books, 1991), 268.

248 **contraband trade worth perhaps 500,000 pounds:** That was the estimate of Henry McCulloh, who was a major speculator in North Carolina land grants and an advocate of the Stamp Act. Quoted in Bullion, *Measure,* 67. See also Charles G. Sellers, Jr. "Private Profits and British Colonial Policy: The Speculations of Henry McCulloh," *William and Mary Quarterly* (3rd Series, 8:4).

248 **at least double that of England:** John J. McCusker and Russell R. Menard, *The Economy of British America, 1607–1789* (Chapel Hill: University of North Carolina Press, 1991), 85. McCusker and Menard calculate that the per capita GNP, a concept that did not then exist to enlighten the British government, would have been between 11 pounds and 12 pounds, 10 shillings.

248 **as Adam Smith pointed out:** *An Inquiry into the Nature and Causes of the Wealth of Nations* (London, 1776; reprint, New York: Modern Library, 1937), 540–541.

248 **risen five times:** M. A. Jones, "American Independence in Its Imperial, Strategic and Diplomatic Aspects," in A. Goodwin (ed.), *The New Cambridge Modern History,* Vol. 8, *The American and French Revolutions, 1763–1793* (Cambridge: Cambridge University Press, 1971), 481–482.

249 **sons of liberty:** Michael J. O'Brien, *The Hidden Phase of American History: Ireland's Part in America's Struggle for Liberty* (Baltimore: Genealogical Publishing, 1973); and Pauline Maier, *From Resistance to Revolution: Colonial Radicals and the Development of American Opposition to Britain, 1765–1776* (London: Routledge and Kegan Paul, 1973), 161.

249 **they advanced two arguments:** Bullion, *Measure,* 155–156.

250 **wrote in 1762:** Quoted in Bullion, *Measure,* 66.

250 **"All colonies," he observed:** Quoted in Bullion, *Measure,* op. cit., 155.

250 **as his lord advised:** John Brewer, *Party Ideology and Popular Politics at the Accession of George III* (Cambridge: Cambridge University Press, 1976), 208.

250 **a century earlier:** Porter, *English Society,* 345.

251 **in both England and America:** *An Appeal to the Justice and Interests of the People of Great Britain in the Present Disputes with America,* published in London in 1774. See Louis W. Potts, *Arthur Lee, A Virtuous Revolutionary* (Baton Rouge: Louisiana State University Press, 1981), 125–127; Michael G. Kammen, *A Rope of Sand* (New York: Vintage, 1974), 213; and Brewer, *Party,* 142–144.

251 **in the American colonies:** Brewer, *Party,* 202.

251 **to the colonies:** Jones, *American Independence,* 482.

251 **stamp tax for colonial usage:** Mack Thompson, "Massachusetts and New York Stamp Acts," *William and Mary Quarterly* (3rd series, 26), 253–258; and "The Massachusetts Stamp Act of 1755," *New England History and Geography Review* (14, 1860), 267.

252 **on private establishments:** Thompson, *Stamp Acts,* 256.

252 **about 60,000 pounds yearly:** Sellers, "Private Profits," 550.

252 **than the duties themselves:** Bullion, *Measure,* 104.

252 **the King and the Counciel:** Kammen, *Rope of Sand,* 5.

252 **outbreak of the Revolutionary war:** Some agents were Americans, but quite a few were Englishmen who either personally sympathized with a given colony or had commercial dealings that brought them into concord with its interests. At least five, including the influential English statesman Edmund Burke, were also members of Parliament. Even more striking, the private secretary of Prime Minister George Grenville was also the agent for Pennsylvania and Connecticut.

253 **intercepting and reading mail:** Intercepting, decoding, and reading the mail of diplomats and others had long been common in London. At the critical period in January 1774, when news of the Boston Tea Party reached London, the letters of Franklin and Arthur Lee were read by the solicitor general in front of the Privy Council, which "pondered whether both men should be charged with treason for attempting to incite Massachusetts against Britain." See Bernard Bailyn, *The Ordeal of Thomas Hutchinson* (Cambridge Mass.: Belknap Press of Harvard University Press, 1974), 254–255; and Pauline Maier, *Resistance,* 89.

253 **had been read:** Charles Chenevix Trench, *Portrait of a Patriot* (Edinburgh: William Blackwood, 1962), 59.

253 **admitted that they could not:** Minutes of the meeting were taken by the Connecticut agent Jared Ingersoll on February 11, 1765. See New Haven Colony Historical Society, *Papers* (New Haven, 1918), Vol. 9, 312–314; quoted in Kammen, *Rope of Sand,* 113.

254 **Contributions in an Enemy's Country:** *No Taxation without Representation: Three Letters of 1754 to Governor William Shirley, with a Preface of 1766,* Benjamin Franklin, *Writings* (New York: Library of America, 1987), 402ff.

255 **workings of the Stamp Act:** The principal provisions are given in Henry Steele Commager (ed.), *Documents of American History* (New York: Crofts, 1947), 53–55.

255 **bled the colonies of specie:** At a time when hard currency was in critically short supply, John Adams observed one incident in which some 18,000 coins were taken from Salem to Boston for shipment. And in a single year, 32,000 ounces of silver were taken out of that area. See Hiller B. Zobel, *The Boston Massacre* (New York: Norton, 1970), 57, 66.

255 **to buy the stamped paper:** A point made to Franklin by an unnamed correspondent in Philadelphia, which Franklin arranged to have printed in the *London Chronicle* of August 17–20, 1765; reprinted in Leonard W. Labaree (ed.), *The Papers of Benjamin Franklin* (New Haven: Yale University Press, 1968), Vol. 12, 186.

255 **dependence and obedience:** John Dickinson, *Letters From a Farmer in Pennsylvania* (Indianapolis: Liberty Fund, 1999), 53.

256 **Virginia House of Burgesses:** The traditional account is given in Moses Coit Tyler, *Patrick Henry* (Boston: Houghton, Mifflin; reprint, 1980, Chelsea House), 68ff.

256 **as treasonable:** Zobel, *Boston Massacre,* 26.

256 **accepted the call:** C. A. Weslager, *The Stamp Act Congress* (Newark: University of Delaware Press, 1976), 11.

256 **with their own consent:** For resolutions of the Stamp Act Congress, see Commager (ed.), *Documents,* 57–58.

256 **nearly burned down his house:** David T. Morgan, *The Devious Dr. Franklin, Colonial Agent: Benjamin Franklin's Years in London* (Macon, Ga.: Mercer University Press, 1996), 106ff. Morgan comments, "The commotion in Philadelphia was precipitated mainly by the proprietary party's efforts to pin the Stamp Act on Franklin and his allies."

256 **never have to be raised:** Kammen, *Rope of Sand,* 119.

256 **impossible to enforce:** M. A. Jones, *American Independence,* 481.

257 **calling themselves "sons of liberty":** The phrase had been coined by Colonel Isaac Barré on February 7, 1765, in a speech before Parliament in which he opposed the Stamp Act.

257 **ever sold in North Carolina:** Blackwell P. Robinson, *The Five Royal Governors of North Carolina, 1729–1775* (Raleigh: State Department of Archives and History, 1968), 51–52.

257 **gathered to demonstrate:** The event is described in Henry B. Dawson, *New York City during the American Revolution, Being a Collection of Original Papers (Now First Published) from the Manuscripts in the Possession of the Mercantile Library Association of New York City* (New York: Privately Printed for the Association, 1861), 25–26, 41.

258 **a long series of riots:** Maier, *Resistance,* 4 ff.

258 **Ebenezer MacIntosh:** Stewart Beach, *Samuel Adams: The Fateful Years, 1764–1776* (New York: Dodd, Mead, 1965), 75–76.

258 **terrorized the neighborhood:** George P. Anderson, "Ebenezer Mackintosh: Stamp Act Rioter," in *Publications of the Colonial Society of Massachusetts* (Boston, 1927), 26.

258 **"a trained mob":** Carl Bridenbaugh, *Cities in Revolt: Urban Life in America 1743–1776* (New York: Capricorn, 1964), 307.

258 **for the town meeting:** Peter Shaw, *American Patriots and the Rituals of*

Revolution (Cambridge, Mass.: Harvard University Press, 1981), 180ff; and Hiller B. Zobel, *The Boston Massacre,* 27.

259 **hanged Oliver in effigy:** The use of effigies was not then a common form of intimidation, but it soon became one; the intent, of course, was intimidation. Since hanging or burning an effigy was merely a threat, it was not legally actionable, but public figures were rarely unmoved by it.

259 **with the British garrison:** Dirk Hoerder, *Crowd Action in Revolutionary Massachusetts, 1765–1780* (New York: Academic, 1977), 101–102.

259 **resigned his commission:** Thomas Hutchinson (who later became governor) was so humiliated by this episode that he skipped lightly over it in his historical writing. See Lawrence Shaw Mayo (ed.), *The History of the Colony and Province of Massachusetts Bay* (Cambridge, Mass.: Harvard University Press, 1936). An excellent, sympathetic account is Bernard Bailyn, *The Ordeal of Thomas Hutchinson,* (Cambridge, Mass.: Belknap Press of Harvard University Press, 1974), from which some of the information in this section is drawn; see esp. Chap. 2.

260 **a flame of fire:** L. H. Butterfield (ed.), *Diary and Autobiography of John Adams* (Cambridge, Mass.: Belknap Press of Harvard University Press, 1961), entry for "1770. Monday Feby. 26 or Thereabouts," Vol. 2, 350.

260 **members of the "loyal nine":** Hoerder, *Crowd Action,* 94–96, fn 32.

260 **one of the effigies:** Quoted ibid., *Crowd Action,* 97.

261 **to eat some Cherries:** Zobel, *The Boston Massacre,* 83.

261 **reluctant to get involved:** Francis Bernard to Secretary at War Lord Barrington on December 132, 1766; quoted in Zobel, *The Boston Massacre,* 61.

264 **Philadelphia Patriotic Society:** Esmond Wright, "American Independence in its American Context . . ." A. Goodwin (ed.), *The New Cambridge Modern History,* Vol. 8, *The American and French Revolutions, 1763–93* (Cambridge: Cambridge University Press, 1971), 519–520.

264 **propagandists and agitators:** Philip Davidson, *Propaganda and the American Revolution* (New York: Norton, 1973; original edition, 1941). See also Bernard Bailyn, *The Ideological Origins of the American Revolution* (Cambridge, Mass.: Belknap Press of Harvard University Press, 1967), 5–6; and Arthur M. Schlesinger, *Prelude to Independence: The Newspaper War on Britain* (New York: Vintage, 1965), 20ff.

264 **continue a free state:** "A List of Infringements and Violations of Rights," *Proceedings of the Town of Boston* (October–November, 1772); in Samuel Eliot Morison, *Sources and Documents Illustrating the American Revolution 1764–1788,* 2nd ed. (New York: Oxford University Press, 1929), 93.

264 ***American* liberty is finished:** Dickinson, *Letters.*

265 **Britain's other great colony, India:** P. J. Marshall, "Bengal: The British Bridgehead, Eastern India 1740–1828," in *The New Cambridge History of India* (Cambridge: Cambridge University Press, 1987), Vol. 2, 2, 84; and Michael Edwardes, *The Nabobs at Home* (London: Constable, 1991), 44ff.

265 **wrote to his prime minister:** W. B. Donne (ed.), *The Correspondence of George III with Lord North,* Letter 246, Kew, September 11, 1774.

CHAPTER 16: FOREIGN FRIENDS AND FELLOW SUFFERERS

270 **by the devil himself:** Leonard W. Labaree (ed.), *The Papers of Benjamin Franklin* (New Haven: Yale University Press, 1968), Vol. 15, 75 and 78.

270 **sold their votes to the highest bidders:** Lewis Namier, *England during the American Revolution* (London: Macmillan, 1961), 3ff. Namier offers a very weak argument that Parliament did represent Englishmen: "Corruption in populous boroughs . . . was a mark of English freedom and independence, for no one bribes where he can bully."

270 **for their money:** Quoted in Roy Porter, *English Society in the Eighteenth Century* (London: Penguin, revised edition, 1991), 111.

271 **perquisites of the state:** Roy Porter, *English Society,* 58.

271 **the landed embracing the loaded:** Porter, *English Society,* 52.

271 **retained a sentimental loyalty:** Bernard Bailyn, *The Ideological Origins of the American Revolution* (Cambridge, Mass.: Belknap Press of Harvard University Press, 1967), 111.

273 **no hero of the Mob but Wilkes:** Letter to Charles O'Hara in Thomas W. Copeland (ed.), *The Correspondence of Edmund Burke* (Chicago: University of Chicago Press, 1958), Vol. 1, 349; quoted in George Rudé, *Wilkes and Liberty: A Social Study of 1763–1774* (London: Oxford University Press, 1962), 46.

274 **on the brink of revolution:** Rudé, *Wilkes,* 56.

274 **their English exemplars:** Hiller B. Zobel, *The Boston Massacre* (New York: Norton, 1970), 179.

274 **must stand or fall together:** William Palfrey to Wilkes, February 21, 1769; quoted in Bailyn, *Ideological Origins,* 112.

275 **your Majesty's colonies:** Rudé, *Wilkes,* 113.

275 **irredeemably corrupt and tyrannical:** Pauline Maier, "John Wilkes and American Disillusionment with Britain," *William and Mary Quarterly* (3rd Series, 20, 1963).

276 **every sentence of Boswell's account:** The quotations and paraphrases are from James Boswell, *The Journal of a Tour to Corsica; and Memoirs of Pascal Paoli* (London: William and Norgate, 1951, reprint of the 1767 edition), 71–73.

276 **Corsican guerrilla leader:** The painting appeared in Boswell's book and is reproduced in Carl and Jessica Bridenbaugh, *Rebels and Gentlemen: Philadelphia in the Age of Franklin* (New York: Oxford University Press, 1965), inset following 176.

276 **between liberty and tyranny:** Franco Venturi, *The End of the Old Regime in Europe: 1768–1776, The First Crisis* (Princeton: Princeton University Press, 1989, translated from the Italian of 1979), xiv–xv.

277 **totally disenfranchised:** Michael J. O'Brien, *A Hidden Phase of American History: Ireland's Part in America's Struggle for Liberty* (Baltimore: Genealogical Publishing, 1973, reprint of the 1919 edition), 10.

277 **Roy Porter has written:** Porter, *English Society,* 34–35. See also S. J. Connolly, "Eighteenth-Century Ireland," in D. George Boyce and Alan O'Day (eds.), *The Making of Modern Irish History* (London: Routledge, 1996), 15. Connolly cavils at Porter's comment but then admits its essential truth.

277 **when he wrote:** Op. cit., letter 10.

277 **as John Dickinson wrote:** Broadside published in Philadelphia on November 27, 1773, and reprinted in a number of newspapers throughout the colonies; quoted in Arthur M. Schlesinger, *Prelude to Independence: The Newspaper War on Britain 1764–1776* (New York: Knopf, 1957, Vintage edition), 170.

278 **seemed clear to Dickinson:** John Dickinson, *Letters from a Farmer in Pennsylvania* (Indianapolis: Liberty Fund, 1999), Letter 10.

278 **all over the earth:** Adam's *Works,* Vol. 3, 452n; quoted in Bailyn, *Ideological Origins,* 140.

278 **to enslave America:** Bailyn, *Ideological Origins,* 119.

278 **as Franco Venturi has written:** Venturi, *Old Regime,* 383.

278 **finally work out her own Salvation:** Quoted in Pauline Maier, *The Old Revolutionaries: Political Lives in the Age of Samuel Adams* (New York: Knopf, 1980), 23.

CHAPTER 17: "AN UNGOVERNABLE PEOPLE"

281 **beat them nearly to death:** Hiller B. Zobel, *The Boston Massacre* (New York: Norton, 1970), 73–75.

281 **never came to trial:** Ibid., 159–160.

283 **very little at present:** Letter of October 31, 1768, to Lord Hillsborough, quoted in Zobel, *The Boston Massacre,* 103.

283 **now entirely obstructed:** Zobel, *The Boston Massacre,* 81.

283 **reasonable Assessments Rates and Taxes:** Quoted in Jack P. Greene (ed.), *Great Britain and the American Colonies, 1606–1763* (Columbia: University of South Carolina Press, 1970), 115–120.

284 **account of their tense interview:** Add. MSS. 35870, ff. 87–91, quoted in Lewis Namier, *England during the American Revolution,* 2nd ed. (London: Macmillan, 1961), 46–47.

284 **not the same legislatures:** Letter of October 2, 1770, to Jacques Barbeau Dubourg. Leonard Labaree (ed.), *The Papers of Benjamin Franklin* (New Haven: Yale University Press), Vol. 17, 233–234.

285 **keeping us down and fleecing us:** Letter to Samuel Cooper, June 8, 1770. Labaree (ed.), *Franklin,* Vol. 17, 161–165.

285 **even the most radical:** Pauline Maier, *From Resistance to Revolution: Colonial*

Radicals and the Development of American Opposition to Britain, 1765–1776 (London: Routledge and Kegan Paul, 1973), 100.

285 **Samuel Adams was quoted as saying:** Deposition of Richard Sylvester, January 23, 1769, in Zobel, *Massacre,* 92.

286 **cut your Masters Throats:** Cited by Zobel, *Massacre,* 102.

287 **General Gage:** On December 19, 1768. *Gage Papers,* quoted in Zobel, *Massacre,* 136.

288 **furnaces of propaganda were ablaze:** Arthur Schlesinger, *Prelude to Independence: The Newspaper War on Britain 1764–1776* (New York: Knopf, 1957, Vintage edition), 16.

288 **circulated in the colonies:** Maier, *Resistance,* 139.

288 **fairer and more free:** Gordon S. Wood, *The Creation of the American Republic, 1776–1787* (New York: Norton, 1972), 13. Wood comments that "this continual talk of desiring nothing new and wishing only to return to the old system and the essentials of the English constitution was only a superficial gloss." Yet this assertion or attitude appears nearly always to be a feature of revolutionary thought.

288 **treating the colonists as subversives:** Maier, *Resistance,* 89.

289 **along the Atlantic coast:** Blackwell P. Robinson: *The Five Royal Governors of North Carolina, 1729–1775* (Raleigh: State Department of Archives and History, 1968), 67.

289 **Benjamin Franklin, Massachusett's agent:** Michael G. Kammen, *A Rope of Sand: The Colonial Agents, British Politics, and the American Revolution* (New York: Vintage, 1968), 289.

290 **buying and storing arms and ammunition:** To give just one example, in May 1775 Mecklenburg County, North Carolina, purchased "300 lb of powder, 600 lb of lead, 1000 flints, for the use of the militia of this county." When the North Carolina Provincial Council met in August, it set up a committee to plan for "internal peace, order and safety." See William R. Polk, *Polk's Folly* (New York: Doubleday, 2000), 115–116.

290 **moratoriums were declared:** Schlesinger, *Newspaper War,* 9.

290 **The *Georgia Gazette* told its readers:** September 6, 1769; quoted in Kenneth Coleman, *The American Revolution in Georgia, 1763–1789* (Athens: University of Georgia Press, 1958), 28–30.

291 **and the rest mongrels:** Quoted in John C. Miller, *Origins of the American Revolution* (Boston: Little, Brown, 1943), 379.

291 **an American peasant:** Franco Venturi, *The End of the Old Regime in Europe, 1768–1776: The First Crisis,* translated from the Italian (Princeton: Princeton University Press, 1989), 429.

291 **as he wrote:** John Dickinson, *Letters from a Farmer in Pennsylvania* (Indianapolis: Liberty Fund, 1999), Letter 3.

292 **William Appleman Williams admits:** William Appleman Williams, "Samuel

Adams: Calvinist, Mercantilist, Revolutionary" *Studies on the Left* (1:2, Winter 1960), 1ff.

292 **quite a different Adams:** Pauline Maier, *The Old Revolutionaries: Political Lives in the Age of Samuel Adams* (New York: Knopf, 1980). While he essentially agrees with her, John K. Alexander regards Adams as America's "first modern politician" using tactics that would not become usual until long after the Revolution. See his *Samuel Adams: America's Revolutionary Politician* (Lanham, Md.: Rowman and Littlefield, 2002), 222.

293 **in favor of his plan:** Benjamin H. Newcomb, *Franklin and Galloway: A Political Partnership* (New Haven: Yale University Press, 1972), 271.

293 **There he published:** Joseph Galloway, *Candid Examination of the Mutual Claims of Great Britain, and the Colonies: With a Plan of Accommodation, on Constitutional Principles* (New York: February 1775).

293 **February 25, 1775:** Benjamin H. Newcomb, *Franklin and Galloway: A Political Partnership* (New Haven: Yale University Press, 1972), 264ff.

294 **to create a community:** But in John Locke, "An Essay Concerning the True Original, Extent, and End of Civil Government" (paragraph 15), Locke argues the contrary and indicates that what he means to emphasize is the transition, i.e. making "themselves Members of some Politick Society." Later (paragraph 87) he makes clear that the distinction he wishes to draw is between men in a "*civil society* one with another" and those who have not joined with their fellows in this way.

295 **Declaration of Independence:** Lewis Namier, "King George III: A Study of Personality," Academy of Arts Lecture, 1953; reprinted in *Personalities and Powers* (London: Hamish Hamilton, 1955), 40–41.

295 **to suppress the rebellion:** John Fortescue (ed.), *The Correspondence of King George the Third from 1760 to December 1783* (London: Macmillan, 1928), number 1508.

295 **subject to this country or independent:** November 18, 1774. See W. Bodham Donne (ed.), *The Correspondence of George III with Lord North* (London: Murray, 1867), Vol. 1, 214–215; quoted in Henry Steele Commager and Richard B. Morris (eds.), *The Spirit of Seventy-Six: The Story of the American Revolution as Told by Participants* (New York: Harper and Row, 1967), 61.

295 **the latest Posterity:** Charles Francis Adams (ed.), *The Works of John Adams* (Boston: Little, Brown, 1856), Vol. 2, 361ff.

296 **emotional context of the congress:** By September 8 it was shown to be an exaggeration. Although General Gage had seized the store of gunpowder at Cambridge, that was merely a "disagreeable circumstance," Adams commented.

297 **narrowly defeated:** John E. Ferling, *The Loyalist Mind: Joseph Galloway and the American Revolution* (University Park: Pennsylvania State University Press, 1977), 26.

CHAPTER 18: CASTING THE DIE

299 **the dye is now cast:** John Fortescue (ed.), *The Correspondence of King George the Third from 1760 to December 1783* (London, Macmillan, 1928), number 1508.

300 **blood and confusion:** Quoted in Esmond Wright, "American Independence in Its American Context: Social and Political Aspects . . ." in A. Goodwin (ed.), *The New Cambridge Modern History,* Vol. 8, *The American and French Revolutions, 1763–1793* (Cambridge: Cambridge University Press, 1971), 513–514.

301 **approximately 2.5 million:** In 1770, the population was estimated at 2,148,000, of whom about 460,000 were black. *Historical Statistics of the United States* (Washington, D.C.: Bureau of the Census, 1960).

302 **nearly a hundred little groups:** I discuss one of these groups, from Mecklenburg County, North Carolina, led by Colonel Thomas Polk, in *Polk's Folly* (New York: Doubleday, 2000), 115–116. See also Pauline Maier, *American Scripture: Making the Declaration of Independence* (New York: Knopf, 1997), 48. Maier has identified some ninety similar groups.

303 **damped down the voice:** They suppressed the publication of the Mecklenburg Resolves. See William Henry Hoyt, *The Mecklenburg Declaration of Independence* (New York: Putnam, 1907), 66ff.

303 **must be annihilated:** Letter of September 26, 1780, quoted in Lewis Namier, "King George III: A Study of Personality," in *Personalities and Powers* (London: Hamish Hamilton, 1955), 45.

304 **be a poor island indeed:** Letter of June 11, 1779; in Lewis Namier, "King George III," 45.

306 **had no arms:** Charles Knowles Bolton, *The Private Soldier under Washington* (New York: Scribner, Sons, 1902), 114.

306 **George Washington called:** Letter to George William Fairfax, dated Williamsburg, June 10, 1774; in Henry Steele Commager and Richard B. Morris (eds.), *The Spirit of 'Seventy-Six: The Story of the American Revolution as Told by Participants* (New York: Harper and Row, 1967), 23–24.

308 **of making firearms:** The most important shop was at Lancaster, Pennsylvania, under William Henry. See Charles Knowles Bolton, *The Private Soldier Under Washington* (New York: Scribner, 1902), 107.

308 **turn saltpeter (sodium nitrate) into gunpowder:** Now dated but still useful is Orlando W. Stephenson, "The Supply of Gunpowder in 1776," *American Historical Review* (30, 1925), 271ff. See also E. Wayne Carp, *To Starve the Army at Pleasure: Continental Army Administration and American Political Culture, 1775–1783* (Chapel Hill: University of North Carolina Press, 1984), 21–22.

308 **by a little Coaxing:** Letter to William Strahan, dated Passy (France), August 19,

1784; in Benjamin Franklin, *Writings* (New York: Library of America, 1987), 1100.

309 **or independent:** November 18, 1774; in W. Bodham Donne (ed.), *Correspondence of George III with Lord North, 1768–1783* (London: Murray, 1867), Vol. 1, 214–215; quoted in Henry Steele Commager and Richard B. Morris (eds.), *Spirit,* 61.

309 **was being born:** As John Adams wrote to Thomas Jefferson much later, on August 24, 1815, the real "revolution" had already occurred. The actual war, he wrote, "was no part of the revolution; it was only an effect and consequence of it."

Index